Praise for *Super Pumped*

"The tale of Uber, the queen of the so-called 'unicorns,' is a parable about power—and the lengths to which some startup founders will go to amass it and hold on to it. Aside from being a delicious read, Mike Isaac's account is also teeming with new revelations that will shock and outrage you."

—John Carreyrou, *New York Times* best-selling author of *Bad Blood*

"The level of detail Isaac brings to [the story of Uber], the way he weaves the cast of characters together, is impressive. It's the history of Uber, yes . . . but it's more. . . . It is the story of the dot-com bubble in the late '90s. . . . It's the story of the convergence of Amazon Web Services, the iPhone and the App Store, and an economic crash that turned the faucet of finance back on again. It's about the ethos of Silicon Valley . . . and the cult of the founder . . . of a business model built on quicksand, and that makes it about the damage Travis Kalanick has done in his pursuit of power."

—Dylan Schleicher, Porchlight Books

"A detailed, unsparing account of entrepreneurial arrogance, breathtaking excess, and cutthroat competition at one of the tech industry's most vaunted, loathed, and socially transformative companies."

—Anna Wiener, *New York Times* best-selling author of *Uncanny Valley*

"[Isaac] spins a compelling yarn. . . . [*Super Pumped*] is no dry business profile but a tale that Isaac has deeply reported yet still made accessible."

—William Nottingham, *Los Angeles Times*

"Rollickingly entertaining."

—Edward Niedermeyer, *Drive*

"Isaac brings his expertise and energy to *Super Pumped*, and helps us better understand one of the most important companies on earth." —Jeremy Hobson, NPR

"*Super Pumped* goes beyond the business profile to reveal something deeper and darker. The Uber of *Super Pumped*— most likely still the Uber of today—is not just a business; it's a Beast." —B. David Zarley, *Paste*

"Wildly entertaining [and] a very important read . . . Isaac shows how Uber's messy inner workings and dramatic power struggles have made a company that, for better and worse, is now part of the fabric of modern life."

—Bethany McLean, author of *The Smartest Guys in the Room*

"A gripping, masterfully reported book that offers an essential window into what can go wrong with Silicon Valley's growth-at-any-cost culture."

—Sheelah Kolhatkar, author of *Black Edge*

"A riveting read about bro culture gone awry, and how one CEO was willing to do anything to win."

—Nick Bilton, special correspondent, *Vanity Fair*

SUPER PUMPED

SUPER PUMPED

THE BATTLE FOR UBER

MIKE ISAAC

W. W. NORTON & COMPANY

Independent Publishers Since 1923

Copyright © 2019 by Mike Isaac

All rights reserved
Printed in the United States of America
First published as a Norton paperback 2020

For information about permission to reproduce selections from this book, write to
Permissions, W. W. Norton & Company, Inc., 500 Fifth Avenue, New York, NY 10110

For information about special discounts for bulk purchases, please contact
W. W. Norton Special Sales at specialsales@wwnorton.com or 800-233-4830

Manufacturing by LSC Communications
Book design by Lovedog Studio
Production manager: Anna Oler

Library of Congress Cataloging-in-Publication Data

Names: Isaac, Mike, author.
Title: Super pumped : the battle for Uber / Mike Isaac.
Description: First edition. | New York, NY : W. W. Norton & Company, Inc., [2019] |
Includes bibliographical references and index.
Identifiers: LCCN 2019025992 | ISBN 9780393652246 (hardcover) |
ISBN 9780393652253 (epub)
Subjects: LCSH: Uber (Firm) | Ridesharing—United States—History.
Classification: LCC HE5620.R53 I83 2019 | DDC 388.4/132120973—dc23
LC record available at https://lccn.loc.gov/2019025992

ISBN 978-0-393-35861-2 pbk.

W. W. Norton & Company, Inc., 500 Fifth Avenue, New York, N.Y. 10110
www.wwnorton.com

W. W. Norton & Company Ltd., 15 Carlisle Street, London W1D 3BS

1 2 3 4 5 6 7 8 9 0

For Sarah and Bruna

*You must know there are two ways of contesting, the one by the law,
the other by force; the first method is proper to men, the second
to beasts; but because the first is frequently not sufficient,
it is necessary to have recourse to the second.*

—NICCOLÒ MACHIAVELLI, 1513

*Being super pumped gives us super powers, turning the hardest
problems into amazing opportunities to do something great.*

—TRAVIS KALANICK, 2015

CONTENTS

PROLOGUE

NO ONE WANTED TO WALK HOME THAT NIGHT.

It was winter in Portland, 2014, cold enough to need a heavy jacket. Downtown traffic was thick with students, commuters, and holiday shoppers buying gifts. It had snowed earlier in the week; the streets were still slick with rain and melted flurries. White, flickering Christmas lights lined the trees along Broadway downtown, a festive backdrop for the holiday season. But it wasn't a good night to be waiting around for a bus. The local transportation officers stood outside in the cold—damp, bored, and annoyed—trying to catch a ride.

The officers weren't looking for a cab home. They worked for the Portland Bureau of Transportation and had a mandate: Find and stop anyone driving for Uber, the fast-growing ride-hailing startup. After months of trying to work with city officials to make the service legal in the city, Uber had thrown negotiations out the window. The service was launching that evening, without the bureau's approval.

For Uber, it was business as usual. Since 2009, the company had faced off against legislators, police officers, taxi operators and owners, transportation unions. In the eyes of Travis Kalanick, Uber's co-founder and chief executive, the entire system was rigged against startups like his. Like many in Silicon Valley, he believed in the transformative power of technology. His service harnessed the incredible powers of code—smartphones, data analysis, real-time GPS readings—to improve people's lives, to make services more efficient, to connect people who wanted to buy things with people who wanted to sell them, to make society a better place. He grew frustrated by people with cautious minds, who wanted to uphold old systems, old structures, old ways of thinking. The corrupt institutions that controlled and upheld the taxi industry had been built in the nineteenth and twentieth centuries, he thought. Uber was here to disrupt their outmoded ideas and usher in the twenty-first. Nevertheless, transportation officials were beholden to legislators, and legislators were beholden to donors and supporters. And those donors often included drivers' unions and Big Taxi, the groups who wanted Uber to fail.

UBER HAD ALREADY TRIED the nice-guy approach in Portland. Twenty-four hours before, Kalanick had dispatched David Plouffe, an expert political strategist, to smooth things over with city transportation officials. Plouffe was a silver-tongued creature of politics. Many believed his mastery had helped Barack Obama clinch the presidency in 2008. Plouffe knew exactly the right notes to hit with local politicians. He called Charlie Hales, Portland's affable mayor, to brief him on Uber's next steps. Hales took the call in an office in City Hall, joined by Steve Novick, his transportation commissioner.

If Hales was a nice guy, Novick was his enforcer. Standing four feet, nine inches tall, with thick glasses and a voice that pitched steadily higher as he got angry, Novick was a bulldog. The son of a waitress and a New Jersey union organizer, Novick was born without a left hand and missing fibula bones in both of his legs, disabilities that enhanced his pugilistic spirit. After graduating from the University of Oregon with his bachelor's degree at eighteen years old, he went on to earn a Harvard law degree by the age of twenty-one. He had a sense of humor, too: in past campaign advertisements, Novick branded himself "The Fighter with the Hard Left Hook"—a reference to the metal hook-shaped prosthesis that capped his left arm.

Plouffe opened talks with a friendly overture, letting the two local politicians know that Uber had waited long enough, and with a folksy, familiar tone in his voice—a classic Plouffe touch—said Uber was planning a launch downtown the next day.

"Well, guys, we're already in a number of suburbs outside of Portland, and there's just so much pent up demand for our service in your great city," Plouffe said. Uber's pitch since Plouffe came aboard was a smart one, populist in tone. The service was a way for individuals to earn money using their own cars, on their own terms, setting their own schedule. It would reduce the number of drunk drivers on the road, improving city safety, and passengers would have another convenient option in places where public transportation wasn't fully mature. "We're really trying to provide a service to your citizens here," he went on.

Novick wasn't having it. "Mr. Plouffe, announcing that you're going to break the law is not civil," he said, his hook digging into the mayor's desk in frustration. "This is not about whether we should have a thoughtful conversation about changing taxi regulations. This is about one company thinking it is above the law."

Novick and Hales had tried to tell Uber for months that the company couldn't just roll into town and set up shop just because it was ready to do so. The taxi union would have a conniption. Furthermore, there were existing regulations that prevented some of Uber's services from operating. And since ride-hailing was such a new phenomenon, much of Portland's existing rules didn't address the practice—laws for Uber just hadn't been written yet. Uber would have to wait.

It wasn't as if Novick and Hales were being inflexible. Hales had promised to overhaul transportation regulations upon entering office. Just a few weeks prior, Portland was one of the first cities in the country to draft rules that allowed Airbnb, the home-sharing startup, to operate legally within the city's confines. And for more than a year, the hope was that such a forward-thinking city could do the same with ride-sharing.

But Portland's good intentions weren't delivering on Kalanick's time frame. Now, the two sides found themselves at an impasse. "Get your fucking company out of our city!" Novick yelled into the speaker phone. Plouffe, the charmer, was silent.

Uber's nice-guy approach hadn't worked. But it wasn't designed to. Over the previous five years, the company had grown from a startup employing a couple of techies in a San Francisco apartment to a burgeoning global behemoth operating in hundreds of cities across the world. It had done so by systematically moving from city to city, sending a strike team of employees to recruit hundreds of drivers, blitz smartphone users with coupons for free rides, and create a marketplace where drivers were picking up passengers faster than the blindsided local authorities could possibly track or control. This was the plan for Portland as well, no matter what the mayor and his enforcer had to say. And Travis Kalanick was tired of waiting.

SIX HUNDRED MILES south of Portland, at 1455 Market Street in San Francisco, Travis Kalanick was power-walking around Uber headquarters.

The thirty-eight-year-old chief executive was a pacer. Pacing was something he had done for as long as any friends of his could remember; his father once remarked that a young Travis had worn a hole in the floor of his bedroom from all the pacing. The habit didn't dissipate with age. As he grew older, Kalanick leaned into it. Pacing became his thing. Occasionally, when taking a business meeting with an unfamiliar face, he'd apologize and stand up—he had to pace.

"You'll have to excuse me, I just gotta get up and move around," Kalanick would say, already out of his chair. Then he would continue the conversation, full of kinetic energy. Everyone inside Uber headquarters was used to Kalanick doing laps around the office. They just made sure to stay out of his way.

Uber headquarters was specifically designed with Kalanick's pacing in mind. The 220,000 square feet of office space in the heart of San Francisco included a quarter mile of indoor, circular track built into the cement floor, which weaved through rows of standing desks and shared conference room tables. The track, he would say, was for "walk and talks." Kalanick liked to boast that during the course of any given week his walk and talks would take him 160 laps around the quarter-mile track, the equivalent of forty miles.

This was not just any walk and talk. Portland officials had been stalling on new transportation regulations for more than a year. Now

Uber was going to launch in the city, without the mayor's consent. They didn't have time for city officials to get their act together and write new laws. "Often regulations fail to keep pace with innovation," an Uber spokeswoman would later tell reporters of the Portland incident. "When Uber launched, no regulations existed for ride-sharing."

The problem wasn't Uber's black car service, which functioned well in a number of cities because it adhered to standard livery and limousine service regulations. The problem was UberX, an ambitious, low-cost model that turned nearly anyone on the road who had a well-conditioned car and could pass a rudimentary background check into a driver for the company. Allowing random citizens to drive other people around for money opened up a slew of problems, most notably that no one had any idea whether or not it was legal. At Uber, no one really cared.

Kalanick didn't think much of the nice-guy approach to dealing with cities. He believed that politicians, when it came down to it, would always act the same way: they would protect the established order. It didn't matter that Uber was transformational, a way for people to catch a ride from a stranger with just a few taps on their iPhone. The new model pissed off the taxi and transit unions, and those people would flood the mayor's office with angry phone calls and emails. Uber, meanwhile, would happily rake in the cash, and do so with a groundswell of public support from locals who loved the ease and simplicity of the service.

Kalanick was done waiting. It was time to go. He gave the word and Uber general managers on the ground in the Pacific Northwest got the message: Protect the drivers, trick the cops, and unleash Uber in Portland.

THE NEXT EVENING, Erich England was waiting in front of a historic venue, the Arlene Schnitzer Concert Hall, along Portland's storied Broadway strip. He was glaring down at his phone, refreshing his Uber app.

England was not a concertgoer: he was there to bust Uber. Posing as a fan of the symphony looking to catch a car home, the Portland Transportation code enforcement officer had opened the app hoping to find a driver seeking new rides.

After the phone call ended with Plouffe, Novick had sent out marching orders to his staff: Go catch the drivers. After an officer like England successfully hailed an Uber, he would write the driver thousands of dollars in civil and criminal penalties—lack of proper insurance, public safety violations, required permits—and threaten to impound the vehicle. Novick knew he might not be able to stop the company, but at least the City of Portland could slow them down a bit by scaring off their drivers. Local press showed up to document the action.

Uber was ready. Whenever it entered a new city, the company used the same, reliable approach. Someone from Uber headquarters would travel to a new city and hire a local "general manager"—usually a fired-up twentysomething, or perhaps someone with a scrappy, startup mentality. That manager would spend weeks flooding Craigslist with want ads for drivers, enticing them with sign-up bonuses and thousands of dollars in cash for hitting milestones. "Let drivers know they get $500 cash when they take their first ride on UberX," the advertisements said. For the most part, the GMs placing these ads had little professional experience, but that wasn't a problem for the company's recruiters. Uber only expected that new field operations staff have ambition, the capacity to work twelve- to fourteen-hour days, and a willingness to evade the rules—even laws—when necessary.

England refreshed his app again. Finally, his request was accepted by a driver. The car was five minutes away.

Until it wasn't. The driver had cancelled on him, and the car had driven past, according to the app. England never saw it go by.

What England didn't know was that Uber's general managers, engineers, and security professionals had developed a sophisticated system, perfected over months, designed to help every city strike team—including the one in Portland—identify would-be regulators, surveil them, and secretly prohibit them from ordering and catching Ubers by deploying a line of code in the app. The effect: Uber's drivers would evade capture as they carried out their duties. Officers like England could not "see" the shady activity, and could never prove it was happening.

England and others in Portland had no idea what they were up against. They considered Uber a group of overzealous young techies, perhaps a bit too enthusiastic about the transformative effects their startup would have on transportation. The staff was presumptuous, even arrogant, but that could be chalked up to the relative youth of the team members.

Behind the scenes, Uber was hardly innocent. Recruiting ex-CIA, NSA, and FBI employees, the company had amassed a high-functioning corporate espionage force. Uber security personnel spied on government officials, looked deep into their digital lives, and at times followed them to their houses.

After zeroing in on problematic individuals, the company would deploy one of its most effective weapons: Greyball. Greyball was a snippet of code affixed to a user's Uber account, a tag that identified that person as a threat to the company. It could be a police officer, a legislative aide or, in England's case, a transportation official.

Having been Greyballed, England and his fellow officers were served up a fake version of the Uber app, populated with ghost cars. They had no chance of ever capturing the rogue drivers. They might not even know if drivers were operating at all.

For the next three years, Uber operated with impunity in Portland. It wasn't until 2017 when the *New York Times* broke the story of how Uber used Greyball to evade the authorities that Portland officials fully understood just how Uber had carried out its subterfuge.

But by 2017 it was too late. Uber was up and running in Portland— legally, even—a fixture of the city, in regular use by citizens who praised its convenience. Kalanick and his team had violated local transportation laws, and instead of being exiled, they had found enormous, game-changing success.

KALANICK AND HIS FORCES had flouted laws in Portland, and in scores of other cities. But ask a typical Uber employee at the time—and even some supporters years later—and they will tell you they didn't see it that way. Greyball was consistent with one of Uber's fourteen company values: Principled Confrontation. Uber was protecting its drivers while confronting what they saw as a "corrupt" taxi industry that had been protected by bureaucracy and outdated regulations. Concepts like "breaking the law" weren't applicable, they believed, when the laws were bullshit in the first place. Kalanick was convinced that once everyone used the service, it would click—they'd understand that the old way was inefficient and expensive, and his way was the right way.

To some extent, he was right. As of this writing, Uber operates worldwide. It is present on almost every continent, with copycats and competitors trying to mimic the growth and power Kalanick achieved in the eight years he was at the helm. Uber has struck deals with local governments to become as ubiquitous as public transit, and is working on a future in which the cars that people request will drive themselves.

And yet, Uber is not always seen as a success story. Uber's rapid rise was nearly undone in 2017, as the company faced the consequences of years of Kalanick's boundary-pushing behavior, unabashed pugnacity and, eventually, the CEO's own personal decline. Kalanick's story is whispered as a cautionary tale for founders and venture capitalists alike, emblematic of both the best and worst of Silicon Valley.

The saga of Uber—which is, essentially, the story of Travis Kalanick—is a tale of hubris and excess set against a technological revolution, with billions of dollars and the future of transportation at stake. It's a story that touches on the major themes of Silicon Valley in the last decade: how rapid developments in technology can crash into long-entrenched labor systems, throw urban development into upheaval, and overturn an entire industry in a matter of years. It is the story of a deeply sexist industry, fueled by gender imbalance and a misguided belief in a tech-supported meritocracy, blind to its own biases. It is the story of the sweeping but poorly understood ways that startups are financed today, and how this can affect the leaders, employees, and customers of fast-growing companies. It is the story of the ugly decisions made around user data and personal information as technology firms seek to exploit consumer data. But most of all, it is a story about how blind worship of startup founders can go wildly wrong, and a cautionary tale that ends in spectacular disaster.

Travis Kalanick and his executive team created a corporate environment that looked like an admixture of Thomas Hobbes, *Animal House*, and *The Wolf of Wall Street*. This toxic startup culture was the result of a young leader surrounded by yes-men and acolytes, being given nearly unlimited financial resources and operating without serious ethical or legal oversight. At war with outsiders and among themselves, the company engaged in spying, backbiting, and litigiousness as it struggled for power and supremacy over a multi-billion-dollar empire.

As a result of Kalanick's actions, Uber's valuation was cut by tens of billions of dollars, competitors that might have been vanquished were strengthened and found new footing around the world, and the company faced a half dozen federal investigations into its sordid history. More than once, investors and employees worried that the entire future of the company was at stake.

As a Bay Area resident and professional journalist during the past decade, I saw Uber rise to power right in front of me. I witnessed how quickly a transformative idea can change the urban fabric of a city, and how strong personalities can have an outsized effect on shaping the way a startup operates.

I began covering Uber for the *New York Times* in 2014. Those were Uber's glory days, when Kalanick's cunning and street-fighting sensibilities helped to outwit competitors, seal billion-dollar financing deals, and make Uber's global conquest seem inevitable.

Just a few years later, Uber was on a collision course with itself, and Kalanick's leadership had grown into a liability: 2017 turned into one of the worst years of sustained crises for any corporation in the history of Silicon Valley, as Uber suffered blow after self-inflicted blow in full view of the public.

MY COVERAGE ULTIMATELY LED ME to becoming part of the story, wrapped up in the twisted saga of lies, betrayal, and deceit that Kalanick and other company leaders used to build and control a tech juggernaut, one of the first unicorns of the mobile era, a company worth billions of dollars that succeeded in changing the way we move through the world, yet nearly destroyed itself in a bonfire of bad behavior, ugly decisions, and greed.

I feel lucky to have been along for the ride.

PART I

Chapter 1

X TO THE X

THE EMAIL BLAST WENT OUT TO EMPLOYEES ACROSS THE WORLD:
Uber had passed another milestone. It was time for Uberettos* to
celebrate.

It was an Uber tradition for Travis Kalanick to take the company
on the road after hitting growth targets. Funded by billions of venture
capital dollars, the trips were conceived as morale boosters, a way to
bring employees closer together. But they were also an excuse for a
week-long bacchanal in some far-flung part of the world. For this celebration, Kalanick had a special city in mind: Las Vegas.

Kalanick would have to be creative in Vegas if he wanted to top
Uber's previous company-wide retreat. In 2013, he had organized a
blowout bash in Miami to celebrate Uber reaching $1 billion in gross
ride bookings—an enormous feat at the time. The trip was memori-
alized internally with the Chinese symbol 九, a character that stood
for the number 9. He claimed it had "internal meaning at Uber," but
was "something we do not discuss externally," according to a letter
he sent to the entire company before the retreat. He went on to advise
his staff not to throw large kegs off of tall buildings, and mandate no
interoffice sex unless co-workers explicitly stated "YES! I will have

---

* Every tech company over the course of its maturation process must create for itself a noun to
describe its collective employees. Google employed Googlers, Twitter employees were called
"tweeps," and for Uber in its early days, the noun was "Uberetto." The exact etymological
origins of the noun—which confused many employees after they joined—are not clear.

sex with you" to one another. He also noted that any puking on hotel grounds would result in a $200 fine. The email set the tone for the rest of the retreat.

Miami would be dwarfed by what Kalanick had in store for Las Vegas. This one was special, a celebration of a key internal metric. Every time the company reached a revenue milestone that corresponded with an exponent of the number ten, Uber celebrated with a party. But as the company swelled in headcount and number of cities served, so did the scope of Uber's celebrations. With every new zero added to the revenue figure, Uber's thousands of employees were rewarded with an all-expenses-paid trip to another global destination.

The ten billion revenue figure was special. Everyone appreciated the significance of such a big, round number, and Kalanick in particular had an affinity for the mathematics of exponential growth. They would call this milestone party "X to the x"—ten to the power of ten. He dispatched an entire team of designers to work on the aesthetics of the trip. The invitations, the signage, even the wristbands all had the same look: A large, white "X," raised to the power of a smaller white "x" against a square, black backdrop. It was very Uber.

High-end party branding aside, some form of celebration was appropriate. By the fall of 2015, Uber employed nearly five thousand people globally, the result of an endless talent poaching campaign across Silicon Valley. Engineers from Amazon, Facebook, Apple, Tesla, and especially Google, came flooding into the company as quickly as recruiters could pluck them from places like the Creamery, the Battery, WeWork—VC–funded startups themselves—that had become the usual haunts of coders in San Francisco.

Engineers had seen the adjectives the press used to describe the company. Uber was "fast-growing," "pugnacious," a "juggernaut." They heard the whispers of staggering revenue growth, and saw the company's surging valuation, which was already well into the billions. They loved how Kalanick brought a hacker-like mentality to the way he built and ran his company. No one wanted to miss their shot at entering the next Google or Facebook on the ground floor.

Recruiters knew exactly how to sell it, tweaking the FOMO* of ambitious engineers. "You don't want to miss this rocket ship," the headhunters said, as they flooded the LinkedIn boxes of engineers across the Valley. Securing equity in a fast-growing company like Uber could one day let them cash out and buy a mini-mansion in San Francisco, the white-hot center of the Bay Area real estate market. Others dreamed of doing four years inside of Uber, then using their vested riches to start new companies of their own.

The Bay Area had seen this before. After the initial public offerings of Google, Twitter, and Facebook, Silicon Valley had absorbed hundreds of newly minted millionaires. And now, for thousands of young engineers who had heard stories from older colleagues about the heady days of the Web 1.0 boom, landing a job at Uber meant they, too, might realize their dreams of tech riches.

Joining Uber in those days was a statement, like driving a Tesla or wearing a Rolex. The anxiety, stress, and crushing schedule of twelve-plus-hour days was all going to be worth it. They were all going to get paid, *big time.*

IN OCTOBER OF 2015, thousands of Uber employees flew into McCarran International Airport in Las Vegas, walked out in the hundred-degree heat, and piled into a line of shuttles and taxicabs,† headed to their hotels on the strip. It was time to party.

Kalanick spared no expense. Uber rented hundreds of rooms up and down the strip at Bally's, the Quad, the Flamingo, and others. Each employee was given Visa prepaid credit cards filled with money for food, fun, and festivities. They didn't always need them; the private parties were stocked with free food and open bars. A company-issued wrist-

---

* "Fear of missing out," naturally.

† Before Uber's 2015 party, the local taxi unions had kept ride-sharing out of Vegas. Uber launched in Las Vegas just one month prior, but the company was barred from picking up passengers from the airport; that was still Big Taxi's turf.

band with the "X" logo gained one access to all of the planned Uber events. Before the trip, engineers quickly spun up an app that served as a personal guide to the week's festivities. Everyone was given small, temporary stick-on tattoos that all read the same: "X to the power of x."

While Kalanick insisted on extravagance, some executives had the foresight to worry about the optics of the celebration. Rachel Whetstone, a former Googler and Uber's top policy and communications executive, sent out memo after internal memo detailing how employees should *not* behave. No Uber T-shirts, no discussion of company numbers or metrics, absolutely no talking to press. Even the small Uber logos on employees' corporate Gmail accounts were removed and replaced with the "X" logos, in case bystanders happened to see an engineer working in public.

Uber already had an aura of arrogance about it in 2015. The pervasive trope of the "tech bro" was the ire of communications representatives across the Valley; young and moneyed, childless, these engineers and salesmen were unburdened by the daily concerns of the baristas, housekeepers, and wait staff they felt existed to serve them. A tech bro's greatest worry was whether or not he was working at that year's hottest "unicorn"—a noun coined in 2013 by a venture capitalist who used it to describe companies valued at more than $1 billion. By the fall of 2015, Uber was the unicorn to end all unicorns; every tech bro *had* to be there.

Uber wasn't alone as a haven for tech bros. Snapchat, once a darling in the Valley for its innovative approach to social networking, was under fire for emails its founder had sent to fraternity brothers during his college days at Stanford. ("Fuck Bitches Get Leid," [sic] one read.) A group of Dropbox and Airbnb employees were filmed trying to kick a group of San Francisco kids off a soccer field to make room for their corporate league game. The clip went viral, and the companies were forced to apologize to an outraged public. Whetstone and other communications team staff cringed at the thought of seeing Uber's Vegas blowout splashed across the pages of Silicon Valley tech blogs—or worse, the *Daily Mail*.

Some took it too far, even by Uber standards. One employee hired a pair of prostitutes to join him in his hotel room. The next morning, he and his roommate woke up with all of their belongings stolen, including their work laptops. Uber management, terrified of company secrets being sold on the black market, fired the employees on the spot and tried to track down the hardware.

A Los Angeles general manager was fired in 2015 for groping the breasts of one of his team members. Managers were doing drugs with their subordinates—cocaine, marijuana, and ecstasy, mostly. Then there was the employee who managed to steal a party transportation shuttle and joyride it with other Uber employees looking for a good time.

Executives designed each night to outdo the previous one. One memorable evening saw employees flood into XS, a club inside the sleek Encore Las Vegas. That night, Kygo and David Guetta—two renowned electronica musicians—played a private set for Uber staff into the early hours of the next morning.

But the crown jewel was the final musical guest. As Uberettos lined the venue inside the Palms hotel, the house lights went dark and the stage filled with smoke. A voice began to sing the first few slow bars of a familiar song. Then she appeared. Wrapped in a blood-red jumpsuit, sequins shimmering against the neon beams behind her, fog machines wrapping her in mist. The words started coming into focus, a hit all the twentysomething employees knew by heart: "Got me looking so crazy right now, your love's got me looking so crazy right now. . . ."

Employees began screaming as the singer stepped into the spotlight. They realized what Kalanick had done: He got Beyoncé.

The night exploded, with employees dancing and singing along to a string of number-one hits. The crowd hushed for a haunting acoustic rendition of "Drunk in Love," a standby. Up in the rows of seats facing the stage sat Beyoncé's husband, Jay-Z, smoking a cigar and smiling.

As Beyoncé's set came to a close, Kalanick stumbled on stage: His employees, giddy with song and free-flowing Cîroc, were loving every moment. They were all celebrities that night.

"I fucking love you! ALL of you!" Kalanick yelled into the microphone, holding Beyoncé's hand, clearly drunk. "I fucking love you back!" one woman screamed back at him.

Then, Kalanick dropped another bomb: Beyoncé and her husband, Jay-Z, were now stockholders in Uber. What he didn't tell them was how the celebrity couple had decided to invest; Kalanick had paid Beyoncé $6 million in Uber restricted stock units for her performance. The stock's value would increase by 50 percent in less than a year.

At the end of the week, Uber's finance team added it all up. The entire "X to the x" celebration cost Uber more than $25 million in cash—more than twice the amount of Uber's Series A round of venture capital funding.

Employees across the company had to appreciate the moment. Many of them were nerds in high school. In college, velvet ropes had kept them out of stylish lounges. Now, they were being ushered into a Las Vegas nightclub with open arms, treated to a private performance by one of the world's biggest musical superstars. The engineers from Stanford, Carnegie Mellon, MIT were suddenly ballers, in business directly with Jay-Z.

The whole ordeal was supposed to be "baller as fuck," as someone put it. And it was.

"X TO THE X" perfectly emblematized a particular moment in Silicon Valley history. After the dot-com-era bust of the early 2000s, a wave of mobile-device innovation quickly swept the world. The unveiling of the iPhone in 2007 put a handheld computer in everyone's pocket. Here, in Las Vegas, Uber employees were celebrating a smartphone app—which they had personally built—that summoned a taxi at the push of a button. Their labor had brought them absurd, unimaginable wealth. Multi-million-dollar mansions, day trips to Napa Valley vineyards, and lakefront properties in Tahoe were within their reach practically overnight.

Their response to this was not to pause and marvel at their astonishing luck to work at a time and place where fortunes came gushing to twentysomethings via smartphone apps. Instead, they imagined taking their few million from Uber and creating a unicorn of their own—for surely their success thus far showed they were destined for even greater success in the years to come.

But for every *WIRED* cover story of a boy genius striking it rich with a smartphone app, there was a mess of secondary effects left in his wake. Many of the next generation of apps catered to the needs and whims of the white, upwardly mobile twentysomething males of Silicon Valley. The press gave significantly less ink to the latent misogyny bubbling up inside of tech companies, and the libertarian view that enabled tech figureheads to unwittingly enable these same biases. The divide between tech's most talented, and the class who waited tables and served them coffee only grew starker by the day. Fast-rising rents pushed wage earners out of San Francisco, while landlords flipped those former apartments to new, wealthier tenants. The "gig economy" unleashed by companies like Uber, Instacart, TaskRabbit, and DoorDash spurred an entirely new class of workers—the blue-collar techno-laborer.

With the rise of Facebook, Google, Instagram, and Snapchat, venture capitalists looked everywhere to fund the next Mark Zuckerberg, Larry Page, or Evan Spiegel—the newest brilliant mind who sought, in the words of Steve Jobs, to "make a dent in the universe." And as more money flowed into the Valley from outside investors—from hedge funds and private equity firms, sovereign wealth funds and Hollywood celebrities—the balance of power shifted from those who held the purse strings to the founders who brought the bright ideas and willingness to execute them. With money easier to come by, founders were able to exact more favorable terms for themselves, wresting control of the companies from the money men—who required diligence, profitability plans, and oversight.

This shift in the funding of American technology businesses would change the way a generation of the most successful startup founders would expect to be treated by their backers—the "cult of the founder" meant celebrating the vision of the founder no matter what, a slavish devotion to the CEO of a company simply because he was the CEO. Twelve-hour workdays and a nonexistent social life became things to

be celebrated, the markers of a "hustle culture" that the tech bro founders embodied. (Of course, these hardworking bros also played hard, at
events like X to the x.) Even when those founders were bending rules
and even laws, they were treated as Platonic philosopher kings. Many
believed the founders were remaking the world, making it smarter,
more logical, meritocratic, efficient, and beautiful—delivering a new
and much improved version: an upgrade on life.

This was the height of tech utopianism. And though Kalanick would
have no way of knowing it until years later, Uber's company trajectory
would closely map that of the tech industry more broadly. Both were
surging faster, higher than anyone could have anticipated. And just
as the country was beginning to cast suspicion on the beneficence of
Facebook's algorithms, consumers reached the limits of what they were
willing to observe with rose-colored lenses. Soon thereafter the world
of uninhibited technological progress came to a screeching halt.

And so did Uber and Travis Kalanick.

THERE WAS ONE OTHER event employees would recall long after leaving
the desert.

After a day of drinking beer in poolside cabanas, Uberettos checked
their apps to find their next destination: Planet Hollywood. They made
their way up the moving staircase, under pink and red shimmering
entrance lights and into the spacious Axis Theater. Large enough for
7,000 people, the Axis was decked in gold and deep purple velvet.

Long-time employees were used to pomp and circumstance on corporate retreats. But this was different. "Ten to the ten" meant something special to Kalanick; he wanted to show everyone how far Uber
had come, and what it meant to him.

As the lights dimmed, a pair of silhouettes wheeled a large, rickety
chalkboard onto the stage, green slate framed with wood, as if they had
robbed a high school science classroom. Onto the stage walked Kalanick, clad in a stark white lab coat and thick-rimmed black glasses.

He became "Professor Kalanick" for the better part of the next three
hours, explaining to his employees his vision for the company. He was

introducing what he called his "philosophy of work," the result of what
he said was hundreds of hours of deliberation and discussion.

The entire presentation was born directly from Kalanick's obsession
with Amazon, the online retailer led by Jeff Bezos, a founder every
young entrepreneur idolized. Bezos's path to success was the stuff of
Kalanick's dreams. The small online bookstore had become a multi-
billion-dollar retail behemoth by skating on razor-thin profit margins,
focusing on long-term growth over short-term gains, and relentlessly
undercutting competitors on prices. Kalanick admired how Bezos rein-
vested profits on future opportunity, to always stay one step ahead of
his competition.

More than any other company, Amazon embodied the type of business he wanted Uber to become. As Kalanick saw it, delivering people
from place to place was only the beginning of Uber's potential; one day,
Uber would match drivers with packages, food, and retail goods, and
solve untold numbers of other logistical problems. Kalanick imagined
he would one day become a direct challenger to Bezos, reshaping the
way people and goods moved major urban centers. Uber wanted to be
the Amazon for the twenty-first century.

Kalanick carefully studied the methods of Bezos and his company,
down to the fourteen core leadership principles posted to Amazon's
website:

1. Customer Obsession
2. Ownership
3. Invent and Simplify
4. Are Right, A Lot
5. Learn and Be Curious
6. Hire and Develop the Best
7. Insist on the Highest Standards
8. Think Big
9. Bias for Action
10. Frugality
11. Earn Trust
12. Dive Deep

13. Have Backbone; Disagree and Commit
14. Deliver Results

Kalanick had a surprise for his employees, inspired by Bezos's leadership, the company he built, and the leadership principles that formed Amazon's culture.

"I want to introduce you to Uber's values," Kalanick said, pointing to the chalkboard on stage. The house lights shone on the blackboard behind Kalanick. Written in white chalk were fourteen bullet points, each a short saying or thought, sprung directly from the brain of the CEO. The audience read the list as Kalanick rattled them off aloud:

1. Always Be Hustlin'
2. Be An Owner, Not Renter
3. Big Bold Bets
4. Celebrate Cities
5. Customer Obsession
6. Inside Out
7. Let Builders Build
8. Make Magic
9. Meritocracy & Toe-Stepping
10. Optimistic Leadership
11. Principled Confrontation
12. Super Pumped
13. Champions Mindset / Winning
14. Be Yourself

Some of the employees in the audience were confused. "Is this a joke?" one twenty-seven-year-old whispered to a colleague sitting next to him. "Is this still part of the whole professor act?"

The list read like Amazon's corporate values run through a bro-speak translation engine. People in Kalanick's world were not happy or sad, they were "super pumped" or "super unpumped." Company brainstorming meetings were "jam sessions." Half the company enjoyed

Kalanick's colorful vocabulary. The other half bit their lips. Kalanick expected everyone to be as "super pumped" about the values as he was.

Over the next two and a half hours, Kalanick explained each value in excruciating detail, carting out a different executive or Uber employee who embodied it. Ryan Graves, head of operations, was brought out to reflect "always be hustlin'," a willingness to move fast into new cities. Austin Geidt, an early intern who eventually rose to become one of Uber's most respected and high-ranking executives, walked onstage to "celebrate cities."

"Customer obsession" came straight from the mind of Bezos. Just like Bezos, Kalanick had an almost single-minded fixation on improving the customer experience of his product. Everything about riding in an Uber—from opening the app to getting out at the destination—should be seamless, easy, enjoyable. To Kalanick, all employee actions should stem from that point of view. "Principled confrontation" rested on the idea that Uber employees wouldn't shy away from conflict or a fight—as long as it came from a place of principle. This value was often used to justify Uber's barging into new cities even when it wasn't lawful or welcomed; Uber knew taxis were corrupt and protectionist. Uber was elbowing its way in for the good of the customers in the city, even though they didn't know it yet.

"Super pumped" was a particular point of pride. In Uber's early years, every employee was evaluated on a list of eight core "Uber competencies," from qualities like "fierceness" to "scale" and "innovation." Scoring low could mean termination, while scoring high influenced pay raises, promotions, and annual bonuses. But it was an employee's level of "super pumpedness" that made all of the difference in a performance review.

"Super pumpedness is all about moving the team forward, working long hours—pretty much a do-whatever-it-takes attitude to move the company in the right direction," as one Uber employee explained the term. If there was one quality Travis Kalanick looked for in a new recruit, it was that they were as super pumped as he was to work for Uber.

Now, six years into the company's history, Kalanick felt Uber was finally coming into its own. With an audience of millions and billions in venture capital in the bank, Uber was unstoppable. It was inevitable—so Kalanick believed—that Uber would one day challenge Amazon as another global tech superpower.

AFTER THE FIRST evening's presentation, Kalanick told employees in the audience he had a special guest to interview: Bill Gurley.

Bill Gurley, a former financial analyst turned legendary Silicon Valley venture capitalist, would prove to be instrumental in Uber's entire arc of success. As a general partner at Benchmark, a top-tier VC firm, Gurley secured a Series A investment in the young company. As a board member and vocal supporter of Uber, Gurley was someone almost everyone in the company looked up to for advice. Two other Uber backers joined the discussion.

Towards the end of the interview, the tone shifted. Kalanick asked what advice the venture capitalists would give to him for the future. Gurley sat back and mulled the question for a moment, furrowing his brow. Then the investors gave it to Travis straight.

One of Uber's greatest strengths was its incredible product focus, drive, and intensity—from every employee, at every level of the company. That ability to strive for greatness drove Uber to global, multibillion dollar heights. "But what I've seen from you, as a leader, is that if you expect people to jump to the ceiling, they'll actually do it," one investor said, as Gurley nodded. "They'll jump so high, they'll smash through the roof with their heads."

But that strength in excess, the investors claimed, was also Uber's greatest weakness. Perhaps Kalanick would do well to help employees take better care of themselves—through wellness, massage, meditation, even yoga, the investor offered.

Some employees were shocked. Uber's backers were telling Travis to take it easy. Even Gurley, one of the most competitive VCs in the Valley, believed it was important for the company. But he was right. Uber employees were always sprinting. They kept working even after they went home, terrified of both their competitors and their bosses. The pace was causing burnout at all levels of the company; some engineers and designers were seeing therapists to deal with the strain.

As the audience of employees applauded at the suggestion, Kalanick smirked, moving into a mock yogic child's pose on the stage in front of his employees. The VC's were right; Kalanick couldn't "run the company under the red line forever."

But Kalanick went on to make clear where he stood—Uber wouldn't be resting on its laurels.

"Make sure we all understand: This is a marathon," he said. "I'm down for that."

Chapter 2

THE MAKING OF A FOUNDER

THE STREETS OF NORTHRIDGE, CALIFORNIA, ARE SECTIONED INTO AN asymmetric grid. There is an order to the layout, a nine-and-a-half-square-mile trapezoid the shape of Utah tucked between the San Fernando and Simi valleys in Greater Los Angeles. Seen from above, Northridge is framed by a near-perfect square of freeways, an emblem of transportation efficiency.

Travis Cordell Kalanick was born on August 6, 1976, in Northridge Hospital to Donald and Bonnie Kalanick, an average, white, middle-class couple who built a comfortable life for themselves in California. Travis spent his formative years in a wood and brick ranch-style home on the corner of a quiet intersection, purchased by his father on a civil engineer's salary. Like the neighborhood of Northridge, even the family driveway was built symmetrically, a stretch of grey cement slabs outlined with red brick.

Bonnie worked at the local paper, the *Los Angeles Daily News*, as an advertising executive. She spent decades selling ad space to small and medium-sized businesses across the San Fernando Valley, a time when the internet was a distant threat and the news business was still lucrative. Bonnie was one part of a normal, Northridge nuclear family, "always happy, upbeat, and never spoke poorly about anybody," Melene Alfonso, a former co-worker, said of her. "Her customers loved her."

She was good at her job, a resilient worker and a charmer. Bonnie had a reputation at the paper for her sales prowess and the charisma to win over clients—a quality that she would pass on to her young son Travis. Though Bonnie's smile was always quick, her co-worker recalled, she possessed an inherent competitive spirit.

But at the end of a day of hard work and constant selling, Bonnie would return home to Donald, Travis, and Travis's little brother, Cory, born just a year later. Bonnie doted on her two boys, spending all her hours away from the newspaper caring for them.

Travis, in particular, was close to his mother, and she was close to him. Later, when Travis rose to power, friends remarked on how terribly proud of her son she was. After he'd left home in his thirties, Travis would return annually to Los Angeles to celebrate Christmas with the family. One friend recalled how Bonnie scurried back and forth between the living room and kitchen, cooking a holiday feast for the family while making sure Travis had enough to eat. Bonnie kept clippings of newspaper articles detailing her son's success, showing them to friends, neighbors, visitors, anyone.

"She wore her heart on her sleeves," Travis later said of her. "And when she walked into a room, her warmth, her smile and her joy would instantly fill it."

Bonnie's dedication to Travis never wavered. He was never the most popular kid, nor did he have overnight success in the startup world. Long before the breakout success of Uber, Kalanick had been seen as an entrepreneurial failure. When pitching new clients on his enterprise products, door after door was closed in his face. When one company was nearly acquired by a tech giant, the opportunity was snatched away at the last minute. And when one of his closest advisors and investors betrayed him early on in his career, it didn't keep Travis from building another new venture shortly thereafter. One friend described him as a pit bull that spent its life getting kicked by its owners—no matter how beaten down Travis was, he never, *ever* gave up.

Later, when an interviewer asked his parents where Travis got his stubbornness, Bonnie raised her hand.

"Working for a newspaper, I was used to sales rejection all the time, so I knew what that was like," she said in an interview in 2014. "But I had hope, since he is very determined and he will not back down when he felt he was right—he's tenacious."

DONALD, WITHOUT A DOUBT, was the left brain of the family. A civil engineer by trade, Donald spent much of his career working for the City of Los Angeles, where he contributed to projects at Los Angeles International Airport, as well as other parts of the city.

Donald's marriage to Bonnie was not his first. He married once before at twenty-seven, to a younger woman, in a pairing he would later call a mismatch. He had two daughters with his first wife, half-siblings to Travis and Cory. Even after remarrying, Donald maintained a positive relationship with his ex-wife. "Peaceful," he'd later note.

Donald considered himself an analytical thinker, a champion of logic, rules, and complex systems. Instead of father and son football games or having a catch, the two bonded by working together on Travis's grade-school science projects. The two once built an electrical transformer together. Travis liked to call him a tinkerer—and he was.

"I liked to build things," Donald later told a reporter. "I thought it'd be nice to be driving by a structure and say 'hey, I had a good part in building that.'" He went to junior college before transferring to receive an engineering degree. He felt at home surrounded by math, by numbers.

Donald was tough on his sons, and had high expectations for them. He also introduced them to the world of computers. Early in Travis's life, his father brought home the family's first computer, giving Travis the ability to practice programming for the first time. He learned to code by the time he was in middle school. Travis ultimately never mastered coding languages—he preferred thinking through product and user-experience issues—but the early connection to technology would stay with him. Travis loved efficiency and hated waste. He appreciated how the rise of software and the internet allowed old, ineffectual, and broken systems to be overturned and rebuilt anew. Code and programming enabled anyone willing to learn and work hard a chance to change the system—to change the world.

TRAVIS TOOK TRAITS FROM both Bonnie and Donald in equal measure. A precocious child, he picked up his father's skills with mathematics, impressing others with his ability to speed through arithmetic in his head where other classmates needed pencil and paper. His mother's sales talent rubbed off on him as well. Travis and Donald were part of the YMCA's Indian Guides youth troop, where Travis was a top seller for the group's annual pancake breakfast fundraiser. Travis spent hours outside his neighborhood grocery store, pitching shoppers on their way inside to donate to his troop's fundraiser. He was charming, persistent, tireless, and competitive; his parents eventually had to drag him home in the evenings.

He maintained that competitive edge as he grew older. At Patrick Henry Middle School—only a half-mile drive from his home in Northridge to Granada Hills—Travis was naturally athletic. Travis ran track, played football, and shot hoops. At eleven, an article in his mother's newspaper praised him for being a basketball player with a 4.0 grade point average. His prize: an enormous trophy—larger than the ones that teams received for winning the regional championship.

"Success in athletics doesn't happen by accident; it requires hard work and discipline," the award presenter said of Travis and his classmates at the time. "When you learn the art of discipline, that's half the battle."

Despite these talents, middle school was not easy for him. Older kids began to pick on the wiry youth for his intelligence, or for not wearing the right clothes or not knowing how to act "cool." The bullying was relentless, in part due to Travis's early lack of emotional intelligence, friends and close ones say. Being a math whiz who could rapidly crunch large numbers in his head scored points with his teachers. But it also put a target on his back; he was a geek. And in Travis's middle school, geeks got bullied.

At some point in middle school, Travis decided he would not take the bullying anymore. He pushed back against his aggressors, and even began bullying others to deflect attention from himself. Fighting came naturally and ultimately his aggression won him a spot in the cool crowd.

In high school, he began to wear the right clothes, got the right girlfriend, and hung out with the right people. Life became much less difficult after he figured out how to fit in; a geek in cool kids' clothing.

His entrepreneurial spirit continued to shine through. As a teenager, Travis began selling Cutco knives door-to-door in his neighborhood. In cold call after cold call he honed his natural sales ability. This would prove indispensable years later when Travis had to raise money pitching startups. That summer back in the '80s, he claimed to sell $20,000 in knives. Cutco salesmen twice his age had trouble meeting those numbers; Kalanick did it with ease, his commissions growing larger with every new knife he sold.

Selling for a big company wasn't enough. At eighteen, Kalanick decided to start his own SAT preparation service with a classmate's father—a strange pairing that ended up working out rather well. The business, which they called "New Way Academy," was a workshop, taught by Kalanick, where he reviewed test-taking strategies and quizzed roomfuls of sixteen-year-olds with sample questions. He saw it as a performance, another way to sell to an audience.

Kalanick himself was no slouch on the SAT. He scored a 1580, just twenty points shy of perfect, and whipped through the math portion of the test with plenty of time to spare.

Friends remember his savant-like math abilities. "We were driving across town in Los Angeles once, and Kalanick saw a street sign that said we were seventeen miles from where we were going," recalled Sean Stanton, a friend and former colleague. "He looked down at the speedometer and saw our average speed, and in a few seconds rattled off how long it would take us to get there so we could make it in time for our meeting. I mean, who does that?"

With his test scores and extracurriculars, Kalanick could have his pick of colleges. He chose to stay close to home and enrolled at the University of California, Los Angeles. It was there he would find his first real opportunity to build a startup.

THE ERA KALANICK MATRICULATED at UCLA would prove to be pivotal in the history of the internet. In 1998, people largely accessed the internet through sluggish modems and dial-up connections. Back then, 28.8 kilobauds per second was considered decently fast; it took minutes to download an image file and a half hour for a three-minute music track, if you were lucky.

College campuses, however, offered young techies like Kalanick an enormous upgrade. By the late 1990s, most major universities provided their on-campus students with access to college networks connected to the internet through so-called "T1" lines. Using fiber optic cables, T1 connections relied on digital signals rather than analog ones used by most telephone lines. A college campus wired with fiber optic cable, delivering 1.5 megabits per second meant a student like Travis could surf the web more than a thousand times faster than he could on his parents' old 28.8-kilobit dial-up connection. Files that used to require hours to download could zip through in seconds.

Kalanick was a double major in computer science and economics, and joined the Computer Science Undergraduate Association, which put him in the middle of a rapidly expanding field.

He and his computer science friends took full advantage of their T1 connection. They battled each other in games like Quake, Doom, and StarCraft. File-sharing parties were common; groups spent hours trading and downloading music, movies, and images, swapping files as if they were baseball cards.

Then it occurred to some of them: "Wouldn't it be cool if we had a page where we could search for some of this stuff directly?" Kalanick would later recall. A central hub, like the internet portals they grew up on, where they could search for any media they wanted and download it. It would make more sense than emailing files between friends; anyone in the world could use it.

What Kalanick was describing, without realizing it, was a proto-version of Napster, the iconic file-sharing network co-founded by Sean Parker, an internet entrepreneur and, later, an early advisor to Mark Zuckerberg at Facebook.

Eventually, Kalanick joined six of his friends to build Scour.net, a Google-like search engine that gave users the ability to "scour" millions of files and then download them, like Napster. Kalanick later claimed to be a co-founder, though his friends disputed this status. Eventually, Kalanick was tasked with Scour's sales and marketing efforts.

By his senior year, Kalanick decided to drop out of UCLA to work on Scour full time, following the example of entrepreneurs like Bill Gates and, later, Mark Zuckerberg. It upset his parents, though they wouldn't tell him as much until years later. He technically lived at home, but spent all his time down the road in a two-bedroom apartment with his six other co-workers, where he "worked, ate and slept." Scour didn't have much of a business model. But Kalanick and friends had absorbed the Silicon Valley maxim that growth was paramount. A path to profit could come later.

Work was everything to Kalanick. He didn't have friends, he didn't have girlfriends. To have a relationship with Kalanick, one former Scour colleague said, you had to be working alongside of him. Beyond his parents, Kalanick had few personal relationships.

All he thought about was building a great company. He wouldn't wash his clothes, leaving piles of laundry accumulating on the floor of his room. He would borrow money from friends and forget to pay them back. He would go weeks at a time without looking at his mail; one person close to Kalanick recalled a stack of unopened letters piling up on Kalanick's bedside table. Work took precedence over everything.

Much like Facebook, Scour grew popular across college campuses with broadband networks, praised for its ability to help students download illegal files quickly. Soon, Scour was competing head-to-head with Napster for file-sharing dominance, though Scour's edge was the ability to search for files other than music.

After a series of articles in local and national newspapers, the Scour team finally captured the attention of investors, a moment that would stick in Kalanick's mind for years to come. "We were running out of money, our server costs were going up, our traffic was going through the roof," Kalanick said. They could run Scour on fumes in the beginning; server costs were free, thanks to the campus network, and they weren't taking salaries. The half dozen members of the founding team pitched close friends and family on a small round of initial investment. But it quickly became clear that Scour was going to need to take on real investors to handle the influx of customers, especially if they wanted it to scale.

Through a friend of a friend, the group was introduced to a pair of investors to help bring Scour to the next level. Their names were Ron Burkle and Michael Ovitz, two venture capitalists who would change the way Kalanick saw VCs for the rest of his life.

Burkle was a billionaire, known for his philanthropy and his private equity and venture firm, The Yucaipa Companies. Ovitz, a legend in the Los Angeles entertainment industry, was a talent agent and co-founder of Creative Artists Agency, or CAA, one of the world's highest-profile sports and entertainment agencies. He also was coming off a gig as president of The Walt Disney Company, where he had been unceremoniously pushed out by then chief executive Michael Eisner.

Burkle and Ovitz offered Scour a term sheet, a detailed charter of investment terms stipulating what percentage of the company the investors would get in exchange for their money. It also included what is called a "no-shop clause," wherein Scour couldn't solicit other investors for money while the company was negotiating the final agreement with Ovitz.

Scour agreed to the term sheet, but found themselves mired in negotiations with Ovitz over the details. Finally, with cash reserves empty and bills piling up, Scour's employees needed to either make the deal or walk away. Kalanick called Ovitz to level with him, and hope he would let the company out of their contract since it didn't look like Ovitz was going to sign an agreement.

"Look, we are running out of money," Kalanick told Ovitz. "It's clear you aren't funding this, and we need to go find money." If Ovitz wasn't going to give them funding soon, Kalanick needed to raise money somewhere else.

Three days later, Ovitz sued Scour for breaking the no-shop clause.

Kalanick was livid. One of his investors—someone who was supposed to have his back and support the company—was suing his own founders for breach of contract.

"We've got this really litigious hardcore dude out of LA suing us," Kalanick later told other entrepreneurs. "Do you think anyone else is going to give us money? No."

Ovitz's tactics worked. To keep Scour from going under, the team agreed to Ovitz's onerous terms; the VC managed to acquire more than half the company for $4 million, wresting control of Scour away from its founders. But the episode—and the lesson in how to negotiate with venture capitalists—would stick with Travis for years to come.

Then Hollywood decided to fight back. In December of 1999, the Recording Industry Association of America, or RIAA, sued Napster for $20 billion. They wanted to send a message: Any entrepreneurs thinking of building their own file-sharing companies would be sued into oblivion. Six months later, the RIAA joined the Motion Picture Association of America and about three dozen other companies in a lawsuit against Scour for $250 billion.

During his long career in entertainment, Ovitz had learned to see around corners. His friends in Hollywood began to glare at the super-agent who was now promoting a file-sharing startup. So Ovitz cleverly distanced himself from Scour using backchannel media connections. The *New York Times* quoted a person familiar with Ovitz saying the mogul was growing "increasingly uncomfortable with his association" with the startup, and further, that months earlier Ovitz had sent letters to Scour's CEO and board "expressing concern about the copyright implications."

The move was a second betrayal. Ovitz hired an investment banker to sell his controlling stake in Scour as soon as the lawsuit hit.

Each of the founders were hurt, but Kalanick took the lawsuit the worst. Scour had been Kalanick's first real attempt at building a company, and he had thrown himself into it completely. He had dropped out of college, forgone a real salary, moved back in with his parents, and abandoned the idea of a romantic relationship.

Moreover, Kalanick had found himself enjoying the startup life. As Scour had grown more popular, he had loved being associated with a cool brand, something hundreds of thousands of people used regularly. He learned how to negotiate deals, strategizing out loud with his partners each of the steps of managing an important client relationship. He loved the hustle of sealing deals, making the Hollywood connections, building and growing.

By the end of the ordeal, he was exhausted and depressed, sleeping fourteen to fifteen hours each day. He watched as Scour—a company they believed could grow into a global destination for media—was sold for parts in bankruptcy court.

Kalanick was devastated. And he swore he would never be played by a man like Ovitz again.

Chapter 3

POST-POP DEPRESSION

DESPITE THE RIAA'S OBLITERATION OF SCOUR AND THE BETRAYAL BY Ovitz, Kalanick walked away from bankruptcy court with some money in his pocket. He had thought Scour was going to be worth millions of dollars—and had he been just a few years older and lived five hundred miles to the north he might have been right.

When Travis Kalanick was still an undergrad, South of Market—SoMa for short—in San Francisco was a dot-com wonderland. In the 1990s, the airy lofts at the corner of Second and Bryant housed dozens of startups with dreams of transforming the web. Companies like Bigwords.com, Macromedia, and Substance were quartered along South Park, a cozy green area tucked between Second and Third streets. *WIRED* magazine covered the rise of the dot-com era in breathless detail from its offices just a block away.

In the days of Scour, Kalanick was just launching his career as an entrepreneur. He watched from the periphery as a culture defined by young startups, rich in venture capital, sprung up around him, bolstered by the promise of the ever-growing internet.

Private estimations of company values soared. Businesses with no revenues and enormous losses were valued at tens of millions of dollars. More than 4,700 companies went public from 1990 to the mid-2000s, many of which had no business doing so. After they hit the public markets, shares in the companies—from Pets.com (dog food delivery) to Webvan (grocery service)—initially skyrocketed. Investors trolled the markets for speculative new internet stocks while bankers cold-called fledgling internet companies to pitch them on going public, since bankers made fees on every IPO.

Some companies were indeed good bets. Amazon, eBay, Priceline, Adobe—a number of the startups formed in the '90s outlived the dotcom era. These companies were able to do something many of their contemporaries weren't: build a sustainable underlying business.

In the 1990s, Silicon Valley in particular was ripe for an economic bubble. Federal interest rates were extremely low at the time, resulting in wide investor access to cheap capital. That cash was injected into a slew of newly formed companies, which in turn used those dollars to purchase things like servers, bandwidth, and other IT products from other dot-coms, creating an artificial bubble of increasing revenues and success. In addition, financial advisors on Wall Street were pumping tech stocks. They encouraged average investors to sink their savings into internet startups, which they described as strong investments with good, long-term growth potential.

An entire ecosystem of companies that catered to dot-com companies sprung up around the Valley (along with the popularity of the shopworn San Francisco adage that it's better to sell shovels during a gold rush than to actually prospect for gold). For a starting price of $25,000, employees at Startups.com would help new companies find an office, pick their furniture, even figure out their payroll software.

In response to this frothing market, and worried about inflation, the Fed raised interest rates several times in quick succession in 1999 and 2000, closing the faucet on free-flowing capital. That, in turn, forced many startups to rely on actual revenues—not those artificially propped up by venture capital dollars—a feat that eluded many. And since so many of the companies purchased products from one another, an economic downturn hurt *all* the companies in the sector. One investor compared it to a collective Wile E. Coyote moment. Startups had run off the edge of a cliff. When they stopped to look down, they realized there was no ground beneath their feet. Hundreds of private companies closed their doors, unable to find further investment. Public companies saw their shares fall to mere pennies on the dollar.

"I remember walking into our office on Dore Street, near Eighth and Townsend, after the bubble burst," recalls Rob Leathern, a former financial analyst at Jupiter Research. "I'd see the empty offices of failed startups all throughout the halls of our building, with weeks' worth of copies of the *Wall Street Journal* piling up in front of their doors, and the same FedEx missed-delivery stickers stuck to their windows for months."

Leathern isn't exaggerating. Billboards lining the stretch of Highway 101 down to Palo Alto were advertising internet companies that no longer existed. A website dedicated to chronicling the startup death march appeared: *Fucked Company*. One fifth of all office space in the SoMa area was vacant in the summer of 2001, an enormous increase from the record lows of .06 percent just eighteen months before. Rents dropped by an average of $300 a month across San Francisco, while Craigslist was flooded with listings for hundreds of computer towers, monitors, servers, and other caches of hardware, some of which had been used for as little as a few weeks.

As companies liquidated themselves, employees moved out of the area—some even left the state—to find other jobs. Some gave up on the industry entirely. Ryan Freitas, who would later become a product design executive at Uber, began working as a line cook (albeit a high-end one) after being laid off from the digital and IT consulting firm Sapient in 2001.

"Anyone trying to start a company in San Francisco back then had to be fucking crazy," Leathern said.

TRAVIS KALANICK APPARENTLY was fucking crazy.

Almost immediately after Scour closed its doors, Kalanick started brainstorming with Michael Todd, one of his Scour co-founders. In relatively short order, the two of them dreamed up what Kalanick called his "revenge business," a way to get back at the RIAA and MPAA, and the other companies who sued the partners and torpedoed Scour. That company was called Red Swoosh.

"We basically took our expertise in peer-to-peer technology, took those thirty-three litigants, and turned them into customers," Kalanick said. The new idea was similar to Scour: Red Swoosh would use connected "peer" computers in a network to transfer files between systems in a more efficient way. This time, however, those files weren't going to be illegal downloads; the media companies were going to supply the files themselves. Kalanick would convince the RIAA and MPAA and others to hire Red Swoosh to deliver multimedia files—videos, music, whatever—to paying customers via set-top boxes on their TVs, or to their home computers.

That kind of efficiency fascinated him—whether transferring computer bits or moving physical atoms in a vehicle. It all came down to one proposition: What is the fastest, simplest way to transfer something from one place to the other?

Launching Red Swoosh required Kalanick to make his first true sojourn to the startup Valhalla: Silicon Valley. Unfortunately for him, he arrived just as the party ended. By the time he was taking meetings in Palo Alto and pitching Red Swoosh to investors in the fall of 2001, the streets were empty.

"Tumbleweeds blowing through," Kalanick said.

He persisted despite aggravating early experiences with potential Red Swoosh investors. VCs who had lost their shirts just months before when the bubble burst mostly laughed him out of the room. Often, he couldn't even get a meeting.

Other investors looked at Red Swoosh and only saw the ghost of Akamai Technologies. A networking software firm, Akamai was the company most similar to Kalanick's startup. Before the bust, Akamai had a $50 billion market capitalization. After the bubble burst, Akamai's shares plunged, and the market cap sank to $160 million. Investing in Kalanick's fledgling startup, if it even had potential, wouldn't yield the outsized returns venture capitalists require.

"It was January of 2001, and I was trying to start a networking software company," Kalanick later said, realizing the futility of the endeavor. "Are you frickin' kidding me?"

They forged ahead regardless. Kalanick set up shop just south of the city, in San Mateo, a few freeway exits down from San Francisco and roughly thirty minutes north of Silicon Valley.

From the start, the team didn't love Kalanick's leadership. His six engineers went months without pay, and he begged them to stay. At one point, when Red Swoosh was running out of money, an employee dipped into the company's payroll tax withholdings—money a company reserves to pay the IRS the taxes it owes—to fund operations. That employee left the company, and Kalanick was stuck with the blame. He was later informed by an advisor that the company might be committing tax fraud. This would stick with Kalanick for years; he felt betrayed, put in legal jeopardy by a colleague.[*] It would form the basis for his difficulty trusting people close to him for years to come.

Red Swoosh barely scraped by, but somehow he kept the place running. Cashflow was a month-to-month adventure. He scored a $150,000 deal from a cable and telecommunications company just two weeks before he was set to go out of business. It was painful and desperate, but Kalanick eventually came to appreciate the experience. It taught him how to negotiate from a position of weakness.

One VC firm made Red Swoosh a promise of a $10 million investment, but it never materialized, eventually falling apart after the venture capitalists couldn't agree on other investors to join the funding round. Once again, Kalanick felt he got screwed by VCs who didn't care about him or his company. It left a bad taste in his mouth. Later, describing the incident and his hatred of venture capitalists, Kalanick would channel the West Coast rap icons of his youth, Snoop Dogg and Dr. Dre: "VCs ain't shit but hos and tricks," he said.

That cycle repeated for the next few years at Red Swoosh. Kalanick would run out of money, then secure a last-minute deal with a larger tech company and keep his business alive for another few months. He'd then find a way to parlay that deal into yet another venture investment, bailing out the company for a year or so longer. "In a weird way it sort of kept me going, because there was always this shiny ball that was just right there," he said. "I could almost taste it, but it kept never happening."

---

[*] Kalanick saw to it that the tax withholdings eventually made their way to the IRS.

One of his most painful episodes happened in Davos, Switzerland, home to the annual elite conference for the world's wealthiest and most powerful people: the World Economic Forum. Kalanick, who had managed to get an invitation to the event, was in the middle of negotiating a $1-million annual revenue deal for Red Swoosh with AOL, a potentially lucrative partner. Before he could close the deal, Kalanick received an email from his last remaining engineer—the one who hadn't quit despite being paid irregularly for months. The engineer said Michael Todd, one of Kalanick's former colleagues from Scour, was recruiting him away to go work for Google.

Losing his last engineer was bad. It went from bad to worse when the news hit the front page of *Fucked Company*, spreading Red Swoosh's embarrassing troubles across Silicon Valley. That, in turn, led to a breakdown in talks between Red Swoosh and AOL.

Finally, in 2005, Kalanick caught a break. Kalanick got into a flame war on a message board with Marc Cuban, the celebrity billionaire investor and owner of the Dallas Mavericks. Kalanick evangelized peer-to-peer tech, while Cuban thought Kalanick was dead wrong. Though Cuban didn't like the tech, he *did* like Kalanick's hustle, the tenacity he saw in Kalanick during the pitch. Cuban sent Kalanick a private message, offering him money to invest $1.8 million in the company. That was a crucial lifeline which eventually led to more contracts with important partners. Another investment from August Capital, a respected Valley firm, pumped even more life into the company.

There was a silver lining to his disappointing trip to Davos: He met the CEO of Akamai Technologies, his largest competitor, and began to make inroads with the company. Finally, after six years of tireless hustling, Kalanick negotiated his best deal yet: he sold Red Swoosh to Akamai for nearly $20 million. After taxes, Travis personally netted roughly $2 million.

After a grueling trudge towards the exit, Kalanick was finally able to take a breath. No longer would he have to work around the clock for peanuts, looking for the next deal while living in his parents' house and eating ramen and other treats from the bargain bin at Safeway.

Four months after the deal closed, he bought a condo in San Francisco's Castro district, set atop one of the tallest hills in the city with a view of the Bay Area. He was able to take some time to relax and enjoy the luxuries that lured global elites to San Francisco. He and his girl-friend, Angie You, could hang out with friends from the startup scene while he let his Akamai shares vest. He could party, chill, and most importantly, figure out his next move.

Aside from the money he made over ten years of dogfights in startup land, Kalanick had gained a great deal of practical experience and emerged with a new understanding of leadership. He now held a siege mentality, one that perceived dangerous enemies all around, and developed a quasi-Darwinian vision of what it takes to survive.

"There are forces all around you when you run a company, . . . ready to take you out," Kalanick said. "The [CEOs] that survive are the ones that are supposed to be there."

But most of all he took a valuable lesson to heart: Never trust a venture capitalist.

"They're all so founder friendly! They exalt founders, put them on pedestals and say 'we're just the measly VCs!'" Kalanick later said to a group of entrepreneurs of his early startup experiences. "It is in the VC's nature to kill a founding CEO. It just is."

Chapter 4

A NEW ECONOMY

TRAVIS KALANICK SOLD RED SWOOSH JUST AS A NATIONAL CRISIS was beginning to unfold.

It was April 2007. For years, American banks had been doling out loans to first-time, "subprime" home buyers, whose financial histories had historically made it impossible for them to secure home loans. But changes in national fiscal policy in the late 1990s led banks to welcome subprime buyers in record high numbers, signing them to seemingly affordable adjustable-rate mortgages, and then packaging these mortgages into derivative products and selling them to other investors.

This practice set the timer on an economic IED. Subprime borrowers who signed up for adjustable rate mortgages soon faced sky-high monthly payments. Wave after wave of home owners defaulted, failures that rippled throughout the economy. It would take years for the country to recover from the catastrophe—and some people never did.

As the great financial crisis came to a boil, the federal government spun up a suite of financial instruments to soften the blow. On September 7, 2008, the Bush administration seized control of Fannie Mae and Freddie Mac, the United States' two largest mortgage financing bodies. Henry Paulson, then secretary of the treasury, pledged billions in bailout money to some of the world's largest financial institutions, including AIG, J.P. Morgan, Wells Fargo, and dozens of others. From September 2007 and onward through the financial crisis, the Federal Reserve Bank cut interest rates from a little over 5 percent to its lowest ever rate, 0.25 percent, by 2009. And that rock-bottom rate is where it would stay for the next seven years.

Through these maneuvers the Treasury Department and the Fed arguably kept the global economy from spiraling further out of control. But during the panic, leaders focused mostly on Wall Street, and not Big Tech. Slashing interest rates to save the banks would have profound effects on technologists and entrepreneurs—particularly on a fifty-mile stretch of Route 101 in Northern California.

IN A WAY, the carnage of the dot-com bust had done the Valley more good than ill.

The bust separated the dot-com poseurs from the actual valuable companies. Led by Larry Page, Sergey Brin, and Mark Zuckerberg, a new generation of entrepreneurs seemed to understand intuitively how to harness the true power of the internet, and turn it into a profitable business.

There were three important ingredients that fueled the new generation of entrepreneurs like Zuckerberg and Page. First, By 2008, more than 75 percent of American households owned computers, and unlike the 1990s and early 2000s, this mass population had access to broadband; more than half of American adults in 2008 purchased a high-speed internet connection for the home. As more and more people connected online, demand for new, internet-enabled services grew by the day.

Second, the hurdles for entrepreneurs who wanted to launch a company were lowering quickly. Amazon Web Services, or AWS, changed the startup game entirely. Amazon started AWS in 2002 as an engineering side project; it would grow to become one of its most successful innovations in Amazon history.

Amazon Web Services powers cloud computing services for coders and entrepreneurs who can't afford to build their own infrastructure or server farms on their own. If a startup is a house, AWS is the electric company, the foundation and the plumbing combined. It keeps the business up and running while the company founders can spend their time focusing on more important things like, say, getting people to come to their house in the first place.

Crucially, AWS was relatively inexpensive. For the first time in computing history, any single programmer with a startup idea and a bit of cash could quickly build a company without having to plow tons of money into infrastructure—they could farm that part out to Amazon, and focus on building the app itself.

But the third and most important ingredient was released just two months after Travis Kalanick sold his startup. It would change the face of computing—and how the world would come to interact with devices—more than anyone could have ever anticipated.

AT THE END OF 2006, two men walked a sunny sidewalk in Palo Alto and talked about the future.

In his signature black turtleneck and faded blue Levi's, Steve Jobs couldn't go anywhere in Silicon Valley without being swarmed by fans. His accomplishments were well known by then; after giving the world the Macintosh, he helped found Pixar, the beloved animation studio. Later he would develop the iPod and iTunes store, a combination that revolutionized the way the world listened to music through digital media. Jobs's legacy was already cemented thrice over.

Biographers were already beginning to sketch that legacy in their heads. Jobs had been diagnosed with a rare form of pancreatic cancer, quickly growing gaunt as the sickness attacked his system.

Beside him was John Doerr, the Intel engineer turned venture capitalist. Doerr, too, was a titan of industry. Doerr was an unassuming man, slight of frame, with wire-rimmed glasses resting atop his pointed nose. He looked like he would be more at home in a laboratory fabricating silicon chips—something he once did back at Intel in the '70s— than zooming around the Valley hosting dinners for Barack Obama.

As a partner at Kleiner Perkins Caufield & Byers, the storied Menlo Park venture firm, Doerr made an early investment in Netscape, a company that eventually became the world's first consumer internet browser. Doerr was early to spot the potential of Amazon, back when Jeff Bezos's operation was selling books in a run-down warehouse in Seattle. And perhaps most famously, in 1999 Doerr invested $12 million in Google, then just a search engine run by a couple of engineers in a garage. Five years later, when Google sold its shares on the public stock markets, that investment was worth more than $3 billion, a return of more than 240 times Doerr's original investment.

But that morning, they were just two friends walking down the sidewalk in Northern California, on the way to their kids' soccer game.

As they chatted about life, family, and the industry, Jobs stopped for a moment and reached into his pocket, pulling out something Doerr had never seen before. It was the first iPhone.

"John, this thing nearly killed our company," Jobs said to Doerr, who stared at the boxy, glass-faced device with wonder. Jobs *never* showed him new products ahead of time, but Doerr—as well as the rest of the technology world—had heard rumors of the iPhone's development. Apple was said to have been working on it for years, a skunkworks project of the highest secrecy. Doerr stayed quiet, not wanting his friend to clam up and put the phone back in his pocket.

"There's so much new technology in it, fitting it all inside was a feat," Jobs went on, beginning to walk again under the valley oak trees that lined the Palo Alto street. "Behind this LCD display we've fit a 412-megahertz processor, a bunch of radios and sensors and enough memory to hold all your songs. We've really done it."

Jobs handed the phone to Doerr, noting that it didn't have all those "fucking ugly buttons" that characterized BlackBerry (the predominant cell phone of the day, used by most professionals). It was touchscreen based—sleek, glossy, gorgeous.

Doerr held the phone as gingerly as he would had he been given a newborn baby. It felt better than the phone he had in his pocket by far. Still staggered that his friend was showing him a new device, he flipped the iPhone over in his hand to look at the back panel. In small, white lettering beneath the iconic Apple logo, Doerr saw a bit of information that intrigued him: It said "8GB," an amount of storage that at the time seemed like more room for files and music than anyone would need.

"What do you need all that storage for?" Doerr asked, watching his friend crack a smile as he took the phone back.

But Doerr already knew. Just as Jobs had trained millions of people to go to the iTunes store and download their music to their computers and iPods, Jobs was going to do the same with music and new program applications—or "apps"—for the iPhone. He knew he was opening up a new way of computing, built for mobility, and would need his pocket computers to do just as many things as his desktop Macintosh computers were able to do. He would eventually call it the App Store.

Doerr knew opportunity when it was in front of him. He tried to seize it.

"Steve, I see what you're doing. I see it. I want to be a part of this," he said. "I want to put together a fund to kickstart this thing."

Doerr was falling back on his VC instincts. Every few years, investors like him would go to their institutional partners to pool millions of dollars in a new fund. Venture capitalists like Doerr would then use that money to purchase stakes in promising startups around the Valley. Like Bill Gates and his era of Windows-based applications, Doerr saw that an iPhone App Store would open a huge new field to programmers—whose startups he could fund.

Jobs chopped the air with his hand. "No. Stop it right there. I don't want a wave of shitty apps from outsiders polluting this phone. Not going to happen."

Doerr dropped the subject and the two men walked on to the soccer game. He knew his friend's mind would be impossible to change once he had made it up, and Apple's approach to software development had rigorously avoided Gates's "come one, come all" approach with Windows third-party apps. But he sensed that Jobs was wrong, that people would be desperate to build things that operated on Jobs's gorgeous new device, and ultimately Apple would let them do it.

PLUCK AN ENTREPRENEUR at random from the streets of Silicon Valley, and you'll likely find an evangelist for Jobs and his vision for the iPhone, of the one product that could be "an iPod, a phone, an internet mobile communicator."

The iPhone radically reimagined what a smartphone was supposed to be. A sleek, glass-faced front with a dazzling array of colorful apps—a rainbow of greens, blues, and yellows. The iPhone took the luxuries of an enterprise-level business device, like email and internet access, and opened mobile computing up to the masses. You didn't need to carry an MP3 player, mobile phone, and bulky laptop around to browse the internet on your commute. You didn't need a separate camera to take photos during an afternoon walk in the park. The iPhone had it all.

Inventing the hardware was laudable enough, and the company would spend the next decade refining it. But the device truly took off when Jobs decided to allow a wave of "shitty apps" into his sandbox. And far from "polluting" the iPhone, they fueled its rise higher and faster than even Jobs could have anticipated.

LATER IN THE SPRING of the next year, a few months after their walk to their kids' soccer game, John Doerr was at his Palo Alto home when he got a phone call from a friend. It was Jobs.

"Remember that thing you pitched me on last year, the fund?" Jobs said.

Doerr immediately knew what his friend was talking about and sat up in his chair. "Yes, yes I do. Did you give it more thought?"

"I did," Jobs said. "I think Kleiner should do it."

The call shocked the investor. Doerr knew how controlling Apple was under Jobs's reign. Everything had to be perfect, from the industrial design led by Jonathan Ive—a dapper British lieutenant and long-time confidant of Jobs—to the software and apps under the direction of Scott Forstall, a fiery and talented executive leading Apple's mobile operating system. Asking Doerr to kickstart a sea of new smartphone apps with a multimillion-dollar fund would create a wave of innovation much messier than Apple was used to dealing with.

But Doerr wasn't going to question an opportunity. He offered to raise $100 million from his limited partners, an unheard-of amount of money—especially one earmarked for funding a new form of program that was unproven and untested. But Doerr believed in Jobs and saw the potential the iPhone could have in the market if the product took off.

To say they were right would be a wild understatement.

UNTIL 2006, computer programmers made their living inside big corporations or software development outfits. To have your software touch millions of people usually required the distribution of major software publishers, ones with sizeable marketing budgets and deals with off-the-shelf, big-box retailers. Places like Best Buy, FuncoLand, and Babbage's had aisles stocked like a grocery store, stuffed with rows of boxes of PC and Mac programs.

The App Store changed the model for software development entirely. All a programmer needed was an idea and facility with Apple's mobile software code. With those two components, anyone could build and distribute their own apps and market them to millions of people instantly. Spin up a server on Amazon Web Services, blast out some code, and submit your app to Apple for review, and your work could be up and running in days.

For people opening the App Store at home, it was like walking the aisles of their local Best Buy. Unfettered access to millions of games and programs on their iPhone required little more than a Wi-Fi connection and a few extra bucks.

Coders across the world looked at the App Store with giant, flashing dollar signs in their eyes. They heard stories from coders like Steve Demeter, an obscure indie developer who with a few friends wrote an app called Trism—a Tetris-like game for which he asked five dollars per download—in a matter of weeks. Two months after its release, Demeter had raked in more than a quarter of a million dollars. Top developers in the early weeks of the App Store were seeing anywhere from five thousand to ten thousand dollars in income from app downloads every single day.

Other Silicon Valley investors followed suit. Venture funds looked at Kleiner Perkins' Doerr and the enormous amount of money being poured into apps and started doing the same thing, scouting around the Bay Area for the best and brightest in app development.

Nearly overnight, the App Store became the Wild West. As had been the case when Jobs and Steve "Woz" Wozniak imagined the first Apple computer in their garage, the next great revolution in computing could come from anywhere, not just big publishers like Microsoft, Adobe—

or even Apple. Hundreds of millions of dollars began to flow outward across San Francisco from the dozens of VC firms that lined the well-known stretch of Palo Alto's Sand Hill Road.

Armchair computer enthusiasts began to look at California with riches in their eyes. Venture funds began throwing money at twenty-somethings, hoping to stumble into funding the next killer app. Doerr called it "the appification of the economy," an era beyond the web and desktop that focused on mobility and independent creation, afforded by all the possibilities the iPhone had to offer.

Those in venture capital, like Doerr, knew how it really worked. There would certainly be meteoric rises of apps built by anonymous coding wunderkinds—the App Store had so many customers, and so much interest in new software, the odds made it a certainty. But the *real* winning apps were backed by top-tier venture capitalists, who made connections to potential partnerships with large companies, built pipe-lines to faster recruiting, offered strategic advice and, of course, tur-bocharged growth and marketing with millions of dollars in funding.

The top-tier firms of the Valley like Sequoia Capital, Kleiner Per-kins, Andreessen Horowitz, Benchmark, and Accel all began hunting for new talent. They wanted young, hungry entrepreneurs whose ideas turned into obsessions. They wanted founders who were willing to push themselves—and the rules—to the limit. They wanted to find the founders who spotted opportunities for innovation in the minor annoyances of daily life.

They particularly liked the idea that occurred to one young entre-preneur, who had already found wealth and fame, but was annoyed that he couldn't seem to catch a cab in downtown San Francisco. That entrepreneur was Garrett Camp.

UPWARDLY IMMOBILE

GARRETT CAMP WAS PISSED OFF.

It was 2008—the twenty-first century—in one of the richest, most forward-thinking cities in the entire world, and he couldn't catch a taxi in under a half hour.

At only seven-by-seven square miles, San Francisco was small enough that one could survive without owning a car, but still large enough for a person to be annoyed they didn't have one.

He could always bike across the city, though a six-speed didn't work so well climbing up steep hills like those on Divisadero Street. And a bike wasn't going to help him get home from a bar at two o'clock in the morning—at least, not without a DUI or a head injury.

There was always BART—Bay Area Rapid Transit—San Francisco's wheezing commuter rail system. But BART was gross, a patchwork of dirty cloth seats and crowded cars, nowhere near large enough for the influx of twentysomethings who had invaded the Bay Area in recent years. And BART didn't run past midnight. Not ideal for a young man pursuing the nightlife.

At first it was an annoyance. Camp, a Canadian by birth and an entrepreneur by heart, had moved to San Francisco after attending business school, with hopes of growing his startup—a Web 2.0 phenom called StumbleUpon. He had high expectations of the City by the Bay, a promised land where young startup founders could strike it big—maybe even invent the next rocket-ship company.

Camp was smart, but he was no Steve Jobs. An introvert by nature, Camp enjoyed tinkering with ideas for startups or solving problems in his head as he walked the sloped streets of San Francisco. Even at thirty years old, Camp still looked like a college student with his close-cropped cut of dirty blond hair and button-down oxford shirts. He was cerebral, a little geeky, able to explain the intricate architecture of the internet, but lacking the polish and showmanship of, say, an Elon Musk. His wide, toothy grin made him look more goofy than dashing—something like "the entrepreneur next door."

Camp was fun to hang out with, though. He enjoyed traveling, loved experimenting with fine dining in the Bay Area. He was always game for a hot tub hang, enjoyed theme parties that obliged one to rent a tux. As he grew further from his Canadian roots and became a Californian, Camp grew his hair past his shoulders, affecting a kind of neo-hippie vibe. He looked as if he'd be just as comfortable hanging out with a surfboard in Long Beach as he would hunched in front of a MacBook Pro at the Creamery. Camp later became an annual regular at Burning Man, the weeks long off-grid bacchanal in the Nevada desert attended by thousands of techies and hippies from all across the West Coast.

StumbleUpon was his claim to fame, a kind of early social network conceived back when he was in college in Calgary, long before the rise of Facebook. The site was perfect for the days of the desktop web; StumbleUpon flicked users between different websites at random, promising to offer surprising and delightful suggestions for users to "stumble upon" and enjoy. It was like a proto-Reddit, a link-aggregation site that at its best delighted users with new, interesting facts, obsessions, and subcultures.

In the early 2000s when Camp created the company, it had been a good idea. But by '07 the site was looking dated, especially with the rise of mobile devices. Suddenly the smart money was on mobile apps. Desktop-centric startup tools like StumbleUpon were growing increasingly irrelevant.

Friends knew him as an obstinate colleague—quick-tempered when challenged directly—and often unwilling to change his mind when convinced he was right. A sense of pigheadedness can often be a virtue for startup founders and CEOs, but only when the idea works. If the idea doesn't work, then a stubborn and pushy CEO ceases to look "exacting" and "visionary" and becomes "difficult."

Nevertheless, StumbleUpon paid off. Camp was able to parlay the buzz around his site into a sale to eBay, the online auction giant, for $75 million—an admirable sum, especially for a small company that had only raised $1.5 million in venture capital. Camp was smart enough to retain a large ownership percentage and the sale made him a rich man. As soon as he signed the deal he had cash in his pocket and startup cred to his name—the young entrepreneur's Silicon Valley dream, realized.

And yet. Camp had all the money in the world and still couldn't get around town. The taxi system was antiquated, the fleet a patchwork of yellow relics, often coming apart at the seams. Taxi base station owners didn't invest in the cars' upkeep. The dispatching system was ancient. Base station dispatchers fielded calls from clients and radioed those requests to taxi drivers circling the streets. But customers had no idea if the cab would actually show up.

The taxi system's unreliability compelled Camp to create hacks and workarounds. One trick he devised was dialing up all the major taxi services in the city, one after the other, to ask for a pickup. He'd take the first cab that arrived and ignore the rest. It was a dick move, but he felt justified; after all, they flaked on *him* most of the time.

The companies caught wise to Camp's tricks. He ghosted them so frequently that they stopped sending cabs to pick him up altogether. "I've been blacklisted," Camp thought. "This is messed up."

The problem plagued him. He tried expensive black car services, but didn't like coordinating all the drop-off spots if he was with more than a couple friends. He would schedule his preferred drivers to pick him up at restaurants later on in the evening, but that was imperfect too, since it sometimes meant he had to rush through a meal when he was supposed to be enjoying it.

Camp had splurged on a new Mercedes-Benz after selling his company, but he didn't want to rely on his car. Parking was always a nightmare; if you found a spot, you'd be lucky if it wasn't on a 35-degree inclined hill.

Camp remained vexed. Getting around San Francisco was a problem, and no one seemed very invested in trying to fix it.

THE SEEDS OF IT first came to him during a Bond flick.

Camp was relaxing at his new luxury apartment in South Park—just yards away from where the idea for Twitter was first conceived, and where Instagram's early offices were located—when he decided to watch a movie. *Casino Royale*, the 2006 reboot starring Daniel Craig, was a favorite, something he watched when he didn't have anything else in mind. There was something about the understated cool of Craig's Bond; perhaps, on some level, Camp liked the idea of the world's greatest spy being a short-tempered, crew-cut blond not dissimilar to himself.

Then he saw it. There was a moment when Bond was driving a Ford through the sunny streets of Nassau, approaching a beachfront resort on the sparkling blue Bahamian seaside.

What caught his eye was a small flourish on Bond's cell phone as he drove through the beachscape. The phone, a boxy, silver Ericsson antiquated by later standards (it still had a numerical push-button keypad!) displayed a GPS-based map on its tiny screen. Bond was watching himself—a small arrow icon gliding across a dark green bitmapped grid—as his car moved across the Nassau landscape, inching toward The Ocean Club.

It was a throwaway scene, something most people would have absorbed passively in the theater. It primarily served to highlight some of Bond's cool gadgetry while plugging an Ericsson product, something the producers got paid to do.

But the image stuck in Garrett's head. The iPhone, in all its glory, had just been released a few months ago and was probably one of the most powerful pieces of handheld technology he had ever seen (far more impressive than Bond's Ericsson). That meant it came with Wi-Fi connectivity, an accelerometer, and future iterations would have GPS capabilities—three key components in determining a user's location on a map.

What if he didn't have to spend his nights dialing for cabbies? What if there was an app for that?

And most importantly: What if, like James Bond, he could look like a total badass using it?

TRAVIS, TOO, was trying to relax after six years of hustling.

The $20-million exit didn't exactly make Kalanick the next Mark Zuckerberg—or even the next Garrett Camp. Both Kalanick and Camp found riches in the arms of an acquiring company, and each sold about a month apart from the other in 2007. But Camp had definitely done better; in a place like Silicon Valley, a hot consumer app like Camp's would *always* fetch a higher price (and a flashier headline) than a peer-to-peer file-sharing infrastructure company.

Still, Kalanick's landing was respectable enough, earning him enough money to stop working and spend time cruising around San Francisco, judging startup events and hopping to parties thrown by early-stage investment funds. For the first time in his life, he was a free agent. He had millions of dollars in his pocket and wanted to act accordingly.

In one of his favorite movies, *Pulp Fiction*, Kalanick was captivated by one character, played by Harvey Keitel. Wearing a thin mustache and a pressed black tuxedo at eight o'clock in the morning, Keitel speeds across the entire city of Los Angeles in a silver Acura NSX in nine minutes and thirty-seven seconds, an impossible feat, to fix the problem of hiding a dead body produced by Travolta and Jackson, whose car is covered in a mess of gore. The character's name was Winston "The Wolf" and his job was to swoop in and fix problems that needed solving.

Kalanick wanted to be a fixer like The Wolf. After buying his hilltop house in the Castro, Kalanick started investing small amounts of money in various startups with the understanding that he'd be available as their own personal fixer, willing to swoop in and solve problems whenever a founder needed his help.

Got a problem with an agitated investor? The Wolf can handle it. Don't know the first thing about hiring new engineers? Just call The

Wolf. Maybe you have late-night thoughts on your company's next move and want to talk it out. Never fear, The Wolf is here.

Kalanick started promoting his investment-portfolio companies on his personal blog, which he called *Swooshing*, an homage to his now-acquired startup. Swooshing featured a blown-out photo of Kalanick as a kind of startup cowboy, complete with pearl snap shirt and ten-gallon hat, atop which he rested his black sunglasses. Self-promotion was a common enough practice among so-called "angel investors," a name for small-time venture capitalists whose five-figure investments and advice to founders earned the angel a slug of early shares in a company that could one day hit it big. For Kalanick, blogging[*] was a way of marketing himself, along with the occasional talk delivered at startup mixers and cocktail parties.

"My people think of me as a funding shepherd," Kalanick once said to a roomful of young engineers at "Startup Mixology," a regular, boozy event for techies in their twenties. Onstage, Kalanick clicked a remote control as a slide behind him flicked into view: Behind him was Jesus Christ, robed, hooded, and holding a shepherd's cane. "I'm really frickin' curious," he noted, hitting the clicker again to showcase a fluffy cat, biting and batting around a toy. "Just think of me as The Wolf."

It was a cheesy, tongue-in-cheek pitch, but Kalanick's swagger and self-assuredness piqued the curiosity of at least a few founders. Eventually, he was able to park personal investments in startups like Expensify—a company that handled workplace expense reports—as well as others like Livefyre (social media management), CrowdFlower (data collection management) and Formspring (social networking), along with about a half dozen others. Kalanick would eventually consider joining Formspring, which looked promising in an age where social media companies were taking off—both with the venture community and the general public.

Kalanick started buying button-downs, less schlubby blue jeans, fun

---

[*] In this modern era, a venture capitalist couldn't just fade into the background, as was historically the case. Now, VCs had to work overtime to become desirable in the eyes of young founders, and took every chance to market themselves.

sneakers, colorful striped socks. He made startup investments like he was buying oil paintings, adorning his online profiles like he would a gallery wall in his apartment. To friends, he called his portfolio his "art collection."

But being a "funding shepherd" wasn't everything. Kalanick still felt he had more to offer. He had paid his dues building and selling Red Swoosh—and then some. He had heard the word "no" a hundred times a day for four years straight, a regimen that would harden any young entrepreneur. Inside him stirred a combatant. He fashioned himself into something like Bruce Banner, the comic book hero who always harbored The Incredible Hulk within him.

At the same time, Kalanick didn't feel fully at ease being a full-time investor. He was angry over the injustices he saw in the venture capital and startup world. "I'm a part of this company where the revenues are shooting through the frickin' roof, we've got an insanely talented senior management team, yet VCs are trying to axe the founder," he told a group of young entrepreneurs. Here was a successful founder defending the founder of his portfolio company. "Why are we getting rid of him? I don't understand. Can you please tell me that?"

VCs, in Kalanick's mind, weren't in the game for the right reasons. They weren't there to change the world like he was, or even to alter it slightly. Venture capitalists cared about one thing: the bottom line.

Over those months, Kalanick perfected his swagger. He delivered pitch after PowerPoint-backed pitch at dozens of startup events. But what he really needed was a place to showcase his talents, somewhere young entrepreneurs could come and riff on new ideas with him. He wanted to create a safe space for young minds eager to change the world through the transformative power of technology. Soon enough, that idea became a reality. The "JamPad"—Kalanick's nickname for the sparsely decorated, million-dollar apartment where he lived at the top of the Castro—opened for business.

Kalanick treated the JamPad like his own personal salon, an informal symposium where technologists could relax, sink into the couch, and talk about the future over beers and a platter of grilled T-bone steaks. (Kalanick hoped for people to call him by the nickname "T-

Bone," securing the Twitter account "@KonaTbone" for his "musings and often controversial aphorisms." His avatar: a bloodied cut of beef.)

But the apartment was hardly flashy. Kalanick barely had any furniture or art decorating the walls, no Ferrari in the garage, no Eames Lounge Chair in the living room. It was poorly lit, making the place look more like a dank cave than a "startup salon." Friends remarked on how drab it was for someone of his status. They expected him to splurge on *some* interesting centerpiece, given how artfully he referred to his collection of startups. It never crossed Kalanick's mind to do so—decor just wasn't something he thought about.

The most memorable parts were the tennis tournaments. Kalanick was a driven tennis player—on the Nintendo Wii. He soon bested all of his friends and most of the global players who scored themselves online. Whipping the white plastic handheld Wii controller back and forth while bouncing around his spartan living room—mostly empty of furniture—he looked like a tech-world McEnroe or Agassi, power serving against hapless competitors.

The JamPad served two primary purposes: a place for Travis Kalanick to crash, and a place for Travis Kalanick and his techie friends to riff on ideas. "Jamming," in Kalanick parlance, was like playing in a jazz quartet or a psychedelic rock band. Kalanick's enthusiasm and support for risk-takers bred around him a small following of devoted friends. It all started, he would say, with a jam sesh.

"It's ad hoc, but eventually it sort of comes together into beautiful music," Kalanick said.

GARRETT CAMP STILL couldn't get the idea out of his head.

The cabs in his city were shit. Worse, since he'd been blackballed from most of the services, he had begun resorting to black car services, and had collected a laundry list of the best private drivers in San Francisco, repeatedly pinging them whenever he needed a ride for a night out.

But even that was imperfect. The money, the complexity of arranging pickups, the confusion of sharing rides with friends. It was too messy. He needed the best kind of cab—one he or any of his friends could hail directly from their iPhones. He needed an ÜberCab.

That became the working title—along with a few other options like "BestCab"—for the imaginary app he had designed in his head. Eventually, he'd drop the umlaut; it was too confusing for American audiences. But Camp wouldn't let the idea go, and brought it up with nearly all of his friends, including one budding entrepreneur and angel investor, fresh off a recent company sale: Travis Kalanick.

Camp joined Kalanick at the JamPad along with a host of other young entrepreneurs, most of whom came from companies Kalanick cared about or had a financial stake in. David Barrett and Lukas Biewald, two of Kalanick's portfolio company CEOs, made regular appearances. Kalanick funded another JamPad friend, Melody McCloskey, who would later go on to found the startup StyleSeat.

And then there was Camp, who wouldn't stop talking about his idea for UberCab. He chattered incessantly to Kalanick about its possibilities. "Did you know taxi medallions can cost, like, a half a million dollars a year?" he'd ask friends. "Have you ever looked into how base stations operate?" he'd go on. And the tech was inefficient, he'd pointed out; yellow, busted old Crown Victorias that barely made sixteen miles to the gallon were reliant on little more than two-way radios and their watchful eyes in order to find fares. There had to be a better way to give people rides.

Camp couldn't leave out the best part. They'd market UberCab to professionals in dense cities—people like themselves—and try to make it feel exclusive, almost like a club. You've got to be a member to use it, guaranteeing a "respectable clientele," and they'd only accept top-of-the-line luxury vehicles. The kind you'd want to be seen in riding around town: Mercedes, BMW, Lincoln. Best-case scenario, Camp believed, is that he created a market leader in private transportation, with the possibility of hundreds of millions in annual revenue. At worst, he'd create a small black car service for executives in San Francisco; basically, an upscale transportation service for himself and all of his friends. Even if he lost, he would win.

He wasn't subtle about pushing the idea. "Uber" replaced "great" in his vocabulary, a way to call things great. Things were "uber" this and "uber" that. That car? Uber cool. Tasty pizza for dinner? Uber slice. He wanted Uber to become something more than a German preposition, a noun synonymous with cool.

Both Camp and Kalanick loved the idea. The problem was, neither Camp nor Kalanick wanted to run it. The thirty-two-year-old Kalanick was still trying to embrace his role as The Wolf, a "funding shepherd," after years of nonstop work at Red Swoosh. And when Kalanick wasn't doling out advice to young CEOs, he was hopping planes to Europe, South America, Southeast Asia, fulfilling a wanderlust he wasn't able to satisfy while hunkered down in the startup cave. Camp also wanted to own the cars and garages used by drivers, something Kalanick was completely uninterested in overseeing. It was a small detail, but one that turned Kalanick off.

But Camp wouldn't drop the idea. And eventually, after letting go of owning the cars and garages, he wore Kalanick down. The two lived together during a trip to Paris for a tech conference. After a series of drunken nights spent doing math over candlelit dinner tables, arguing over how much they could make per car, or whether they should own the vehicles, Kalanick and Camp returned home inspired. It would take Kalanick a few more months to sign on full-time, but Camp had finally convinced him.

Uber needed a fighter to lead it, one who could take on the cutthroat world of venture-funded competitors while battling the entrenched taxi cartels. They both knew that Kalanick was the right man for the job.

PART II

Chapter 6

"LET BUILDERS BUILD"

BEING THE RIGHT MAN FOR THE JOB DOESN'T MAKE THE JOB ANY easier.

Building a startup is very, very hard. To create actual software, a founder must first convince engineers to take a pay cut in exchange for company stock, then do the same with marketers, salespeople, and the rest of a lean staff. A founder needs to figure out payroll, finances, and taxes—and perhaps rent an office, if the founder doesn't own a garage.

A founder needs to be able to wear any number of hats, from human resources one day to conference speaker and PR manager the next. As optimist, cheerleader, therapist, and problem solver, the founder must balance the needs of a growing company with those of each individual on staff, without neglecting their own spouse or children. And when the bank accounts start getting low, a founder needs to get back down to Silicon Valley and start hustling for more dollars. With money in the bank, a founder must then juggle the demands of the backers, who expect nonstop growth.

Even if a founder masters all of these things, there's no guarantee the company is going to work. The timing may be wrong. The company might run out of money before the idea can flourish. Or maybe the idea and cash flow are both solid, but the product itself isn't resonating. Having a good idea is important. Executing on that idea is paramount. Silicon Valley is teeming with people with big ideas and empty bank accounts. It is a town where being first to an idea doesn't always mean you'll end up the winner.

Neither Camp nor Kalanick wanted to take on the founder's challenge for their on-demand, app-based fleet of luxury black cars. So they tweeted out a call to arms.

On January 5, 2010, Kalanick posted: "Looking 4 entrepreneurial product mgr/biz-dev killer 4 a location based service. . pre-launch, BIG equity, big peeps involved--ANY TIPS??" he tweeted.

Just then, a twenty-six-year-old intern named Ryan Graves happened to be looking at Twitter, and spotted Kalanick's request. He was interested, but didn't want to come off as too desperate. Three minutes later, he tweeted back at Kalanick with a cheeky response: "heres a tip. email me. :) graves.ryan[at]gmail.com."

Though Graves didn't know it at the time, that tweet would eventually net him more than a billion dollars. It proved to be the luckiest decision he'd make in his life.

But at the beginning of 2010, Graves was still an aimless twenty-something, one of many trying to hit it big in the startup world. Taking a chance on a gig at UberCab seemed like a cool thing to do.

Graves looked like the captain of the football team. He was 6'3" with dirty blond hair, a strong jaw and a toned, athletic build. "Surfer bro" might have been another apt nickname. Graves grew up in San Diego near the beach, paddling through the Pacific swells. On any given Saturday, you'd probably have found him near Ocean Beach or Tourmaline Surfing Park. When Graves left home to go to college in Ohio, he traded surfing for water polo and pledged Beta Theta Pi. His warm demeanor put people at ease, rare in the tech world. Friends loved to say Graves had a high "EQ," or emotional intelligence, atypical of many engineers and analytical types who inhabit positions of power in the Valley. Friends and co-workers invariably described him the same way; Ryan Graves was "a good dude."

Graves caught the entrepreneurial bug early. He worshiped entrepreneurs like Steve Jobs, Larry Page, and Sergey Brin, idolizing the way they built something enormously successful out of nothing but an idea and a computer. Graves's Tumblr was filled with photos of Jeff Bezos, quotes from Albert Einstein, articles about Elon Musk. One personal favorite was an iconic quote from Shawn Carter, better known by hip-hop fans as Jay-Z: "I'm not a businessman. I'm a business, man."

In 2009, he was bored of his job as a database admin at GE's health care unit in Chicago. He wanted a *cool* job, perhaps at one of the start-ups whose apps populated his iPhone home screen. One of those was Foursquare, a buzzy, location-based mobile check-in startup that had cache among the Valley elite. He tried applying through the front door, but was quickly turned down; Foursquare was inundated with offers from eager would-be tech workers. Instead of giving up, Graves had a better idea. On nights and weekends, he started calling around bars and restaurants in Chicago, pitching owners and managers to sign their businesses up for the Foursquare app. By pretending that he actually worked for the company, Graves managed to sign up thirty new customers in the Chicago area. So Graves tried again, sending that list of new customers to Foursquare and some of its investors.

Managers at Foursquare were immediately impressed. Self-starters like Graves tended to excel in startup-land. They kept Graves on as an intern doing business development work for the company, based in Chicago.

During his time at Foursquare, Graves posted a picture of a small, metallic statue of an ape-man wearing a backwards baseball cap, waving a bone over its head while perched atop a pile of broken electronics. (The image was plucked from *2001: A Space Odyssey*, a film more than twice as old as most Foursquare interns.) The hideous trophy was a Crunchie award, a prize given to Foursquare for having the best mobile app in Silicon Valley that year. It was the Oscar statuette of the tech industry, and Graves wanted one of his own.

Graves went to startup networking events and happy hours. He read *TechCrunch, VentureBeat,* the *Times,* the *Journal, Techmeme*—feeling the pulse of all things tech. His eyes were glued to his Twitter feed, where he followed all the venture capitalists, tech CEOs, and founders. One day, Graves hoped, he would star in an article by Michael Arrington, the Valley-famous lawyer turned *TechCrunch* founder whose stories could make or break a startup. All he needed was a shot. So when Graves saw Kalanick's tweet, Graves seized the opportunity and replied.

The two took to one another almost immediately. Graves liked

Kalanick's worldliness and "funding shepherd" machismo. Kalanick appreciated Graves's audacity, hustle, and energy. Graves was game for anything. Shortly thereafter, the twenty-six-year-old Ryan Graves became UberCab's first full-time hire.

"I'll be at the ground floor of a startup that has the opportunity to change the world," Graves posted to his Facebook as he prepared to leave the Midwest. "The world of no health insurance, jamming late nights, endless responsibility, and some of the most fun I've ever had are ahead of me and I'm so stoked."

Graves and his new bride, Molly, packed their truck, pulled away from their Chicago apartment and headed west to San Francisco.

SINCE NEITHER OF THEM wanted to do the job, Camp and Kalanick decided that Graves, young and full of hustle, should be the company's first chief executive. Graves was ecstatic; he finally had his chance to prove he could make it at a startup.

It didn't last long. Graves's friends have always considered him an "A-plus guy," but he turned out to be a B-minus chief executive. During the company's early fundraising days, he'd walk into important meetings with venture capitalists and fumble stats or other talking points. Despite his confidence, he could never deliver a convincing enough pitch to seal the deal. Graves didn't have company-building experience, like Camp, or the ability to rapidly crunch numbers, like Kalanick. Graves was a charmer and a hard worker, but those qualities only went so far. Investors were interested in the idea, but didn't think Graves had what it took to make it big.

There's a familiar line of thinking among the technorati: Good ideas are important, but venture capital is all about making the right bet on the right person at the right time. When sizing up a founder, a venture capitalist asks: *Will this guy*—and in the sexist tech industry, it was almost always a guy—*be the one to take a startup from a handful of hard-working kids to a Fortune 500 company someday? Will this guy stick around when the shit hits the fan? Is this a guy I'm willing to bet millions of dollars on?* People liked Graves. But for most of the VCs who met him, the answer to those questions was no.

During the early days under Graves's CEO tenure, co-founder Camp began tweeting cryptically about UberCab. They hadn't announced anything about their new venture yet, but the three men teased their "stealth startup," a commonly used phrase to build allure (whether a project deserved it or not).

Rob Hayes, a partner at First Round Capital, saw Camp's Twitter schtick and was intrigued. He sent an email, met the company, and quickly cut a check for nearly half a million dollars in the company's first "seed" round of funding. Chris Sacca, a friend from Kalanick's "JamPad" days, also threw in a chunk of capital, along with a handful of other close acquaintances who became "advisors"—a glorified title for early supporters. Of the early group of seed investors, though, Hayes and Sacca were the most hands-on, offering advice and strategy. Hayes and Sacca's seed investments would one day be worth hundreds of millions of dollars.

That first seed round gave UberCab enough runway to build the essentials of a real startup. After working out of Hayes's office at First Round Capital for months, the UberCab team rented desks in a shared workspace and started bringing on early team members.

Hayes, Sacca and others agreed: Graves was a great guy, but he was no CEO. He had to go. In an early meeting with Kalanick, Camp, and Hayes, they tried to break the news to Graves as gently as possible. Graves's pride was hurt, but he took it well enough and accepted a position as general manager and vice president of business operations.

Kalanick took the opportunity to seize control. Upon agreeing to the CEO role, he insisted on being given a larger ownership stake in the company. It was important, Kalanick believed, that the leader of UberCab have complete say over his company's path forward, which meant he should hold majority control. Kalanick didn't care about his salary; he already had a taste of wealth after selling Red Swoosh. What he wanted was power.

He got it. Camp and Graves signed a chunk of their shares over to

Kalanick as compensation for his new position, a move that would tie
Kalanick permanently to the company's outcome whether it emerged
as a success or, more likely, an embarrassing failure.

During the reorganization, Graves achieved his long-simmering
wish. On December 22, 2010, Uber's first full-time employee was the
subject of a *TechCrunch* article. It just wasn't the one he had hoped.
"Uber CEO 'Super Pumped' About Being Replaced By Founder," the
headline read, an emasculating take on Graves being punted down-
stairs. (Behind the scenes, he was decidedly less "pumped.")

Kalanick didn't have to pretend. His enthusiasm was real: "I'm
frickin' pumped to be on board full-time with Uber!" Kalanick said to
the journalist Michael Arrington. Though Arrington saw the startup's
potential early on, he couldn't keep a straight face.

"People are seriously *pumped* about this change," Arrington wrote.

THE FIRST VERSION of UberCab was not an app. Users logged in to
a desktop computer browser, navigated to UberCab.com, requested a
black car and, in theory, would receive a ride within ten minutes or
less for only one and a half times the price of a yellow cab. It was more
expensive, yes, but the idea was people would pay more for the reliability and convenience of on-demand service. Soon enough, the company
farmed out development, and contract programmers hacked together
a rudimentary version of an UberCab iPhone app. It was buggy and
slow, but it worked.

Camp, a sucker for luxury, focused on branding. He was fixated on
maintaining a fleet of high-end black cars like Lincolns, Suburbans,
and Escalades. Even the initial launch motto—"Everyone's Private
Driver"—was supposed to convey a sense of exclusivity, an upscale
way to get around town. Camp believed everything about the brand
should exude coolness.

In the early days, that meant cold-calling hundreds of limo drivers
around San Francisco and convincing them they should drive for the
new service. That grunt work largely fell to Graves, who would Google

black car services across San Francisco, show up to their garages, and pitch the bemused fleet staff on driving for UberCab.

The company struck an early deal with AT&T, wherein they bought thousands of iPhones in bulk at a discounted price. These they would hand out for free to drivers, pre-programmed to run UberCab's software. The AT&T deal brought Luddite drivers onto the network as quickly as possible. Tens of thousands of dollars in iPhones lined the walls of UberCab's offices, stacked like white bricks. They piled atop one another faster than staff could give them away. Matt Sweeney, an early employee, posed for an Instagram snapshot of himself splayed across a pallet of iPhone 4s with his eyes closed, a bed of shrink-wrapped handsets in pristine, minimalist Apple packaging.

The tactic worked. New UberCab drivers flooded the market in San Francisco as the handful of early employees began to promote the app to anyone who would listen. The app shot up in the App Store rankings, especially after it began receiving glowing initial reviews from the press. *TechCrunch*, now the company's favorite industry blog, hailed UberCab's model as innovative and disruptive, something akin to "Airbnb for cars." Ironically, in just a few years startups would begin to describe themselves as the "Uber for x."

"Choose your car, driver and price and get exactly what you pay for," as one *TechCrunch* article by Arrington said. "Help break the back of the taxi medallion evil empire." Uber couldn't have phrased it better itself.

Word of mouth spread across San Francisco. Those who tried Uber-Cab swore by it. For everyone who had ever been stranded in Potrero Hill beyond the reach of Muni, or stuck out in the Sunset district; for people who got stuck in the city after BART stopped running at midnight—UberCab was exactly the thing San Franciscans had been waiting for.

The app pleased its users because Kalanick and Camp had spent a great deal of time thinking about user experience, "UX" in tech industry parlance. They believed every part of an UberCab ride, from hailing the driver to exiting the car, should be as easy and enjoyable as possible. A "frictionless" experience, as Kalanick put it, was crucial to making the "UX sing."

For instance, often when people called for a traditional taxi, they didn't know whether it'd be there in a matter of minutes or if it wouldn't show up at all. When a user ordered an UberCab, she could watch the car's journey, pixel by pixel, across the map on the screen of their iPhone. San Francisco's aging taxicabs were grimy, their seats sticky and torn. UberCab's private black cars would show up spotless, with slick black leather interiors and comfortable air conditioning, replete with wintergreen breath mints and chilled bottles of Aquafina.

One of the most important parts of the UberCab experience was paying for the ride. Kalanick was insistent that payment was something people shouldn't even have to think about. With UberCab, the ride would simply be charged to a credit card stored on your account. Ending the trip was as simple as opening the door and stepping out onto the curb. No tips, no change, no hassles.

Soon enough, startup CEOs and venture capitalists started expensing their UberCab rides. Having the Uber app—knowing to order an Uber rather than take your chances with a taxi—became a status symbol. UberCab employees printed out dozens of promotional gift cards, handing them out to influential Twitter users and other high-profile members of the Bay Area's tech elite, encouraging them to talk and Tweet about it.

Within months, Kalanick and Camp's startup was the talk of Silicon Valley.

TO PROVE THE COMPANY could scale, however, Kalanick needed to replicate UberCab's success outside the Bay Area. San Francisco felt like kind of a "gimme," a tech-friendly haven where a sizeable population of young people with money to blow enjoyed early-adopting new ideas. If your consumer-tech iPhone app doesn't flourish in San Francisco, you might as well pack up and go home.

Twenty-four-year-old Austin Geidt was tasked with figuring this out. In 2010, Geidt had just graduated with a degree in English from the University of California, Berkeley, and no idea what to do with her life. She had never worked a full-time job outside of the retail industry. The day Geidt applied for an intern position at UberCab, she had been turned down for a barista gig at a Peet's Coffee shop in downtown Mill Valley, one of the richest parts of Northern California per capita— home to many of the people Uber would eventually wish to court for its service.

Geidt scored an internship with UberCab before it had a real office or much of a customer base. With no marketable skills and very little idea of what she was doing, she ended up doing some of everything. She'd ring up limo companies across San Francisco, convincing them to join the service. She'd post countless Craigslist ads and blanket the city side-walks with want ads and flyers. It was scut work, but Geidt was grateful for the job, and exhibited "hustle," a favorite characteristic of Kalanick's.

She was Uber's first city launcher, a made-up job that involved para-chuting into new markets, setting up shop, and launching the service. She planned the earliest city launches meticulously, from finding office space and forging relationships with local black car companies, to items as granular as "buy a sheet cake for our team's launch party."

She quickly found that major metropolitan areas are filled with small businesses that provide black car and limousine rides, mostly for occasions like bachelor parties, weekend charters to tourist desti-nations, or ferrying rich customers to the airport. But drivers suffered through long lax periods, waiting around in garages or on side streets for the next call from the radio dispatcher.

Geidt would offer a solution. "We're going to give your drivers a free iPhone with an app on it, courtesy of our company," Geidt said. "When they have a bit of downtime between their usual gigs, they can turn on the app and make a chunk of extra change on the side." Meanwhile, Uber takes a 20 to 30 percent cut of every ride for providing the net-work that connects riders to drivers.

"Everybody wins," Geidt said.

"It was honestly pretty much a no-brainer for the livery company operators, since the cars were just sitting there otherwise," one early employee said. To kickstart demand, UberCab would dole out incentives to both drivers and riders, a method that proved to be one of the company's most enduring marketing techniques. Riders, for instance, would get a free first trip upon signing up for the app. Drivers were promised hundreds of dollars in bonuses if they completed a minimum number of trips during the week. And to incentivize customers to return, future fares would be discounted anywhere from 20 to 50 percent, and sometimes given away completely—UberCab footed the bill, paying drivers the difference for those rides.

The strategy was pricey, since the company lost money on each subsidized ride. But it paid off after people started using the service more and more. "As the company operators saw how much business they got from Uber, they eventually started buying new cars and hiring more full-time drivers to handle all the extra business," one employee said.

In each new city, Geidt established a team to continue operations after she moved on. Communications managers handled marketing, messaging, and drumming up rider and driver interest. She would hire a few MBA types to handle what they called "driver operations," which meant spreadsheet work managing supply and demand among a population of riders and drivers in continuous flux. General managers were at the top, and acted as the boss of the individual city.

Geidt finally felt like she had found her professional footing. Bringing UberCab into new cities became a routine. She systematized the approach on an internal company Wikipedia-like page, creating a playbook for city launches. Send in a launch team to Seattle, San Antonio, Chicago—wherever—have them follow the playbook, and watch the demand flywheel begin to spin. She became extremely efficient at launching local operations, and would spend the next eight years of her life on airplanes, replicating what she had done in San Francisco in other cities all over the world.

AS GEIDT WAS PERFECTING the playbook in the United States, the idea of launching UberCab in foreign countries seemed unimaginable. But before they could even spread outside of California, the group faced an existential crisis.

On October 20, 2010, just days after Graves had agreed to officially step aside as Uber's CEO, transportation officials showed up at the offices of the young startup. They hadn't read *TechCrunch* and asked to see Graves. UberCab, they said, had been served with a cease and desist order; the company was breaking the law by skirting existing transportation regulations, the San Francisco Municipal Transportation Agency said. Every day UberCab was in operation, the company faced fines of up to $5,000 *per trip.*

The potential fines were enough to put the company out of business. UberCab was already completing hundreds of trips per day in San Francisco. Moreover, Graves, Kalanick, and other employees faced up to ninety days in jail for each day the company remained in operation beyond October 20.

Graves, Geidt, Kalanick, and board member Rob Hayes were in a cramped room together at their shared workspace office when they got the cease and desist order. They scanned the letter in disbelief.

Graves was scared. "What are we supposed to do here?" he said aloud, reading his name on a piece of paper that said he could be going to jail. Hayes, the venture capitalist, wasn't sure what to say. He was used to investing in consumer tech companies, but rarely (if ever) did they run afoul of the law. Geidt, just a few months out of college, stood quiet and nervous, too. This was her first foray into the professional world. Now she was looking at jail time.

Kalanick didn't miss a beat. "We ignore it," he said to the room.

The others looked at Kalanick like he had grown horns. "What do you mean 'ignore it?'" Graves said. The ex-CEO looked at Hayes for advice, since the VC at least had some experience managing startups. Hayes shrugged back at him.

"We ignore it," Kalanick repeated. "We'll drop 'Cab' from our name," he said, something his lawyers claimed gave the company greater legal exposure to false advertising claims.

UberCab was now known as "Uber," and it was staying open for business.

Chapter 7

THE TALLEST MAN IN VENTURE CAPITAL

BILL GURLEY NEEDED TO GET IN ON THIS DEAL.

Over his decade-plus of venture investing, Gurley had watched enough startups succeed and fail to know that this one—Uber, "everyone's private driver"—was special. Not only was the company growing fast, but it was perfect for the iPhone, the device that was changing the world.

Unlike Camp and Kalanick, Gurley wasn't drawn to visions of luxury or the idea of being a "baller." Nor did he have much difficulty getting around; Gurley owned a car and lived in a suburb near Woodside, an extremely wealthy area between San Francisco and Silicon Valley.

What Gurley admired was the potential for scale. Most startups took a business that already existed and tried to make it slightly better or more efficient. Uber promised to upend an entire industry, one that had seen little innovation in decades. The sheer size of the taxi market could make Uber worth billions if the company continued its growth trajectory. And best of all, this new entity, potentially worth billions, had been created out of thin air. It could theoretically drag the entire transportation industry out of the analog world and into the digital one practically overnight. Best of all, whoever did the dragging would set the terms for the entire marketplace.

By downloading the Uber app, riders gave themselves the power and freedom to summon a car instantly, to any location, at any time. And drivers didn't need to spend hundreds of dollars installing some cumbersome box in their dashboard to connect to these customers. Maybe they'd have to spend ten bucks on a dashboard smartphone caddy— Uber would give them the phone for free.

"It's magic," Gurley said.

Uber popped up on Gurley's radar at the exact right time. Throughout his career Gurley had been enamored with what he called "marketplaces," a category of business that neither made new products nor sold others, but merely matched the desires of one side of a market with the products of the other side, and took a cut as the middleman.

By the time Gurley arrived at Benchmark, the venture capital firm where he had worked for the past seven years, marketplaces had consumed him. eBay, one of Benchmark's most successful investments, was a natural marketplace, matching millions of buyers to sellers, all enabled by the rising power of the internet. So was Zillow, an eBay for real estate. OpenTable, one of Gurley's earliest investments, matched people to restaurant reservations. Grubhub, similarly, connected people to food delivery. DogVacay—Airbnb for pooches—was self-explanatory.

Nearly every one of Gurley's investments relied on one basic thesis: the internet had brought with it a profound capability to meet the desires of existing, real-world people for experiences, places, and things. Whereas before a Beanie Baby enthusiast might have had to search high and low for a particular plush giraffe, the web could put that person in touch with someone who had stockpiled a warehouse of them. There were endless combinations of buyers and sellers, and hundreds of potential marketplaces bubbling up from the minds of young entrepreneurs, waiting to be brought to life with Benchmark's blessing—and capital.

Before Gurley arrived, eBay was Benchmark's crown jewel investment. In 1997, the small, tight-knit VC firm had invested $6.7 million in eBay. Two years later, Benchmark's position was valued at more than $5 billion.

Gurley came to Benchmark with a good track record. At Hummer Winblad Venture Partners, his home before joining Benchmark, the firm's first $50-million fund returned $250 million to its institutional investors. And after he joined Benchmark in the middle of '99—a few years shy of the impending tech bubble burst—Gurley had done a number of very successful investments.

But he still wanted a home run of his own. He needed to get into this deal.

JOHN WILLIAM "BILL" GURLEY was born on May 10, 1966, in the small town of Dickinson, Texas, population 7,000. Tourists would pass through the Houston suburb on their way to Galveston on the East Texas shore. In the 1920s, Dickinson was known best for gambling establishments run by the Maceo crime family. Today Dickinson is better known for its annual crawfish festival, "Red, White and Bayou."

John Gurley, Bill's father, was an early NASA aeronautics engineer who worked at the Johnson Space Center in Houston. John had a particular facility with numbers and analysis, both of which he passed to his son. Bill's mother, Lucia, was driven as well. Besides her job as a substitute teacher for the town's schools, she was a city councilwoman for eleven years, volunteered at the local library, and raised thousands in grants for the city's public schools. Lucia spent her spare time working for Dickinson's beautification program, helping to clean up the streets. Bill loved his mother, but more than that, he admired her—her work ethic, her loyalty, and sense of duty to her community.

Enrolled in Dickinson's public school system, Bill soon caught the computing bug; in 1981, Gurley got a Commodore VIC-20 desktop for $299, or about $850 in today's dollars—one of the first relatively inexpensive home color computers. By ninth grade, Gurley began coding his own programs, working from templates he found at the back of computer magazines.

From a young age Gurley stood out for his height. In grade school and at Dickinson High he towered above his classmates. He was different and knew it—and he didn't always like it. But his height played to his advantage in college. A few years into college studies in Mississippi, Gurley transferred to the University of Florida at Gainesville, playing as a walk-on to the team and later received a Division 1 scholarship. Though the Gators played in the SEC, Gurley's time with the team was hardly glamorous; he mostly rode the bench. He played for one minute of one game, missing the only shot he took, during the Gators' blowout loss to Michigan in the NCAA tournament. Still, he managed to pick up a degree in computer engineering.

Gurley continued with computers after college and landed a job at Compaq in Houston, down the road from his hometown. In 1989, Compaq was a growing powerhouse of computer manufacturing, and Gurley was lucky to score a job debugging software for the company. It helped that his sister, an electrical engineering major, was employee number 63.

When he wasn't spotting problems with software at work, he tracked technological advancements closely. He traded stocks on his Prodigy internet personal account. He devoured tech magazines and plowed through dense, finance-heavy analyst reports on up-and-coming tech companies. He couldn't help himself; he was infatuated. Gurley saw an intoxicating, transformative power in technology. He wanted to get closer.

After a stint at the University of Texas at Austin, where he earned his MBA, Gurley found a marketing position with Advanced Micro Devices—a computer chip company—but he quickly grew dissatisfied with the job. He wanted to do something bigger, something in emerging tech that employed his facility with analysis and numbers.

In business school he had caught a glimpse of a field that attracted him: venture capital. The enterprise made perfect sense to him. Crunching numbers and picking emerging tech trends was what Gurley already did for fun. Getting paid to do it—that was the dream. But it wasn't as simple as walking into a venture firm with a resume; several venture investors in Austin turned him down for being too young and inexperienced. So Gurley decided to try his luck on Wall Street instead.

The Wall Street mindset of the 1990s was the photonegative of the one in Silicon Valley. In the Valley, VCs were looking for moonshots—the big, dent-in-the-universe ideas that founders spent years chasing for low or little upfront pay. Wall Street thought in three-month increments.

As a Texan, sitting dead center in between the coasts, Gurley took aspects of both mindsets to heart. He appreciated the audacity of tech founders and their brazen disregard for short-term profits. But Gurley was also a pragmatist; companies that spent all their time dreaming of projects of the future rather than watching their balance sheet could find themselves out of luck—and cash—long before realizing those dreams.

The newly minted MBA started cold-calling brand name firms. Preppy East Coast businessmen found themselves interviewing a giant Texan, wide-eyed and awkward, asking for a job picking tech companies. But in 1993, Gurley finally got his wish. He scored a job at Credit Suisse First Boston as a sell-side analyst, a big break for a twenty-seven-year-old kid with no real analyst or trading experience. Still, the job was perfect for him; Gurley was responsible for synthesizing research and analyzing the personal computing industry. Other firms would use Gurley's reports to decide whether to buy and sell millions of dollars in equities. He saw older, experienced analysts at his firm—Charlie Wolf, David Course, Dan Benton, smart thinkers on the PC industry at the time—being quoted by newspaper reporters and doing stand-up interviews on television. He wanted that glory, and that wealth, too. It was challenging work, but more than that, to Gurley it was fun. The idea of being asked to opine on tech—to be paid for it, even—thrilled him.

Gurley quickly became a star of the Street. He moved up the chain rapidly as his older colleagues cycled out of the industry. They shared their financial models with the young Gurley, imparting years of valuable insights. One colleague, Charlie Wolf, helped Gurley get into Agenda, a famous annual conference of the tech elite in San Francisco. Gurley wandered the conference starstruck, the former benchwarmer trying to imagine himself belonging in a crowd that contained people like Bill Gates, Larry Ellison and Michael Dell—some of the biggest names in the history of computing.

His success wasn't due only to helpful mentors. He quickly forged his reputation by making the right calls on technology stocks and market trends. So much so that he impressed one of Credit Suisse's big-shots, Frank Quattrone, a legendary Silicon Valley investment banker involved in some of the highest profile technology company deals in history. The two men would grow close while at Credit Suisse, and eventually work together again at another firm, Deutsche Bank. Early on, Quattrone recognized that Gurley, an engineer by training and analyst by trade, possessed keen insight into the world he was covering.

The executives at the companies Gurley covered saw it too. As Amazon worked on its initial public offering to the stock markets in

1997, Jeff Bezos and his team of executives didn't pick one of the two high-profile investment banking firms—Morgan Stanley and Goldman Sachs—to lead its IPO. Instead, Bezos's firm picked Deutsche Bank, a less prominent but still excellent firm, to take Amazon public. It was the combo of Deutsche Bank's star banker and his lead analyst that sealed the deal: Frank Quattrone and Bill Gurley. The duo wowed Bezos and his board with their knowledge of the online bookseller and its underlying business. Morgan Stanley and Goldman Sachs had the glitzy name, but Deutsche Bank had Quattrone and Gurley.

Gurley became the go-to analyst on what was then one of the world's largest online booksellers. Gurley saw, very early on, the opportunity for Amazon to become much, much more than a bookstore.

Gurley's biggest talent was his willingness to be a contrarian. In the heady days of the late nineties, when tech analysts like Gurley were often seen as glorified internet stock boosters, Gurley cut a path for himself by bucking against trends. His most notorious call at Deutsche Bank was his infamous report on Netscape, the first web browser and early internet pioneer. Most analysts rated Netscape positively, even as Microsoft readied its Internet Explorer browser for market—and promised to distribute it free of charge. Gurley saw this as a threat to Netscape's browser dominance, and, unlike other analysts, worried about how Netscape would execute its business decisions under pressure from Microsoft. He thought Netscape's shares were overvalued, and downgraded the stock. Netscape shares plunged nearly 20 percent the next day. Netscape never fully recovered.[*]

Despite his success, Gurley told Quattrone he wanted to stop being just an analyst and start making actual investments. Quattrone made it happen. He helped get Gurley a job with a respected venture firm,

---

[*] That call also earned Gurley the ire of a young entrepreneur who would one day become another influential venture capitalist—Marc Andreessen. Andreessen was a co-founder of Netscape and is credited with helping to invent the consumer internet. Though Netscape eventually floundered and sold itself to AOL, Andreessen never forgot Gurley's report. Years later, after both men had achieved personal success and enormous wealth, the two still carry a grudge. In an interview with the *New Yorker* years later, Andreessen remarked of Gurley: "I can't stand him. If you've seen *Seinfeld*, Bill Gurley is my Newman."

Hummer Winblad, but he was soon headed to the big leagues. After just eighteen months, he was recruited by a top-tier firm called Benchmark Capital.

The courtship between Gurley and the firm was long, but necessarily so. Benchmark operated as a small, tight-knit unit, with each partner involved in the decision-making and advising for all the companies in the portfolio. Any new partner would have to be in sync with the others.

Kevin Harvey, a founding partner at Benchmark, took Gurley hunting. While in the woods together, Harvey got to see Gurley's analytical mind at work. But what stuck out most for Harvey was Gurley's tenacity.

"He's kind of an animal," Harvey told his partners. As the two sat in the bush, Harvey watched Gurley spring to his feet, jump over a steep cliff, and scramble down a hill after a wild boar they were tracking, something that Harvey wasn't willing to do. "He thought I was kind of lazy 'cause I didn't want to."

In 1999, Benchmark Capital had five venture partners. Gurley became the sixth. Each was exceptionally tall. They looked remarkably like the starting lineup of a college basketball team. Over time, partners would cycle in and out, but Gurley remained a constant.

Bill Gurley would always remain the tallest.

EVEN NOW, nearly twenty years into a phenomenal career in venture capital, the first thing anyone notices upon meeting Bill Gurley is that he is enormous.

Save for professional basketball players, the six-foot-nine Gurley towers above most everyone he meets. Men in Gurley's position might have used such an outsized stature to their advantage, perhaps to intimidate competitors, a physical manifestation of VC swagger.

Not Bill Gurley. He is painfully aware of his size, and often goes to great lengths to avoid flaunting it. Gurley is more comfortable standing in the back of a room, trying to blend into the curtains at a dinner party. (It never works; friends, reporters, entrepreneurs all flock to Gurley as soon as they see him.) Gurley seems unused to inhabiting his body, visibly calculating how to maneuver his gangly legs and thick frame. One close friend said he wouldn't be surprised if one day, like in a scene from *Men in Black*, Gurley's head opened up to reveal a tiny intergalactic space traveler struggling with the controls of his Gurley-shaped spaceship.

When there is a lull in a conversation or an onstage presentation, Gurley won't fill it with idle chatter. He'll remain quiet. Sometimes after someone says something important, he'll take a step backwards in the room as if physically absorbing the comment.

That's Gurley thinking, analyzing what's happening, what's been said, what *will* be said. Or it's him just being awkward, because he *is* awkward. In a habitat like Silicon Valley, awkwardness is ignored or encouraged, and the only thing that matters is whether you have the brains to back up your ideas.

Brains, and one other thing: zeal. Like so many of his peers in the Valley, Gurley truly believes in the transformative power of technology and innovation. He appreciates the positive impact that a young founder with a big idea and a few million dollars can make in the world. The tech press loves to fixate on his negative comments about Silicon Valley, but Gurley insists he is an optimist.

Even in some of the industry's most dire moments, Gurley didn't shy away from the venture business. He was there during the dot-com bust at the turn of the century, looking for promising founders. And when the financial crisis rocked the foundation of the global economy in 2008, he doubled down on startups.

"Environments like this tend to sort out the true entrepreneurs from the pretenders," Gurley wrote during the height of the crisis. "When money is easy in Silicon Valley, it tends to attract short-term opportunists looking to make a fast-buck rather than build a lasting company. Only the best entrepreneurs set sail in rough seas like this."

Chapter 8

PAS DE DEUX

VENTURE CAPITAL ISN'T AS MUCH A PROFESSION AS IT IS A BRAWL.
If it were a sport, it would be like rugby without the mouthguards.
There are no *real* rules, except that players should do whatever they
need to do to seal a deal.

It doesn't seem like a hard job. All you do is give away other people's
money. But it is. A VC's calendar is packed with daily meetings—with
founders, with their financial backers, with industry analysts, with
journalists. VCs spend time talking to the CEOs of large, established
companies about market trends and recruiting practices. They talk
to investment bankers about private companies and public markets.
They have to fend off hordes of eager founders seeking their favor. Even
while relaxing at the bar in the Rosewood—the luxury hotel that has
long acted as the social hub of tech money in Palo Alto—they're likely
to be interrupted by an awkward elevator pitch.

A venture capitalist's job is to cut through all the noise and find
the startups that will deliver outsized returns for the pension funds,
endowments, family offices, even other high-net-worth individuals who
have invested their money as limited partners, or LPs, in the VC firm.
The lifecycle of a VC fund is typically ten years, by the end of which
these LPs expect returns of at least 20 to 30 percent on their initial
investments.

Venture capital is risky. Roughly one-third of VC investments will
fail. But a heightened "risk profile" comes with the territory. If institutional investors prefer lower-risk investments, they can stick to reliable

municipal bonds or money market funds. With low risk comes low returns.

To compensate for such high failure rates, VCs tend to spread their investments across a number of different industries and sectors. One grand slam investment with a return of ten, twenty, even fifty times the amount of the investment can make up for an entire investment portfolio of losses or weakly performing startups. In venture capital, so-called "moonshot" companies—run by entrepreneurs who aim to remake and dominate entire industries—are the most sought after, the ones that bring the greatest glory.

The investment equation is simple: a venture capital firm provides money to a startup in exchange for an equity stake in the company. For founders who decide to take on venture capital,* a company begins raising its first round of funding early in its life cycle. This "seed" round typically involves modest investments in the tens of thousands to hundreds of thousands of dollars. After that, venture rounds continue by letter: Series A round, Series B round, and so on. Those funding rounds continue until either the company:

A. Dies. This is the most likely scenario.
B. Is acquired by another larger company.
C. Holds an initial public offering of its shares, allowing outside investors to purchase shares in the company through a public stock exchange.

For venture capitalists and founders alike, the goal is to guide the company to either B or C rounds, or "liquidity events." Those are when a VC can finally convert shares in a company into cash.

Each round has a certain kind of politics, and conveys a different kind of status. Typically, the earlier a venture firm invests in a hot

---

* Not every startup decides to take on venture capital. These companies are said to be "bootstrapped," or entirely self-funded. Bootstrapping founders keep all the equity in the company and reap all the rewards if the startup succeeds. Their founders also go broke when they fail.

company, the more prestigious it is for the firm. The firm benefits, ret-roactively, by being seen as having the foresight and skill to invest in a lucrative startup years before it grew into a powerhouse. David Sze, of Greylock Partners, will always be known for his early investments in both Facebook and LinkedIn, when their valuations were still in the millions, not billions. Besides his seed investment in Uber, Chris Sacca made early bets on Twitter and Instagram, each of which have since made him a billionaire.

The other reason a firm wants to invest early is simple: the earlier you invest in a company, the greater share of equity the firm gets for a smaller amount of money.

The hardest part of a VC's job isn't even necessarily about finding the right company, the right idea, or even the right industry to park their next investment. It is about finding the right person to run the company: the founder.

THE MOST VAUNTED TITLE in Silicon Valley is, has been, and ever will be "founder."

It's less of a title than a statement. "I made this," the founder pro-claims. "I invented it out of nothing. I conjured it into being." Travis Kala-nick frequently compared building a startup to parenting a young child.

A good founder lives and breathes the startup. As Mark Zuckerberg said, a founder moves fast and breaks things. The founder embraces the spirit of "the hacker way"; he is captain of the pirate ship. A good founder will work harder tomorrow than he did today. A good founder will sleep when he is dead (or after returning from a week at Burning Man). Like Kalanick at Red Swoosh, a good founder shepherds his company through difficult funding environments, but chooses his bene-factors wisely. A good founder takes credit for his company's successes, and faces the blame for its shortcomings. A good idea for a company, even if it lands at the right time and in the right place, is still only as good as the founder who runs it. Most important of all, there can only ever be one real founder.

If this sounds messianic, that's because it is. Founder culture—or more accurately, founder *worship*—emerged as bedrock faith in Silicon Valley from several strains of quasi-religious philosophy. Sixties-era San Francisco embraced a sexual, chemical, hippie-led revolution inspired by dreams of liberated consciousness and utopian social structures. This antiestablishment counterculture mixed well with emerging ideas about the efficiency of individual greed and the gospel of creative destruction.

Out of those two strands, technologists began building a different kind of counterculture, one that would uproot entrenched power structures and create innovative new ways for society to function. Founders saw inefficiencies in city infrastructure, payment systems, and living quarters. Using the tools of modern capitalism, they created software companies to improve our lives, while simultaneously wresting power away from lazy elites. The founders became the philosopher kings, the rugged individuals who would save society from bureaucratic, unfair, and outmoded systems.

Marc Andreessen famously said, "Software is eating the world." Back then, technologists thought this was a good thing. Until recently, most of the rest of the world agreed. Venture deals increased by 73 percent from the early 2000s into the 2010s. The amount of global venture capital invested soared from tens of billions in 2005 into the hundreds of billions invested post-2010. San Francisco emerged as the world's epicenter of such deals.

But then the balance of power began to shift. As startups upended global infrastructure at an unprecedented pace, entrepreneurs found that old power centers had eroded and been replaced in some cases by the upstarts that sprung up around them. Clayton Christensen's "Innovator's Dilemma" articulated the perils that awaited any company that grew so large that it no longer saw threats coming from more nimble competitors. The venture-backed startups became the new establishment.

Something else happened: Founders realized they liked being in control. They wanted freedom from meddling by outsiders like shareholders, investors, or the general public. Over time, founders discovered ways to protect their power. They used their visionary status to convince investors to cede control to the founders themselves.

Larry Page and Sergey Brin, the co-founders of Google, cemented and institutionalized this practice. In a cramped garage in 1998, Page and Brin founded a search engine to perform a task that sounded bonkers; "to organize the world's information and make it universally accessible and useful." It was the exact type of moonshot thinking venture capitalists encouraged.

But while the Google founders were excited to change the world, they didn't want to make decisions based on what the money men wanted. The motto "Don't be evil"* became synonymous with Google's founders and their approach, the message being "even though we're growing into a mature company, we won't be doing terrible things for money."

In 2004, when Google undertook its IPO, it used a controversial financial instrument called a "dual-class stock structure." Google sold "Class A" shares to the public, while its founders held onto "Class B" shares. The two classes held the same monetary value, but Class B shares came with special privileges; every Class B share represented ten "votes," or ten individual chances to yea or nay company leadership decisions. Class A shares, on the other hand, held only one vote per share. Page and Brin made sure that over the years, they had held onto enough stock in their company—and more importantly, were issued enough Class B shares at the time of the IPO—to maintain majority control.

Page and Brin didn't actually *want* to go public. For the founders, listing their stock on the Nasdaq meant opening Google up to oversight from annoying people who knew nothing about tech. Investors would want to skim cash from Google. And when those investors felt revenue growth wasn't strong enough, they'd try to change the company by imposing their collective will upon the two co-founders.

As one investor told it, Brin and Page agreed to go public only after meeting Warren Buffett, the legendary American business mogul, who introduced the two young founders to the dual-class stock structure.

---

* Google removed the "Don't be evil" mantra from the preface of its corporate code of conduct in 2018.

"We are creating a corporate structure that is designed for stability over long time horizons," Page wrote in a letter cheekily titled "An Owner's Manual For Google Investors." "By investing in Google, you are placing an unusual long term bet on the team, especially Sergey and me, and on our innovative approach. . . . New investors will fully share in Google's long term economic future but will have little ability to influence its strategic decisions through their voting rights."

Many founders followed this same playbook. "Larry and Sergey did it, why shouldn't we?" young entrepreneurs asked themselves. Mark Zuckerberg was considered crazy when he spurned a $1-billion acquisition offer from Microsoft. After Facebook went public in 2012, Zuckerberg maintained outsized influence due to a dual-class share structure and faced no board resistance when he pivoted the entire company to focus on building for mobile devices, an enormous gamble that paid off handsomely.[*]

Facebook was followed by Internet 2.0 companies like LinkedIn, Zynga, and Groupon, all of which mimicked the dual-class structure. Snap Inc., helmed by another tech wunderkind, Evan Spiegel, famously declined a $3.5-billion acquisition from Facebook in 2013. When the company went public, in 2015, the twenty-six-year-old Spiegel became the world's youngest billionaire.

Only in a place like Silicon Valley, where founders are celebrated above all, could an executive like Spiegel spurn such an offer and be celebrated for his bravery. Where nonbelievers might consider such a choice irrational, the "cult of the founder" suggests that no matter what the chief executive may decide, he was probably right because he was the right guy to begin with.

---

[*] Founder worship of Zuckerberg evaporated after 2016, when news coverage of events ranging from the presidential election in the United States to reported ethnic cleansing in Myanmar suggested that Facebook lacked oversight of its platform. Even the boy genius himself, pundits said, was not aware of how powerful—and vulnerable—his own software could be.

AS THE BALANCE OF POWER shifted to founders in 2010, venture capitalists had to fight—*hard*—to beat out their competitors to invest in the best young companies. They hosted parties for entrepreneurs, wined and dined them at chic eateries like Nopa, Bar Crudo, and Spruce. Sometimes a flashier approach worked; chartering a Learjet 31 to bring a group of twentysomething techies to SXSW showed founders that a VC firm could travel in class. Nothing was more "baller" than a private jet.

Gurley didn't rely on expensive gimmicks alone. He gave impeccable guidance, answering calls from a founder at 11:30 at night, after his kids were asleep and he was near dozing himself, to talk strategy or walk a young entrepreneur off some panicked ledge. Gurley competed for the most important deals. And more often than not, he won.

Benchmark had been looking for a ride-hailing or taxi-based business to invest in for some time. Gurley had already been meeting with companies like Cabulous, Taxi Magic, and a handful of other San Francisco–based ride-hailing companies. A popular ride-hailing company could quickly produce what technologists call a "network effect"—a shorter way of saying "the more people that use a service, the more beneficial it is to everyone else over time." And Uber's growing popularity in San Francisco meant it was creating strong network effects among its two-sided marketplace of both riders and drivers.

Just a few months after Uber had raised its seed round with investors like Chris Sacca and Rob Hayes, Kalanick was already out hunting for investors for Uber's Series A. The next round of funding would supply millions for new growth. Gurley had approached his partners about investing in Uber's seed round, but wasn't able to win over everyone at the firm. He wouldn't let it happen again; he *had* to invest in Uber. The opportunity was too big to pass up.

What Gurley didn't know is that Kalanick wanted to strike a deal with Benchmark as much as Benchmark wanted Uber. (Benchmark's reputation preceded it.) He also liked the idea of adding a coveted figure like Gurley to Uber's board, where he could open doors for the company while participating in key decisions. And Kalanick knew Benchmark had made great bets on companies for years.

Benchmark was a blue-chip firm, venerable, even, among the glut of venture capitalists entering the valley after the crash. Kalanick wanted the best, but the *right kind* of "best." Sequoia Capital, for example, was one of the most prestigious firms in tech investing. Kalanick had made repeated overtures to Sequoia for funding, but was repeatedly rebuffed over multiple early rounds.

Gurley had grown famous for his personal blog, *Above the Crowd*, where he would occasionally post investment treatises and thoughts on the state of technology investing. (Aside from the grandiosity of the title, it was also a sly recognition of Gurley's height.) He started it as a fax newsletter during his analyst days, long before the ubiquity of the consumer internet. It grew much larger when Gurley launched it as a public blog in 1996. He would mull over the content of a single 3,000-word blog post for months, vetting his thoughts with friends and colleagues, before putting it out into the world. And when Gurley updated his blog, people read it. A single post could lead Valley chatter for weeks—something Kalanick appreciated.

As the two courted each other, Kalanick—who was living at his hilltop home in the Castro—called Gurley on a Sunday night in 2011 at around eleven o'clock, wanting the VC to make the forty-minute drive from Gurley's home in the suburbs to meet Kalanick and talk through some ideas.

Gurley didn't think twice. He jumped into his car and drove the thirty miles north to meet Kalanick at the W Hotel, one of the only upscale bars in the city that stayed open late on a Sunday evening. The two "jammed" together on ideas for Uber over cold beers at the bar, batting product thoughts and long-term strategic goals back and forth for hours. In the early morning hours, as Gurley's family was asleep in their beds, the VC and the founder sealed an investment in Uber. In a handshake deal, the two valued Uber at about $50 million, and Benchmark would own just under 20 percent of the young company.

The next day, Benchmark got going on the paperwork, and shortly thereafter, the venture capital firm delivered the $11 million investment to Kalanick. Gurley also took a seat on Uber's board, which at the moment had only three members: Garrett Camp, Ryan Graves, and

Travis himself. In Kalanick, Gurley knew he was investing in a dogged CEO; though just ten years Gurley's junior, Kalanick was still in his thirties, and as tenacious as any founder Benchmark had invested in. That tenacity would give Kalanick the courage to challenge entrenched transportation interests worldwide. And though neither of them would know it, it would make Kalanick more powerful and uncontrollable than any entrepreneur Gurley had ever met.

But in that moment, Gurley wasn't thinking about any of these things. At last, over beers at a bar after last call, Gurley had bagged a transportation networking startup, his wild boar.

He was in.

Chapter 9

CHAMPION'S MINDSET

IN KALANICK'S VIEW, ENTREPRENEURS WERE WORTHY OF THE PRAISE they received.

Founders like him spent every day hustling to keep their companies running. They put their reputations, finances, and well-being on the line. Venture capitalists, on the other hand, only risked OPM—"other people's money." VCs anticipate company failures in their portfolio of investments; it's why they diversify their approach and spread cash around to multiple sectors. If a young Uber failed, it was no skin off the investor's back. It was Kalanick and his staff who would bear the brunt. So Travis Kalanick was suiting up for war.

As Uber prepared to expand throughout the country, Kalanick swore that this time, things would be different. He had learned from his last two startups. At Scour, he had left *far* too much control in the hands of investors—investors who, when Scour was under attack, saved themselves and fed him to the wolves. At Red Swoosh, he had survived, but the timing was terrible, and the product less compelling.

Now, with Uber, Kalanick was selling a winning product that was landing at an ideal time. Above all else, Kalanick was in complete control. Everything about Uber—from the design of the app to the raucous, take-no-prisoners culture—was his. He saw himself locked in an existential battle with corrupt, entrenched taxi operators and the politicians they paid to protect them. Kalanick was the general on the front lines.

Aware that war metaphors could seem overblown, Kalanick often compared the ongoing battle to a political campaign. "The candidate is

Uber and the opponent is an asshole named Taxi," Kalanick once said onstage at a tech industry conference. "Nobody likes him, he's not a nice character, but he's so woven into the political machinery and fabric that a lot of people owe him favors."

But this was window dressing. Kalanick had designed Uber for battle. If government decided to push back in any individual city, Kalanick quickly weaponized his users against City Hall. Uber would blast emails out to riders, asking them to contact their local representatives and voice their frustration with anti-Uber crackdowns. Uber city teams would send mass text messages to drivers, urging them to stay on the road even if they were ticketed or their cars were towed by law enforcement.

"There's been so much corruption and so much cronyism in the taxi industry and so much regulatory capture that if you ask for permission upfront for something that's already legal, you'll never get it," Kalanick once told a reporter. Kalanick evidently believed there was no way Uber could win if it played by the rules—his competition certainly wouldn't.

The founder's instinct proved correct. Uber's guerilla tactics far outmatched the resources and technical acumen of government workers or taxi operators. In Seattle, for instance, Austin Geidt dropped in like a paratrooper, quickly hiring ground support staff to drum up interest from riders and drivers. Ryan Graves then swooped in and made the pitch to town car companies: "We're giving your drivers a way to earn extra money." In a matter of weeks, Uber was able to grow its ridership before the city even knew what had happened. By the time regulators had arrived, Uber was too popular with citizens to try and shut it down. Once Uber hit critical mass, transportation authorities lacked the manpower to stop the fleet.

To Kalanick, Uber wasn't doing anything wrong. After all, these were official limo and town car drivers, operating well-maintained, insured vehicles and using Uber's service to make extra money during inefficient downtime. Everyone working for Uber was a licensed, professional driver—period. (This was before UberX allowed anyone with a car to become a driver.) As Uber's footprint spread across the United States—Seattle, New York, Los Angeles, Chicago—it became more popular and thus more difficult for cities to block the company.

Kalanick never revealed stats, but offered a bro-speak narrative of wild success. "The best metric I can give you is that Uber is killing it in San Francisco and we're crushing it in New York," Kalanick told a reporter in the early days after launching in Seattle.

Kalanick hired throngs of ambitious twentysomethings, fresh out of college and starry-eyed about Kalanick's pitch. He spun stories of Uber's eventual ubiquity, providing "transportation as reliable as running water."* It wasn't uncommon for a new hire to enter Uber's headquarters having never managed anything more than a Starbucks, and be sent out to take over a new city.

Kalanick trusted his employees with significant power. Each city's general manager became a quasi-chief executive, given the autonomy to make significant financial decisions. Everyone was responsible for "owning" their position. Empowering his workers, Kalanick believed, was better than trying to micromanage every city. Later, when Uber had billions of dollars in the bank, city managers were given the latitude to spend millions of dollars in driver and rider "incentives"—freebies to get people to use the service—in order to spur demand and, later, to lure riders away from other ride-hailing competitors. Those employees rarely had to check in with headquarters. Top managers in Uber's San Francisco office barely knew employees in, say, Chicago or Philadelphia. And they had little oversight over the money. Local managers were greenlighting seven-figure promotional campaigns based on little more than a hunch and data from their personal spreadsheets.

In many ways, Kalanick's approach was brilliant. A local employee in Miami would be better prepared to fit Uber to their own city than,

---

* Kalanick and other executives said this regularly to inspire employees. That much of the world doesn't actually have access to running water, and might want that need met first, was a detail that the Uber CEO and his peers never addressed.

say, a new hire from San Francisco who knew nothing about the people and institutions that make up a locale.

There were drawbacks. Give *too* much autonomy to a legion of twentysomethings, and you'll occasionally empower a battalion of douchebags. In France, one local promotion boasted "free rides from incredibly hot chicks." The New York office was infamous for its bro-culture. Helmed by Josh Mohrer, a former frat boy turned MBA graduate, the bravado and aggression of management led to resignations and allegations of harassment. Every city office had its own cultural microclimate, for better or worse.

But that sense of freewheeling autonomy made employees lionize Kalanick's leadership. It was as if Kalanick had hired a private army of mini-entrepreneurs and given them one mandate: Conquer. Everyone was a founder of their own city-level fiefdom. Everyone got to live the startup, hacker ethos, something Kalanick cherished and never wanted his company to lose, even as Uber spread like wildfire. With a slap on the back, Kalanick would send his officers out into the field to build their own infantry and fight for Uber. "Always be hustlin'," he'd say.

Kalanick envisioned his company becoming a new Silicon Valley institution, a juggernaut that encouraged a spirit of entrepreneurialism. He wanted "ex-Uber" to carry a certain Valley cultural capital, much like being ex-Facebook or ex-Google. After their time at Uber, Kalanick wanted his troops to go off and start their own companies. He saw the very act of founding a company as a virtue unto itself.

Kalanick may have been an Ayn Rand–esque libertarian spouting cheesy startup platitudes. He may have pushed his team to work to the brink of exhaustion. But it mattered to employees that Kalanick had their backs. They were in this fight together.

They couldn't have asked for a better founder.

UBER HAD PERFECTED viral growth.

After Seattle and New York came Chicago, Washington DC, Los Angeles. But Kalanick's ambitions were much larger; he wanted to go global. By 2012 they were in Paris, with expansion to London, Sydney,

Melbourne, Milan, and dozens of other cities to follow. Guerilla marketing campaigns spread notice of Uber to new passengers, and word of mouth from users brought others flocking organically to the service.

Kalanick's and Camp's vision of on-demand black cars—the baller vision—was lucrative. Uber was already minting money in San Francisco, where every VC and startup founder delighted in having a private car service you called with your phone. But what flipped the switch to enormous growth was moving beyond the luxury model.

Sunil Paul, a serial entrepreneur and longtime transportation geek, was experimenting with a different way of offering rides to people with his San Francisco–based startup, Sidecar. Paul saw what Uber was doing and appreciated their intensity and aggression. But Paul realized there was a much larger market opportunity in what he called "peer-to-peer ride-sharing." That is, instead of focusing on professional limo drivers, Paul wanted to convince normal, everyday people who owned cars to become part-time drivers themselves. The way Paul saw it, the roads were already packed with underutilized vehicles, four- and six-seated vehicles that only contained one person driving the car. It was a glut of capacity, otherwise wasted space.

Paul was first to the thought, and more prescient than he could have known at the time. But in Silicon Valley, being first doesn't matter—being the best does.

As Paul tried to transform his peer-to-peer vision into a reality, another startup was mulling the same approach. Zimride, a carpooling startup co-founded in the Bay Area by a transportation enthusiast and an ex-Lehman Brothers employee (who escaped the firm just three months before its 2008 bankruptcy), was considering a pivot of its own. Until then, Zimride focused mostly on long-distance carpooling between college campuses, something co-founder Logan Green had been obsessed with since his days at UC Santa Barbara. But despite working long hours with his partner, John Zimmer, Zimride was mired in the doldrums. Peer-to-peer sharing—the kind Sunil Paul was pursuing at Sidecar—presented an interesting opportunity.

Kalanick was growing nervous. Across town at Uber's headquarters, he had heard about Zimride's plans, and he had heard whispers about

Sunil Paul's escapades, too. Kalanick considered Mark Zuckerberg a friend—or at least a familiar acquaintance—and the Facebook CEO had given Kalanick a heads up. Facebook employees were going *crazy* for Sidecar, Zuckerberg told him. Zuckerberg warned Kalanick that he might want to keep an eye on the company.

Soon after, Green and Zimmer announced their pivot. Zimride would abandon its long-distance carpooling program and launch a new service called Lyft; the plan was to make casual ride-sharing a fun, friendly experience, asking passengers to ride shotgun next to their drivers and strike up friendships while joyriding to their destination. The cherry on top was a cutesy pink mustache. Lyft sent all of its drivers giant, whimsical, plush hood ornaments to affix to the front of their cars.* It was an instant hit.

Kalanick sprang into action. He told his lieutenants, Ryan Graves and Austin Geidt, to take care of Lyft before it grew into a real threat.

Graves, Geidt, and especially Kalanick weren't above playing dirty. They started booking secret meetings with regulators in San Francisco and encouraging them to go after Lyft and Sidecar. Where once Uber had scoffed at City Hall, now they implored city officials to shut the other companies down. "They're breaking the law!" Geidt and Graves said to the indifferent regulators. Though Sunil Paul's efforts with Sidecar weren't taking off, Lyft was gaining traction quickly. People loved the stupid pink mustaches.

In theory, regulators were against Lyft's antics; after all, the company *was* breaking rules. Uber had been recruiting drivers for some time, but within limits; all of Uber's drivers were licensed livery vehicle operators registered with local transportation offices. Lyft turned that on its head. The mustachioed startup invited anyone with a car and an ordinary Class C driver's license to start driving for Lyft.

But as one Uber employee competing with Lyft at the time said,

---

* Ridiculously enough, the idea for Lyft's pink mustache sprung out of one employee's recognition of the popularity of "truck nutz," literally a pair of fake testicles that drivers could affix to their cars' bumpers. For some reason, both were wildly successful with the public.

"The law isn't what is written. It's what is enforced." To Kalanick's dismay, SF transit authorities weren't enforcing a damn thing. For all his bluster about ignoring regulators and disrupting an industry, Kalanick hadn't actually gone as far as Lyft and Sidecar. Up until then he hadn't been willing to cross the line into extreme ride-sharing.

But he was wrong to hesitate. After Kalanick took his first Sidecar, it clicked. There was an enormous potential market in peer-to-peer ride-hailing with everyday drivers. Kalanick needed to build the same thing for Uber.

From the sidelines, what Gurley saw struck him like a lightning bolt. Uber wasn't just fighting for a piece of the taxi and limousine market. It was competing against every mode of transportation in existence.

"Could Uber reach a point in terms of price and convenience that it becomes a preferable alternative to owning a car?" Gurley later wrote on his blog.

Uber decided to go all in. In a policy paper published to the company's website, Uber announced that it had created a low-cost option, "UberX," that allowed for ride-sharing. Uber was going head-to-head with Lyft.

"We could have chosen to use regulation to thwart our competitors," Kalanick wrote, disingenuously, upon flipping the switch to launch UberX. "Instead, we chose the path that reflects our company's core: we chose to compete."

MOST PEOPLE WHO KNOW Travis Kalanick remark on one thing: in every game he plays, every race he enters, in *anything* where he's asked to compete against others, he seeks nothing less than utter domination.

Friends who grew up with Kalanick said he was obsessed with being the best, be it running track in middle school against the teams from across the Central Valley or participating in debate competitions—something he did for fun—all in order to win.

"He used to give some of his teachers nervous breakdowns," his mother, Bonnie, once said of Kalanick's tenacity. Debate was particularly stimulating for him. He enjoyed finding logical pathways forward in an argument and exposing weaknesses in his opposition. (Even decades later, little excited Kalanick more than discovering an opponent's vulnerability and exploiting it.)

It wasn't just that he liked to win. Kalanick *needed* to win. Winning was the only option, his only goal. If you weren't going to go home with the gold medal at the end of the day, why even show up to the game?

At Uber, winning meant the obliteration of any opponent. There wasn't enough room for Uber and Lyft to coexist, he believed. The game was zero-sum. Every single ride-hailing car on the road in every single important market should have an Uber driver behind the wheel. Nothing less than a complete monopoly would suffice.

Kalanick enjoyed the fight. At first he began to needle John Zimmer, Lyft's co-founder, on Twitter. In playful jabs, he would troll Zimmer by asking about Lyft's insurance policies, business practices, and other seemingly esoteric shoptalk. Then he would start picking Zimmer and Lyft's business apart.

"You've got a lot of catching up to do," Kalanick would tweet at Zimmer. He loved adding the hashtag "#clone" to his tweets, insinuating that Lyft was an Uber copycat. Zimmer tried to take the high road when he responded, but Kalanick was pissing him off.

"He wasn't satisfied with winning," one former Uber executive said of Kalanick's drive. "He needed to rub your nose in it. Like a master training a dog to submit. It was intense."

Every time Kalanick drew blood, he pushed further. Zimmer spent months on the road as Lyft began to gain traction, soliciting Silicon Valley venture capital firms, hedge fund managers, and private equity outfits for funding to grow their business. Whenever Zimmer walked out of a meeting with a new potential investor, however, Kalanick would undermine him. Somehow Kalanick always knew where Zimmer had been.

"We knew that Lyft was going to raise a ton of money," Kalanick once admitted on the record, bragging about his desire to cripple his competitor. Kalanick would make sure investors knew that, between the two companies, they could only invest in one. His primary concern was information sharing. He would tell potential investors, "Just so you know, we're going to be fundraising after this, so before you decide whether you want to invest in them, just make sure you know that we are going to be fund-raising immediately after."

The tactic worked. Zimmer would soon get a call from the investor, apologizing and backing out of Lyft's latest series.

Wherever Lyft went, Uber showed up to harass them. One of Lyft's most effective grassroots tactics was holding what they called "driver events," small parties for a hundred people that Lyft was trying to court as drivers. These events—replete with booze, pizza, cakes, and party games—often endeared the drivers to Lyft; people who attended them felt like the company actually cared about them.

Kalanick made sure to ruin those for Lyft, too. He'd send his own employees to the events, where they would show up in jet black T-shirts—Uber's signature color—carrying plates filled with cookies, each with the word "Uber" written in icing. Each Uber employee had a referral code printed on the back of their T-shirt. The codes were for Lyft drivers to enter when they signed up for Uber, earning them a bonus.

Even when they weren't crashing Lyft parties, Uber found ways to mess with Lyft. All around San Francisco, Uber bought street signs and billboards targeting Lyft. Each billboard showed a large, black disposable razor blade with "Uber" printed on the handle, poised above one of Lyft's pink, cuddly trademark. In the text beside the graphic, Uber made its message clear: "Shave the 'Stache."

Beyond the pranks and Twitter trash-talk, Kalanick figured out a much more effective way of killing off his competitors.

As he once put it to his employees, quoting Puff Daddy: "It was all about the Benjamins."

UBER HAD DISCOVERED a winning formula to expansion. But each new city required capital, an upfront investment to kickstart what they called the demand "flywheel." Drivers wouldn't work for Uber unless there was enough demand from riders. And new riders wouldn't sign up or return unless there was a critical mass of available drivers. It was a classic chicken-and-egg problem.

"Uber solved that problem by straight-up buying the chicken," Ilya Abyzov, an early Uber manager in San Francisco, told friends of the strategy. Uber began torching hundreds of thousands of dollars, giving away the money as driver subsidies. They paid bonus cash when a driver completed a certain number of rides or drove for a certain number of days. Uber would also flood the rider side of the market with cash, doling out thousands of dollars in free rides to new customers. Their theory was, if we can get people to use our service, they'll see how amazing it is and won't want to stop.

And they were right. Once Uber hit a new city, word of mouth spread quickly; customers loved the novelty of seeing their ride wind its way towards their location on the app. People loved how shockingly cheap the (subsidized) rides were. They enjoyed not fumbling for cash, or having to tip drivers when they left the car. Uber appeared out of nowhere, and it was magical.

But making the magic cheap for users required cash. Kalanick knew Uber had to grow quickly, in hundreds of cities, before competitors and regulators could stop them. To do that, Kalanick knew what he needed: a war chest.

Kalanick was good at putting on a show for venture capitalists. Even as a child, he had always been a talented showman. He had already spent years giving pep talks and advice to young entrepreneurs during his period of angel investing. Now, preparing his funding talks, he'd spend hours preparing a slick PowerPoint slide deck with eye-popping financial statistics. He'd rehearse his presentation by himself, over and over, making sure he clicked the remote control for the next slide at the exact right moment in his speech; timing was crucial.

When he was on, Kalanick was *on*. He was a force of nature with investors, a Jobsian tech wizard crossed with the hard-charging motivational speaker played by Alec Baldwin in *Glengarry Glen Ross*. "A-B-C," Kalanick chanted to himself, repeating Baldwin's words in his head. "A-Always, B-Be, C-Closing. Always be closing. Always be closing!" Kalanick didn't fuck around; he knew how to close a deal.

The first few rounds brought Uber tens of millions in venture capital. But Kalanick needed more. A lot more. The company was entering the big leagues of fundraising, where Uber wouldn't be asking for an errant five to ten million dollars from a rich tech enthusiast.

Uber needed billions.

Chapter 10

THE HOMESHOW

IT WAS GURLEY WHO CONNECTED KALANICK TO HIS SECRET FUNDRAIS-
ing weapon.

A good venture capitalist helps a startup recruit. Gurley wanted to find Kalanick a funding wingman, and he had the exact right person in mind: a talented dealmaker from Tellme Networks, a telecommunications software outfit that had been around since the late nineties. Tellme powered telephone-enabled apps, like voice-based personal assistants, or the automated software airlines use to answer the calls of irate customers with flight delays.

Emil Michael was the deals guy for Tellme. Though his sense of humor was brash, Michael presented to clients as a polished entrepreneur; he knew how to glad-hand MBA types around the Valley. Tellme had survived the dot-com bust in part because Michael had struck partnerships with big corporations like AT&T, Southwestern Bell, Fandango, and Merrill Lynch. Even after the bubble burst and Tellme needed to lay off staff and retrench its operations, the company was able to parlay its assets and talent into a home-run sale to Microsoft in 2007 for more than $800 million.* Michael knew how to get a deal done.

---

* Ex-Tellme staff would have wide influence the next generation of internet development. Mike McCue, its CEO, went on to found Flipboard, while others like Alfred Lin would work on Zappos and eventually land at Sequoia Capital. Hadi and Ali Partovi, well-respected entrepreneurs and brothers, founded Code.org. Others went on to join Stripe, Facebook, Amazon, and others. Emil Michael was in good company.

A first-generation immigrant from Egypt, Michael grew up as the son of a pharmacist father and a chemist mother in Westchester County, New York. Michael was a child of the New Rochelle suburbs, a family of color in a largely working-class neighborhood. To fit in, Michael networked. He was gregarious from a young age, chatting with customers more than twice his age while working behind the counter for his father at the small-town pharmacy. People knew the Michaels, and young Emil and his family knew them in turn.

Michael earned good grades and secured a spot at Harvard for undergrad. There, he studied government and then went on to Stanford Law School, which brought Michael to the heart of Silicon Valley. He graduated with top marks, eventually landing a job at Goldman Sachs in the communications, media and technology group. There, Michael cut his teeth in the dealmaking world, watching his co-workers buy and sell companies every day. They fought like gladiators, recapitalizing firms, flipping them, stripping them for parts. But it was the burgeoning technology group that whet his appetite for eventually landing inside a Silicon Valley startup.

After Goldman, Michael began his nine-year run at Tellme, and then went to Washington to take a job as a White House fellow in the Obama administration, working as a special assistant to the secretary of defense. That rounded out his skill set; a deals guy with Washington contacts could go far in the private sector, where Michael felt most at home. The government job lasted a few years before Michael went to Klout, a social influence measurement company. Using proprietary algorithms, Klout scored the amount of reach a given consumer might have across sites like Facebook, Twitter, and Tumblr. Users with high Klout scores were rewarded with perks at partner companies; an influential Klout user might get an upgrade on a Virgin America flight, or a free breakfast at the Palms hotel in Las Vegas. Michael was the guy who struck all the deals with partner companies. Klout executives loved his hustle.

Bill Gurley had connected the two men in 2011, but they wouldn't truly sync up until Gurley called Emil Michael in 2013, asking if he wanted to hear about an enormous opportunity. "We need you over here," Gurley said. Working for Uber early in the company's life cycle, Gurley said, was too good an opportunity to pass up. Gurley loved how aggressive Kalanick was, but he knew the CEO needed a counterbalance, someone to check Kalanick's baser tendencies. Michael, Gurley thought, could be the grown-up in the room.*

Besides babysitting Kalanick, Michael would do for Uber what he had done for Tellme and Klout: strike lucrative deals with partner companies. Gurley had never seen a more talented dealmaker than Michael, whose silver tongue and affable personality could charm business development types.

The pharmacist's son had something Kalanick lacked; he had the emotional intelligence to adapt to any situation. Kalanick could be pigheaded. With slicked-back black hair and dark features, Michael would shake your hand and meet you with a wide smile, putting you at ease even as he sized you up. Every interaction was a negotiation, every opening a potential weakness. He spoke the language of the Street and, having spent more than a decade around technology companies, could adequately sell the abilities of Uber.

But the qualities that made him such an effective dealmaker had their flip side. Michael tended to mirror his partners, taking on the qualities of the group in order to fit in. It was his instinct from his youth in Westchester; to avoid being an outsider, he would become the ultimate insider. At his best, he was a great drinking buddy and even better new friend. At his worst, he was an enabler—a partner who not only helped mastermind the plan, but conspired in the coverup as well.

Kalanick took to Michael immediately. Michael was hired on as Kalanick's second in command of sorts, officially Uber's "Chief Business Officer." The title was akin to a chief operating officer, though in practice he became "dealmaker in chief."

His real job would eventually enmesh itself with his other job; best friend to Travis Kalanick. Michael and Kalanick became inseparable, chatting strategy and business together throughout the day

---

* Gurley thought wrong. The relationship would go spectacularly awry.

while spending evenings and weekends hanging out. They would dine together, take road trips to speak to partner companies together—they'd eventually begin vacationing together, a foursome composed of Kalanick, Michael, and their respective girlfriends. They took trips to places like Ibiza and Greece together—boundaries between the two melted away, personal and professional merged. They were "bros" and acted accordingly, spending lavishly at nightclubs and upscale dinners that befitted the lifestyle Kalanick imagined appropriate for himself and his close friends.

Where Kalanick and Michael really shined, however, was in raising money. The two perfected their technique through sheer force of repetition. In a Wall Street IPO, for instance, a company puts on what is called a "roadshow," in which bankers representing the startup travel from city to city pitching investment firms on their company. Kalanick, however, had no intention of going public (at least, not any time soon). So, he and Michael developed their own method, affectionately titled "the Homeshow." There was enough interest in Uber that the two flipped the power dynamic, forcing investors to come to Uber's San Francisco headquarters, fighting to get in on *their* dance card and waltzing to *their* tune.

Kalanick and Michael created a system built around scarcity. Uber would hold only three meetings with bankers per day for the span of a week, and the investment firms would have to jockey for a time slot.

They called Kalanick "the showman," and he was. He had the poise, timing, and "wow-factor" needed to pique the interest of bankers, VCs, and hedge funds who sat through hundreds of startup pitches a year. Kalanick brought a meticulously composed slide deck, larded with cherry-picked numbers that showed Uber's enormous "hockey stick potential"—a term that referred to the shape of the growth curve every entrepreneur and venture capitalist wants to see when building a company. And he didn't have to work hard to get those numbers. Uber had what was called "negative churn"—a term often used to describe software as a service, or SaaS, companies. Having negative churn meant that once customers used the product, they were more likely to keep using it regularly thereafter. "It means that customer accounts are like high-yield savings accounts," a venture capitalist once wrote of the term. "Every month, more money comes in, without much effort."

Kalanick's data showed that by the time a customer used Uber an average of 2.7 times, they became a customer for life. The product was *just that good.*

Kalanick modeled his approach after his idols: Steve Jobs, Mark Zuckerberg, Larry and Sergey. He positioned Uber among the famous world-changing tech companies, and implicitly put himself among those legendary founders. His performance in the boardroom convinced each set of new executives that he might be right.

Then, after Kalanick had wowed the room, Emil Michael was the closer. As Kalanick whirled through his pitch deck, Michael kept an eye around the boardroom table for body language. Who was leaning in? Whose eyes lit up at the sight of our growth numbers? Who couldn't wait to make an offer? Investment firms would send enthusiastic follow-up notes, but Michael would wait to respond, making them sweat. A week later, potential investors would receive an Excel spreadsheet to fill out, asking them how much money they'd be willing to put into the company and at what valuation. Kalanick set them up, Michael knocked 'em down. The entire process, soup to nuts, took the duo three weeks. It was a dance they would repeat, time and again, over the next five years.

Kalanick and Michael had another advantage, due mostly to luck and timing. From the earliest days of Silicon Valley, the funding ecosystem had been a relatively small one. Local VCs invested in local startups. Venture firms had investment partners with technical chops, those who could appreciate the complexity and logic of their portfolio companies. VCs picked their companies wisely—or at least due to a kind of logic and overarching investment thesis. This dynamic persisted through generations of boom-and-bust cycles.

But the rise of technology companies attracted a different kind of bankroll. Outsiders were starting to have "FOMO"—fear of missing out—as tiny startups began to yield outsized returns. Over two years beginning in 2005, YouTube raised about $10 million in VC capital; by 2006, Google acquired the startup for more than 150 times that amount. Mark Zuckerberg spent $1 billion on Instagram when the company had just thirteen employees. No one wanted to miss the waves of tech money flooding in.

Mutual funds, investment bankers, overseas sovereign wealth funds, and foreign governments noticed the enormous wealth being created by IPOs held by Google, Twitter, and Facebook in Silicon Valley. And they saw that the most obscene wealth accrued to the early investors who bought before the companies went public.

Traditionally, hedge funds stuck to markets they knew and invested across a range of publicly traded companies. But slowly, institutional investors from these funds—the T. Rowe Prices and Fidelity Investments of the world—started to trickle over to Silicon Valley. Hedge fund portfolio managers who oversaw hundreds of millions of dollars *knew* they had to be invested in tech, lest they miss the boom. And of all the private companies raising money in the Valley, Uber was the most important. Uber was the unicorn to end all unicorns, and investors were desperate to land a share.

Kalanick took advantage of that demand. He still harbored wounds from his early experience with Michael Ovitz, the venture investor who betrayed him when Scour was sued by the entertainment industry. After that experience, Kalanick never trusted investors again. So as a condition of allowing them to offer him money, Kalanick offered them miserable terms. Private companies aren't obligated to make their internal statistics public, but investors with a significant ownership stake are generally given insight into the company's financials. Kalanick, however, over time stripped some major investors of all "information rights," and limited the degree of detail offered to others. Moreover, investors had to agree that Kalanick would continue to hold his supervoting shares while newcomers only received shares with weaker voting power. Every supervoting share Kalanick held counted as ten votes in the company, whereas every common share only counted as one vote. Kalanick also had the allegiance of Garrett Camp and Ryan Graves—his two early co-founders and strong allies who also held their own cache of supervoting shares.

In effect, Kalanick had created a powerful cabal that supported his power as chief executive. No investors could meddle in how he spent Uber's money, no shareholders could tell him who to hire, who to fire, and so on.

Uber was Travis Kalanick's company—and if you were lucky, he would let you invest.

GOOGLE VENTURES HUSTLED to be invited to buy into Uber. But at every step, Travis Kalanick kept asking for more.

David Krane, a longtime Google employee turned venture partner, had spent months stalking this deal. He had heard Uber was raising capital again. Krane just needed a chance to get himself in front of Kalanick to charm the entrepreneur into taking his money.

Krane made headway whenever he could. In early 2013, the investor spotted Kalanick at the annual TED Conference at the Long Beach Performing Arts Center. Kalanick was sharing a laugh with Cameron Diaz, starstruck at the actress's fame. Sensing an opening, Krane sidled up to the conversation and politely nudged Diaz aside, inserting himself in front of Kalanick. Krane had done some big deals for GV in the past—including Nest, the smart thermostat company, and Blue Bottle, the boutique coffee chain—but Uber was the whale he dreamed of landing.

The TED moment had left an impression on Kalanick; he liked the idea that a company as vaunted as Google was pursuing him. Later that year, Krane and the other top partners at Google Ventures spent months courting Kalanick, hoping to get a piece of Uber's Series C round of venture financing. As the two teams sized each other up, Kalanick sent over the usual, blunt demands: "Your firm comes to our building," Kalanick said, "and you present an investment proposal to us. Then, we decide if we want to let you in."

The Google guys weren't used to this. Getting an investment from Google Ventures was a *privilege*, not something an entrepreneur had to think about. Though Google Ventures hadn't been around as long as storied institutions like Kleiner Perkins or Sequoia, an investment from GV was a strong signal that your company was legit.

Krane and his partners made a first-class presentation to Kalanick at Uber headquarters. They vowed to give Uber all kinds of support, be it by helping to recruit talented executives from GV's vast network, or offering GV's deep strategic experience. All of that, plus a yachtful of money. Krane and his partners did well. They convinced Uber to do the deal.

Then Kalanick gave them the numbers: Uber, he said, wanted to raise $250 million from a single investor, valuing the company at a whopping $3.5 billion.

The Google guys bristled. This was a staggering amount of money, even for venture capital. Google Ventures hadn't typically been doing investment rounds of that size, usually opting to focus on early and "growth-stage" rounds. They felt more comfortable writing smaller checks to companies, in the realm of single- and low-double-digit millions. And GV usually invested earlier in a company's history for a larger amount of equity ownership. Seed investing meant taking on more risk. It also meant more reward if the company turned out to be a home run.

But this Uber scenario was different. Google Ventures was being asked to write a quarter-billion-dollar check—a substantial chunk of the capital *in the entire fund*—to just one company. And it was being told it should be grateful for the chance. Krane and his partners were not used to this.

After a prolonged back and forth, Krane convinced his partners to bite the bullet. They cut the single biggest check GV has ever written to a portfolio company—and they were treated just like everyone else.

As he had done with previous investors, Kalanick shut Google Ventures off from receiving regular, detailed information about Uber's progress. The sheer size of the check only managed to buy an observer seat on Uber's board of directors—a much coveted, if limited, spot for such a high-profile company. Normally, the guy who leads a large investment like this one gets a seat with voting power. But Kalanick brushed Krane aside for someone higher in the Google hierarchy: David Drummond.[*]

---

[*] Krane negotiated a "board observer" seat, which allowed him to attend meetings, but without the ability to vote. It was irksome, given that Krane had sourced the deal himself, but better that than be cut out of the room entirely.

Kalanick gave Drummond a proper seat on the board, which was no small matter. Drummond had started working with Google in its infancy. He had been a partner at Wilson Sonsini Goodrich & Rosati, the high-profile Valley law firm, when he first met Larry Page and Sergey Brin. After helping Google raise some of the company's first investment rounds and bonding with the co-founders, Drummond joined Google full-time in 2002 to help lead the company through its eventual IPO. Having arrived early, Drummond rose to become one of Larry and Sergey's most trusted lieutenants, eventually landing such lofty titles as "Senior Vice President of Business Development" and "Chief Legal Officer" at Google. He was also given purview over Google Ventures' investments, as well as investments through Google's other major investment arm, Google Capital. In short, Drummond was what they referred to in the Valley as a "BFD"—a big freakin' deal. He was strategic, well-connected, and highly prominent. To Kalanick, having David Drummond join Uber's board would signal that the company enjoyed the full strategic power of Google. Drummond acquiesced, agreeing to join the board.

Kalanick also managed to sneak in another last-minute surprise for Krane. Up until the day the deal would close, Krane knew Google Ventures had been competing with another, unidentified, firm for the investment spot. And Krane was led to believe that Uber ended up choosing GV over that other firm. But at the eleventh hour, after Krane and Kalanick had spent weeks hammering out the deal terms, Kalanick informed Krane he wanted to include another investor in the round: TPG Capital.

Krane was pissed. TPG Capital was one of the world's preeminent private equity firms, having participated in some of the most high-profile leveraged buyouts of companies in corporate history. In 2007, TPG partnered with Goldman Sachs to buy out Alltel, then the world's fifth-largest cellular carrier, for roughly $27.5 billion. At the time, it was the largest leveraged buyout in the telecommunications industry. Google Ventures was one of the biggest fish in the Bay Area, but Kalanick wanted the swagger and global connections that came with taking money from TPG Capital. The firm also offered him a trip on its

corporate jet, a creature comfort that came with only top-tier firms.*
Though he worked with David Trujillo, a TPG partner, to put the
deal together, Kalanick wanted a bigger name from TPG on the Uber
board—David Bonderman. A legend in private equity, Bonderman was
a founding partner at TPG who harbored connections with celebrities,
executives, regulators, and heads of state around the world. Just as
Drummond's name sent a certain message to the tech world, Bonderman's participation telegraphed Uber's importance to the broader business community. And in the end, Kalanick just wanted the investment.

In the end, despite investing some $258 million in the round, Google
Ventures and Krane had to acquiesce. TPG ended up purchasing $88
million in shares directly from Garrett Camp, who was willing to sell
some of his stock.† There was nothing Krane could do to stop the sale.

And Kalanick wasn't done asking for things. He then demanded the
most coveted sugar-plum status symbol in Silicon Valley: a meeting
with Larry Page.

---

* If it was meant to dazzle Kalanick, it didn't work; his TPG jet flight was a one-way trip to
Beijing. On the way back home, he had to fly commercial.

† Camp, unbelievably, used the money to continue funding StumbleUpon, which he still
believed could become a dominant force in social networking. After years of flailing, Camp
finally pulled the plug in June 2018.

PART III

Chapter 11

BIG BROTHER AND LITTLE BROTHER

THE FOUR SEASONS IN PALO ALTO IS A TOWERING SEMICIRCLE, sheathed entirely in reflective glass, rising above US-101. The windows shimmer silver in the midday sun, as if the building were a giant processor chip lodged in the heart of Silicon Valley.

It was also a ten-minute drive from Google's global headquarters in Mountain View. Somehow Krane had been able to accommodate Kalanick's final demand and lock in a meeting with Larry Page and David Drummond. Kalanick and his dealmaker, Emil Michael, were invited to a 9:00 a.m. breakfast at the Googleplex.

But Kalanick was a night owl, accustomed to working until 11:00 p.m. followed by a nightcap with other entrepreneurs nearby. There was no *way* Kalanick could make a nine o'clock meeting in Mountain View. So, Krane booked Kalanick a suite at the Four Seasons.

Krane had a surprise in store for the CEO that morning. Kalanick walked out of the hotel doors to his idling Uber. He tossed his backpack in the backseat and prepared to head south.

Before his Uber driver could pull away, another car showed up—one that looked nothing like the Porsches and Teslas idling near the valet stand. Krane had talked the engineers in Google's "X" division into loaning him one of Google's famed self-driving cars. The white Lexus SUV pulled up, sporting Google's logo beneath an array of lasers and cameras. Kalanick's unmanned chariot had arrived.

Krane's stunt worked. Kalanick was stunned, giddy like a teenager. He cancelled his Uber, hopped into the back of the Lexus, and accepted his ride down south towards the future. (He was so excited, in fact, that he left his backpack in the other car.)

The meeting proved exactly as wonderful as Kalanick had imagined. The group—Larry, Drummond, Kalanick, Michael, and Bill Maris, the managing partner at Google Ventures—spoke like old friends, musing together on what fruits the partnership would bring to the world.

Meeting Page, in particular, thrilled Kalanick. Page was the type of founder Kalanick had worshipped from a young age. He was a self-made man who had engineered an elegant solution to an insanely difficult problem, organizing the entire world's information using search algorithms. Kalanick loved efficiency—Scour and Red Swoosh, his first two startups, were predicated entirely on the notion of being efficient—and Google was the most efficient search engine ever built. Kalanick felt that with Page as his mentor Uber would be unstoppable.

"It was like big brother and little brother," Kalanick later said about that first meeting.

Travis's impression of the meeting didn't exactly sync with reality. Anyone who has met Larry Page knows that he is *not* a personable guy, and the furthest thing from anyone's "big brother." Page is an engineer's engineer; socially awkward, disinclined to take meetings with anyone outside of his inner circle, and obsessive about the incredibly complex problems he wishes to solve.

For Page, the investment was strategic. The Google co-founder harbored a deep interest in transportation. He kickstarted Google's self-driving car research long before other tech and automobile companies thought it was possible. He poured millions of his personal wealth into researching flying cars. Larry Page didn't care about Travis Kalanick; he cared about the future of transportation.

Moreover, Kalanick never internalized Page's philosophy on in-house competition. Larry Page gave his divisions a large degree of autonomy. Google Ventures, in particular, told outsiders that it was a separate entity from Google proper, meaning it didn't necessarily report back to the mothership. Ostensibly, it was also true that just because you were getting a GV investment didn't mean you had Google's support.

Despite Page's lack of visible affection during the breakfast, Kalanick believed he was making a crucial ally. The group talked about potential ways they could collaborate. Perhaps the companies could improve Google Maps through Uber's millions of daily trips. (Uber, in turn, was powered navigationally by Google Maps.)

Page stayed for a short time to talk about these partnerships, and then excused himself to wander his sprawling campus. Kalanick couldn't wait to meet him again for their next "jam sesh."

AS KALANICK WAS HAMMERING out an exciting future for Uber with top brass at the Googleplex, Anthony Levandowski sat frustrated just a few buildings away.

Levandowski had dedicated his life to technology and robotics. Born in Brussels, Levandowski immigrated to the United States as a teenager and landed in Marin County, just across the Golden Gate Bridge from San Francisco. From a young age he was obsessed with maps and vehicles. And he loved building and tinkering. He spent his undergrad years in the East Bay, at the University of California, Berkeley, where as an industrial engineer he built one of his first robots— a Lego-constructed machine that could pick up and sort Monopoly money. Soon he convinced his classmates to enter the DARPA Robotics Challenge with him, a program put on by the Department of Defense in which competitors would build autonomous cars and race them across the Mojave Desert. They entered with high hopes, but the autonomous vehicle they built, a motorcycle nicknamed "Ghostrider," ended up crashing within seconds of beginning the race.[*] The loss deflated him; Levandowski liked to win almost as much as he liked building his robots.

He scored a post-collegiate job at Google working on the company's Street View project. Levandowski was the exact type of engineer

---

[*] Levandowski and his team still ended up winning, in a way; Ghostrider is on display at the Smithsonian Institution.

Google loved to hire; curious, brilliant, and harboring a wide array of interests outside of his main duties at Google.

Levandowski became an unmistakable presence on campus, in part because of his enormous height; he was six feet, seven inches tall. But his personality loomed just as large. He was gregarious, engaging, sharp, and messianic about tech—especially his own projects.

At the time Levandowski worked there, Google encouraged employees to embrace the company's "20 percent time" initiative. That meant 80 percent of your work at Google was meant to focus on your job, but you could spend 20 percent of your time working on other interests.

For Levandowski, that meant building robots. He formed a startup outside of Google named 510 Systems—a nod to the Berkeley telephone area code—and with a group of other employees began building tech that could one day prove useful to Google. That included sensors and other software specifically for self-driving cars. Unbeknownst to Google, the search giant was soon buying much of its tech for the street-mapping project from one of its own employees, Levandowski, who sold the gear via a middleman.

Google eventually found out about Levandowski's ruse. Instead of firing him, Google decided to buy Levandowski's startup for $20 million.

Side hustles like 510 Systems defined Levandowski. He liked money, but what he liked more was finding hacks and work-arounds. Levandowski may have labored at a giant corporation, but he was still a scrappy startup guy at heart. Building a business and selling it back to Google was validating; he had found a hole in the 20 percent time system, exploited it, and won. The $20 million windfall was good, too.

But he was after more than just money and hacking. For years, Levandowski believed that humans moved around the world in a way that made no sense. Tens of thousands of people were killed in automobile accidents annually. Traffic in major urban areas, especially the San Francisco Bay Area, was abysmal. People clogged the streets with inefficiently operated cars. One person to every vehicle on the road was inefficient and wasteful. A fleet of self-driving cars, used only when necessary, would be far cleaner and more cost-effective.

Once Google had bought Levandowski's startup, he dove headlong into mapping and self-driving tech for his superiors, joining the secretive Google X division. Colleagues said Levandowski deserved much of the credit for convincing Google's top brass, especially Larry Page, to pour millions into self-driving research. And by virtue of working on a project dear to the CEO's heart, Levandowski began to develop a special relationship with Page.

But he was also shrewd. When Google bought 510 Systems, Levandowski sold it for just under the amount that would have required him to share the profits with the fifty or so employees under him, depriving dozens of his colleagues of a rich payday. Even worse, Google hired less than half of 510 Systems' staff. The rest had little to show for their time spent working on Levandowski's robots.

Levandowski should have been ecstatic. Instead, a few years later, as Page was hashing out terms with Kalanick, Levandowski felt handcuffed. He had come to Google to build self-driving cars and upend the world of transportation. But Google, for all its foresight, was proving skittish.

Google was terrified to approve what Levandowski really wanted; true, open-road testing of autonomous vehicles. Aside from the ever-present concern about negative public opinion, the nonsensical design of San Francisco's traffic-clogged grid presented an absurdly thorny engineering problem. The smallest error risked a dangerous accident. Naysayers imagined a video of a Google-branded SUV wrapped around the mangled chassis of another car—or worse, the mangled body of a pedestrian.

But Levandowski knew Google needed real-world testing to get autonomous vehicles out of the conceptual phase. Levandowski imagined a future without automobile deaths or congestion, where carpooling was automatic and simple. And here was Google, dragging its feet because it was too scared to break a few rules.

Levandowski's leadership style often irritated other Googlers. He had sharp elbows, was pushy, tough on people, sneering when someone disagreed with him. While Google was careful and methodical, employees saw Levandowski as corner-cutting and occasionally reckless. Without telling his bosses, Levandowski hired an outside lobbyist in Nevada to write a new law that allowed autonomous vehicles to operate in the state without a backup safety driver. Google executives were furious, yet the law passed statewide in 2011.

Levandowski's divisive methods earned him enemies. When he made a play to become leader of the Google X autonomous vehicle unit, a group of employees staged a mutiny, requiring Page himself to step in and name Chris Urmson, a rival of Levandowski's, the head of the self-driving division. Levandowski was crushed and made no attempt to hide it; at one point, he stopped coming into work entirely.

Levandowski was in agony; he worked for the company with the most advanced tech in self-driving vehicles, yet seemed happy to let some other, more aggressive competitor take the lead. There had to be another way.

Chapter 12

GROWTH

GROW OR DIE.

It is the maxim by which every entrepreneur in Silicon Valley lives. From the moment a founder signs their first term sheet from investors, they've made a pledge to fight to keep the startup alive and growing, growing, growing.

Growth became Kalanick's mantra. Each morning, he would crack open his MacBook and skim through progress reports from his lieutenants in the field. He tracked new users in each city. He tracked "supply"— the name Kalanick used for the workforce of human drivers. He lived by the numbers. One day, people would be able to open the Uber app and get an array of items—from diapers to iPhone chargers—delivered anytime, anywhere on earth. Uber would be the logistics company that moved people and things across the planet. Amazon on steroids.

Since he was working all hours of the day, Kalanick expected the same of his employees. A job at Uber wasn't just a job, after all—it was a mission, a calling. If you weren't ready to stay late at the office and work nights and weekends, you shouldn't be working at Uber. Company-wide dinner service—a perk that most large Silicon Valley companies offered for those who worked after hours—wasn't served until 8:15 p.m. That meant you couldn't work an extra hour after five o'clock and strategically grab a free meal at, say, six on your way out the door. You'd have to work an extra 3.25 hours to get your meal.

And there was always work to be done. Each time Uber entered a new city, the company's "hockey-stick growth" attracted attention and competition. That meant employees would have to put in overtime to beat back their opponents, who were most often city regulators and taxi owners and operators, or if not them, then the local city councilman who served them. This dynamic—with Uber entering cities and taxi workers bitterly fighting back—would affect Kalanick. He began to act as though he were under siege.

To Kalanick, local "laws" were hypocritical rules that city officials put together at the behest of transportation groups. The way Kalanick saw it, Uber was engaged in a crusade; the company needed to win over consumers while battling the underhanded, street-fighting tactics of the entrenched interests—from the city council to the governor's office—who were colluding to keep taxi service bad and overpriced. He thought the taxi industry was run by "cartels." The whole system was corrupt.

BUT THE "CARTELS" weren't messing around.

Taxi owners knew they had to stop Uber. In some major cities, taxi owners had paid hundreds of thousands of dollars to purchase "medallions," taxi-service permits required by the local government. Medallions could be absurdly expensive, upwards of a million dollars in peak markets like New York City. Drivers and dispatchers took out huge mortgages to buy them. The limited number of medallions created an artificially constrained market, which meant cab drivers and taxi company owners could charge enough to earn a decent living (and pay for the medallion.)

Then Uber showed up. The medallion system—a market based entirely on scarcity and exclusivity—was threatened to its core. With UberX, the company's peer-to-peer service, anyone with a car could drive for Uber. That simple concept destroyed Big Taxi's barrier-to-entry system, sending the price of medallions plummeting. In 2011, medallions in Manhattan were going for $1 million apiece; six years later, one fire-sale auction of forty-six medallions in Queens fetched an average price of $186,000 per medallion. Overnight, taxi drivers whose entire livelihoods were tied up in paying off an expensive medallion went underwater.

Cabbies were aghast. Doug Schifter, a livery driver from Manhattan, faced financial ruin after the rise of Uber wrecked his income driving for traditional car services. Schifter drove to City Hall in Lower Manhattan on a cold Monday morning in February 2018, put a shotgun to his head, and pulled the trigger.

"When the industry started in 1981, I averaged 40–50 hours," Schifter wrote in a final post to his Facebook page. "I cannot survive any longer with working 120 hours! I am not a Slave and I refuse to be one." From Uber's early days up through 2018, more than a dozen other taxi drivers in New York and other major metropolitan areas also took their own lives.

Taxi drivers who didn't give in to despair, however, fought back. Some tried to beat Uber at its own game by forming taxi alliances and creating their own apps like iRide, Arro, Curb, and others. But taxi operators soon found the best way to fight back wasn't to compete with an app. It was to protect the turf they already had.

When Uber launched in a new city, taxi operators would often lean on their local transit agencies and taxi authorities, who would dispatch an official to Uber's local headquarters. Armed with a thick rulebook and a scowl, officials in New York, Nevada, Oregon, Illinois, Pennsylvania, and other states would point out the rules and laws the company was breaking. The "Weights and Measures" book was supposed to tabulate the cost of a ride, they would say, not some complicated algorithm inside the Uber app. When that didn't work, local legislators were known to dispatch city and state agencies to shut Uber down.

If none of these things worked, there was always good old-fashioned skull cracking. Taxi cartels in areas like Las Vegas and elsewhere had deep ties to organized crime, which meant serious and sometimes violent retaliation. Cars were stolen. Sometimes taxi owners would assault drivers and set their cars aflame.

In Italy, Benedetta Lucini faced pushback from local taxi thugs. As general manager of Uber's Milan office, she worked overtime to convince Uber drivers to stay on the road, even as taxi operators would hail Ubers to their location, pull drivers out of their cars, and beat them.

Eventually, taxi drivers targeted Lucini. They plastered posters with a photo of her face on taxi stands across the city, along with the phrase "I love to steal." On another occasion, cabbies threw eggs at her during a press conference. And one night when Lucini was returning home from work, she found a sign hung from a power line not far from her apartment. On the sign was Lucini's home address, and a message calling her a prostitute who provided her "services" to Milan's transportation chief.

And yet, under Kalanick, Uber didn't flinch. As they struggled with local officials, Uber teams devised a playbook for evading crackdowns. Deregulation—a pure, free market, untouched by the corrupt hands of government and Big Taxi—was the ultimate goal for Uber in every city.

Barging into a market first gave Uber a major advantage. In Philadelphia, for example, Uber pushed headlong into the market illegally, much to the consternation of the local Public Utility Commission. The city would levy a $12-million fine on Uber for its 120,000 violations of the transit code. (The company settled the matter for $3.5 million.) By then it didn't matter. Uber was up and running, having courted more than 12,000 new drivers and spinning up the demand for rides among consumers.

If transit authorities began policing transportation laws, local managers would blast emails and text messages to their driver corps, telling them Uber had their back. Kalanick viewed fines and tickets as just another cost of doing business. Text messages, like the one below, often promised full restitution from Uber if you happened to, say, have your car impounded by the police:

UBERX: REMINDER: If you are ticketed by the PPA, CALL US at XXX-XXX-XXXX. You have 100% of our support anytime you are on the road using Uber—we are here for you, and we will get you home safe. All costs associated will be covered by us. Thank you for committing to providing safe, reliable rides to the citizens of Philadelphia. Uber-ON!

At the same time, local managers were given millions in "incentives" to kickstart demand. Everyone had a smartphone, everyone was fed up with their local metro and taxi services, and everyone loved the free rides.

Uber made it as easy as possible for drivers to sign up. The company used a background check system that moved new recruits through the system quickly. Taxi and livery services used fingerprint testing, which offers a thorough history of a driver's past, but often took weeks to complete. Uber used an outside firm, Hirease, which boasted an average turnaround time of "less than 36 hours." Hirease did not require fingerprint tests.

Waiting weeks for a background check was intolerable for Uber. A week was a year; a month was eternity. After perfecting the quick background check process, Uber's political machine went to work. In states where fingerprint-based background checks were legally required, Uber hired lobbyists to get laws rewritten that mandated drivers undergo the traditional checks.

Uber spared no expense on local lobbying campaigns. The company regularly topped the list of biggest spenders across states like New York, Texas, and Colorado—and dozens of others where they faced legislative opposition—throwing down tens of millions of dollars annually to sway lawmakers. David Plouffe, a former Obama administration political operative, was a major hire who knew how to influence city-level as well as national politics. In Portland, Uber hired Mark Weiner, one of the most powerful political consultants in the city. In Austin, Uber and Lyft paid $50,000 to the former Democratic mayor to lead their campaign against regulation. Later, as Uber matured, the company's staff swelled to include nearly four hundred paid lobbyists across forty-four states; the number of ride-hailing lobbyists outnumbered the paid lobbying staffs of Amazon, Microsoft, and Walmart combined.

The money was well spent. Uber was able to sway legislation in many states. As a result, legislators rarely, if ever, raised the issue of Uber's employment liability for its "driver-partners." That meant Uber could define Uber drivers as contract workers—designated 1099 in tax code parlance—which allowed Uber to skirt paying for benefits like unemployment tax, insurance, and health care. Avoiding these normal employment expenses saved enormous amounts of money, and wildly decreased Uber's liability for drivers' actions.

Lobbying wasn't always a silver bullet; sometimes Uber had to play hardball. Kalanick would order lieutenants to threaten to withhold service to customers or cease operations entirely if it looked like city legislators weren't going to cave to Uber's demands on issues like fingerprint background checks or driver caps.

Uber treated each market less like a negotiation and more like a hostage situation. Kalanick had no problem pulling out of a market entirely—as it did in Austin—especially after they had been operating for a number of months beforehand. The company had leverage: people loved using the service. In nearly every major metropolitan area, the "product-market fit"—a tech industry term to describe how well a given service may do with the public—was near perfect. People hated taxis, and loved ordering a car with their phone. To have such a service taken away roused public anger.

Kalanick saw that weakness and did what he did best: he exploited it. Uber general managers would run entire campaigns harnessing public frustration and telling people to direct their anger to their local lawmakers and elected officials.

In New York in 2015, when Mayor Bill de Blasio threatened to cap the number of cars on the road, Uber tweaked the software inside of its app for New York based riders to show what it called "De Blasio's Uber." That option showed fewer animated cars driving around on the mini-map inside the Uber app, with approximate wait times of up to a half hour—five to six times longer than people usually had to wait for a ride. "This is what Uber will look like in NYC if Mayor de Blasio's Uber Cap Bill passes," said the text inside a small, pop-up notification. Users were invited to "take action," and were presented with a button inside the app that emailed the mayor and the city council directly with a form letter prewritten by Uber. By the end of the campaign, the mayor's office had received thousands of letters from upset users protesting the potential ban. De Blasio ended up shelving the proposal.[*]

---

[*] De Blasio got his revenge and imposed a cap in 2019.

The strategy was extremely effective. So effective, in fact, that Uber decided to systematize and weaponize it across the company. To this end, Uber hired Ben Metcalfe, a caustic, outspoken British engineer who described his job on LinkedIn as building "custom tools to support citizen engagement across legislative matters" to drive "social good and social change." Metcalfe and his team built automated tools that the company used to spam lawmakers and rally users. With easy, in-app buttons, users could send emails, texts, and phone calls to elected officials whenever an important legislative matter was up for debate. By 2015, more than half a million drivers and riders had signed petitions supporting the company across dozens of states. After Uber sent out a mass text message asking for support, petitions began gaining new signatures rapidly, in some cases as many as seven per second.

If all else failed, theatrics and pageantry worked, too. After the Metropolitan Taxicab Commission blocked Uber from operating in St. Louis, Sagar Shah, Uber's local general manager, called local television news stations and print reporters to the MTC offices, where a line of Uber employees marched up with nine white, 15-by-12-inch file-folder storage boxes labeled "1,000 PETITIONS." After stacking the boxes high against the front door of the MTC, Shah delivered a short, lofty speech on democratic ideals and "listening to the voices" of the people supporting Uber.

After the cameras were turned off and Uber officials had left the scene, a reporter decided to look inside one of the boxes the company had left. It was filled with six-packs of plastic seventeen-ounce water bottles, as were the eight other boxes that accompanied it.

On another occasion in New York City, Josh Mohrer, the brash and contentious general manager who led Uber's Manhattan office, organized a rally on the steps of City Hall to take a stand against Mayor de Blasio. Mohrer's team had pushed out alerts to drivers and riders in the days prior, asking them to show up on a sweltering June day to "make your voice heard to your elected leaders."

Not many drivers or riders showed up to the protest. To make it look like Uber had grassroots support, Mohrer ordered his employees to rush from Uber's Chelsea office to City Hall, where Mohrer led them in a chanting protest. Mohrer never let on to reporters or city officials that the protesters, sweating in black, Uber-branded T-shirts, were paid employees of the company.

It didn't matter. In both St. Louis and in New York, Uber's tactics worked. The lawmakers backed down.

Chapter 13

THE CHARM OFFENSIVE

TRAVIS KALANICK COULDN'T FIGURE OUT WHY EVERYONE HATED his guts.

Feelings had no place in the business world. Being cutthroat was a quality to be celebrated, not hidden, in a CEO. When it came to describing an executive, "pugnacious" was never meant to be an insult.

Kalanick had proven himself to all his doubters. By 2014, Uber was a transportation behemoth, backed by the best of the best in venture capital and expanding globally. His company was growing so fast his rivals could barely compete.

And yet every time he looked at his mentions on Twitter, he'd read at least two or three tweets from random people calling him a jerk. Two technology reporters in particular—Sarah Lacy and Paul Carr—seemed to be on jihad against Kalanick, blaming him for the "asshole culture" spreading throughout Silicon Valley. *GQ* had made him look like a caricature of a "bro," a dirty word in techland. The opening sentence of a *Vanity Fair* profile—which he had hoped would be balanced—said he had a "face like a fist."

"What the fuck?" Kalanick wondered. He didn't think the public perception of him matched up to reality.

Every time someone cited Uber's belligerence, they cited Kalanick's attitude toward Lyft, Uber's closest US competitors. Reports that Uber employees were hailing Lyfts and then trying to recruit the driver were met with disgust—something that confused Kalanick and Uber employees. Business, they thought, was supposed to be a competition.

Logan Green, Lyft's CEO, was a good tactician. But Kalanick out-maneuvered his rivals every single time. And he felt fine trouncing his competition.

One prime example: Kalanick's network of spies in the Valley—mostly made up of other tech workers and venture capitalists—picked up early rumors of Lyft's new carpooling service. To get the jump on Lyft, Kalanick tasked his chief product officer, Jeff Holden, to drop everything and copy the carpooling feature immediately. Uber announced the impending launch of "Uberpool," a carpooling feature, mere hours before Lyft announced the product it had invented. By the time Green and Zimmer hit the publish button on their corporate blog, they looked like also-rans. Kalanick had scooped his competition, but his glee at upstaging his rival outraged the public.

Kalanick knew he had made some unforced errors. In the midst of the *GQ* profile, he let it slip that his newfound tech celebrity, and the attendant riches, made attracting women much easier now than it was when, say, he was living with his parents while building Red Swoosh. On-demand women, he joked, wasn't that far off.

"We call that boob-er," Kalanick told the reporter.

Suddenly, Kalanick wasn't just a grown man-child in readers' eyes, he was a blatant misogynist. One particularly cringe-worthy para-graph in the *GQ* story had Kalanick quoting the infamous Charlie Sheen, describing Uber's potential success as "hashtag winning." He name-dropped boutique hotels in Miami like the Shore Club and SLS as places he'd rather be than hustling at Uber. He was trying to be honest—and perhaps a little bit cool—but to the public he sounded like an enormous douchebag.

More than just a douche: Kalanick checked all the boxes of what people *imagined* cocky tech founders were like. He imagined himself as the hero in his own narrative—to the point that his Twitter avatar was the cover of Ayn Rand's *The Fountainhead*, a book espoused by libertarians for its celebration of self-reliance and disdain of government.

When other people looked at Kalanick, they saw another rich white guy riding the wave of venture capital while putting hard-working, blue-collar taxi drivers out of their jobs. Worse, he was living large—women, wine, song—and flouting it.

Kalanick didn't understand. He was hardly the first CEO to enjoy the fruits of his success. He knew how Mark Zuckerberg and Sean Parker had partied after their first few big rounds of venture capital. Larry and Sergey were literally jumping out of airplanes and burning millions of dollars building robots.

"And yet, *I* am the asshole?" Kalanick wondered aloud, pacing around Uber's headquarters. He'd take the anger home with him, obsessing over it while complaining to his girlfriend, continuing to pace in his living room.

Every time Uber got a bad run of press—which was happening more and more frequently—Kalanick fumed that reporters were out to get him. They couldn't appreciate the success of Uber. They were jealous of the company he had built. "Perception versus reality," he said to employees who worried about the company's image. "Their perception of us is nowhere near matching with our reality." This became his refrain. He needed to believe it. It was that, or believe the torrent of negative press, or the daily cascade of vitriolic tweets.

"Exploitative piece of shit."

"You don't give a fuck about drivers."

"DOUCHE."

The random commenters didn't bother him. But Sarah Lacy got under his skin. Lacy, a long-time tech journalist who made her name at *Bloomberg Businessweek* and *Time*, frequently bashed Kalanick. While other journalists were writing about the eye-popping amounts of money Uber was raising, Lacy focused on Uber's cult of staff "bros." "It troubles me that Uber is so OK with lying," Lacy tweeted, referring to some of Uber's lobbying practices. "Uber driver hits, kills 6-year-old girl. Is 'Not our problem' still an appropriate response?" she said, referring to a tragic accident and subsequent tone-deaf response from the company. "The horrific trickle down of Asshole culture: Why I've just deleted Uber from my phone," read the headline of one of her popular articles. According to those close to Kalanick, the CEO felt like Lacy was dragging him for no reason.

"How would they like it if we did it to them?" Kalanick asked Emil Michael, his second in command.

Kalanick's bad boy image was starting to get in the way.

WHILE KALANICK STEWED about the press, Bill Gurley was growing annoyed with his founder.

In the early days of their relationship, he and Kalanick were a dynamic duo. Gurley had a keycard tied to his belt loop granting him full access to Uber's 1455 Market Street headquarters. Gurley would walk past the glass-paned, street-level front doors, ride the elevator up to the fifth floor and scan himself through security uninterrupted, never breaking stride. Everyone knew Gurley—the beanpole Texan was unmistakable.

That was back when Travis cared what Gurley thought, when the founder still looked up to Gurley for advice. And Gurley was no softy. He encouraged Kalanick to be competitive. Kalanick and Gurley shared a frustration with existing legislation, and the older man appreciated the way Kalanick exploited cities' weaknesses. He saw how easy it would be to replicate the playbook worldwide, and cheered Kalanick at every step.

But by the end of 2014, things had changed. Kalanick had begun to sour on Gurley. In public, Gurley was still Kalanick's biggest cheerleader. In private, he expressed doubts. Kalanick grew tired of Gurley's concerns that the company was spending too much money trying to expand across the world into every possible market, or that Kalanick was doing end-runs around his own chief financial officer.

Worst of all, Gurley worried about Kalanick's obsession with China, the El Dorado of Western capitalism, a market few tech companies had yet to successfully crack. Kalanick wanted to fight his way inside and take on Didi Dache,[*] the "Uber of China." Gurley wasn't as keen. In China, Gurley saw a market he didn't fully understand, a set of cul-

---

[*] Didi Dache would eventually merge with a Chinese competitor, changing its name to Didi Chuxing.

tural norms unfamiliar to Uber employees, and a protectionist government hostile to most American businesses. When Gurley looked to the region, all he could see was red ink.

After years of thinking of Gurley as his personal cheerleader, Kalanick started to see Gurley as a gadfly, always harassing Kalanick and poking holes in Kalanick's ideas. Where Kalanick saw opportunity, Gurley started to see problems.

When Travis Kalanick decides he likes someone, they might as well be his best friend. People close to Kalanick describe it as a fragile infatuation, a platonic mini-affair where Kalanick thinks you can do no wrong. Gurley, when the two first met, was the object of Kalanick's infatuation.

When Travis Kalanick decides he *doesn't* like someone, they might as well be dead. If someone challenges Kalanick—in the wrong way, not via "principled confrontation"—they get iced out. If someone doesn't live up to Travis's lofty expectations? Iced out. Or, in Gurley's case, nitpicking Travis with questions and doubts. Gurley got iced out.

Kalanick rarely told someone directly to get lost. It was a slow, subtle frost. The person's name would start to drop off the email list for important strategy and planning meetings. Maybe they wouldn't get invited on as many walk and talks. Suddenly, the person wouldn't be on the "A-Team" anymore—Kalanick's cadre of top lieutenants. When "TK" fell out of love with someone, everyone knew it.

Gurley recognized what was happening, but had few options to rein Kalanick in. When the firms had been itching to invest in Uber, Kalanick had made sure to gut their investor rights. They had board seats but limited power. So Gurley couldn't leverage a board vote to influence Kalanick—at least not by himself.

He started using other means. Gurley began a whisper campaign to try and influence Kalanick, reaching out to people Kalanick trusted for advice. At times, Gurley was on the phone on a near daily basis with Emil Michael.

"He needs to recognize his fiduciary duty to shareholders," Gurley said. "This is crazy." Of all Kalanick's transgressions, his pushing out of Brent Callinicos, Uber's chief financial officer, peeved Gurley the

most. Kalanick believed Callinicos was unnecessary, and felt that most
of the position's duties could be executed by Uber's head of finance.
Gurley suspected that Kalanick didn't want a CFO watching how he
spent Uber's money.

Finance wasn't Gurley's only worry. He knew Uber didn't have a
particularly strong legal department in place, partially by design. Salle
Yoo, Uber's chief legal officer, was someone Kalanick felt he could control. She would push back on Kalanick occasionally, but her fear of
being "iced out" kept her from getting in Kalanick's face about every
legal concern she had.

In almost every other area of her life, Yoo was a leader; she sat on
the council of the Asian Art Museum of San Francisco, was a member of the Council of Korean Americans, and worked as the secretary,
director, and chair of the judiciary committee for the Asian American
Bar Association of the Greater Bay Area. Later that year, Yoo would be
named one of San Francisco Business Times' Most Influential Women
in Bay Area Business. And yet Yoo was often unable—and at times
reluctant—to influence her boss. When she did decide to raise an issue
with Kalanick, Travis regularly treated her concerns as just another
annoyance, especially when they had to do with legal compliance.

Uber's compliance division was marginal. Compliance is one of the
most important safeguards a company can have, as it ensures a company acts within the law. But when a company actively seeks out legal
"grey areas" during rapid expansion, compliance, by definition, is not
a priority. By the end of 2014, Uber was operating in hundreds of cities
across dozens of countries around the world. Even if she had the tools
to do so, there was no way Yoo's team could keep track of what each
city manager was doing.

In one meeting with top general managers, Ryan Graves—who was
by then head of all operations—made it clear where Uber stood on
compliance. While legal claimed it wanted employees to follow the rule
of law, Graves cared more about getting things done.

In effect, GMs had free rein. Kalanick wanted to keep Uber from
feeling too "big company," like a Google or an Apple. That meant protecting his employees from corporate bureaucracy. He wanted them to

ignore all rules except Kalanick's beloved fourteen principles. As Kalanick looked upon his empire, he was proud of what he saw—dozens of young, hungry entrepreneurs, autonomous and improvising as each situation required.

What Gurley saw was a sprawling mess. He was trying and failing to convince Kalanick to hire another CFO. Talking to Michael got Gurley nowhere. Kalanick had no intention of slowing down his spending. And every time Gurley brought up concerns about finances in board meetings, Kalanick would find a way to dodge the issue, or reassure everyone he knew what he was doing.

So Gurley decided to do what he often done over the years with thorny problems: he blogged about it. He had always been a contrarian, warning founders and venture capitalists about the pitfalls of their unpredictable industry. But 2014 and '15 brought out a new version of Gurley. Over a series of entries on his personal blog, *Above the Crowd*, Gurley slowly morphed himself into Silicon Valley's Cassandra.

Like the mythical Greek figure, Gurley forecast a collapse of apocalyptic proportions. Gurley howled about the impending downturn in venture capital, exacerbated by the waves of new money. Savvy investors in the Valley assumed this was Gurley playing the game; the more he scared off late-stage, institutional funds from investing in tech companies, the better the chance the landscape would return to the old model; startups would go public at a normal time in their life cycle rather than deferring, and investors would see their paydays much sooner.

But these blog posts were privately aimed at Kalanick. Gurley was telegraphing his concerns for Uber, his pride and joy company. "We are in a *risk* bubble," Gurley wrote. "Companies are taking on huge burn rates to justify spending the capital they are raising in these enormous financings, putting their long-term viability in jeopardy."

The apotheosis came in front of thousands of people at South by Southwest, the annual springtime music, film, and technology festival in Austin. Dressed in extra-long blue jeans, a pair of brown leather boots and a white University of Texas pullover stamped with the school's burnt orange Longhorn mascot, Gurley took to the stage in a crowded auditorium for an hour of questioning from the writer Malcolm Gladwell.

He launched into his usual bit. "There is no fear in Silicon Valley right now," he said to Gladwell. He noted that there were more than a hundred "unicorns" running loose in the Valley—in his view an insane amount. A unicorn earned the name by being unspeakably rare. Practically overnight, dozens of consumer startups had been valued well into the billions, many with little revenue to speak of. A hundred unicorns suggested to Gurley that some would turn out to be ponies with papier-mâché horns.

"I do think you'll see some dead unicorns this year," he told Gladwell.

Seventeen hundred miles away back at Uber headquarters in San Francisco, Kalanick and Michael scoffed at their overbearing investor, who always thought the sky was falling. They had a nickname for Gurley: "Chicken Little."

THE WAVERLY INN was a good place to woo the East Coast media elite.

Tucked away on Bank Street, a quiet, tree-lined road in hip Greenwich Village, The Waverly Inn was a storied institution for New York media, made famous by Graydon Carter, the longtime head of *Vanity Fair*, who used the restaurant to host exclusive evenings with Manhattan society. On summer evenings passersby would notice celebrities dining outside on the ivy-lined front patio. Dinner at the Waverly *meant* something.

For Kalanick, it meant expensing a meal to ingratiate himself to East Coast reporters who hated him. That week, he had come to New York to check on the Manhattan office and meet with bankers. Nairi Hourdajian, his head of communications, thought they could kill two birds with one stone. Hourdajian bet that if the reporters got to know Kalanick in person, they might realize he wasn't such a bad guy.

Hourdajian had gone through the same process herself. Hourdajian, a proud Armenian-American who breezed through her government studies at Georgetown and Harvard, came from the world of politics. She was familiar with schmoozers and phony executives. Though she knew her boss had rough edges, she had come to believe that inside, Travis Kalanick was a good person.

Hourdajian worked alongside Kalanick through some of Uber's earliest, toughest days. He trusted her to build out the communications team from scratch, and then run it. And when Uber was up against its nastiest early opponents—including taxi operators and government officials—Hourdajian and Kalanick fought side by side in the trenches. She knew Kalanick would never change. But perhaps, if she got reporters closer to him, they would see Kalanick the way she did.

Hourdajian set up a meet-and-greet that Friday afternoon with reporters at the Gramercy Park Hotel, a swank destination in Manhattan's Flatiron District. In a private room, sitting on leather sofas over plates of brie and mini-muffins, Kalanick made the case to reporters that he wasn't a monster, and suggested that Uber wanted to build a relationship.[*]

They had handed off the job of organizing dinner to Ian Osborne, a well-connected British media fixer, whose job it was to pair important members of the business community with equally important members of the press and Hollywood.

Guests were seated in a private room in the back of the Waverly, away from the common dining area. After cocktails, the diners were asked to sit at a long, skinny wood table—almost too skinny to eat over. Guests were uncomfortably close to one another. Kalanick sat at the head, flanked by Arianna Huffington, the media mogul and celebrity who had built influence in politics and publishing. Huffington and Kalanick had grown close in recent years, meeting for the first time at a technology conference in 2012.

Down the table from Huffington was Leigh Gallagher, a senior editor at *Fortune* who oversaw the "40 Under 40" list of influential leaders in the business world. On Travis's other side sat Hourdajian, then Osborne, Uber chief business officer Emil Michael, a handful of other

---

[*] I was at the meeting and agreed to its off-the-record terms, which restricts me from recounting the events in great detail. *BuzzFeed*, however, published a story on the meeting; that story informs the descriptions here. I did not attend the media dinner later that evening.

influential New York media writers, and Edward Norton, the actor turned Uber investor. Norton, who had become pals with Kalanick, was Uber's first official rider when the company launched in Los Angeles.

As Kalanick settled in to schmooze with magazine writers at one end of the table, Emil Michael, his deputy, was cozying up to the media writer Michael Wolff on the other end. Wolff had brought along Ben Smith, editor in chief at *BuzzFeed*.

Smith's bubbly personality made him a wonderful dinner guest; his affable manner often disarmed the people he reported on. But those qualities belied a pugnacious streak. Smith had become known in Washington DC for never backing down from a fight. As a reporter at *Politico* he often sparred on Twitter with those he covered and those he competed with for scoops. When he moved to *BuzzFeed* in 2012, his mandate was to turn the outlet, long famous for its lolcats and list-oriented viral articles, into a respectable, hard-hitting news organization. Smith rebranded his division as *BuzzFeed News*, and soon built a serious outfit whose reporting standards and aggressive pursuit of scoops rivaled that of the most traditional newsrooms.

Smith was thus shocked when he, a member of the media, found himself sitting across from an Uber executive who was so openly disdainful about Uber's relationship with the press. As the group dug into seared halibut and ribeye, Emil Michael, thinking he was amongst a room full of Uber sympathizers and friends, had gone off on a rant about how Uber was unfairly targeted by the press, and had been a victim of its own success.

As the dinner progressed, Smith noticed Michael's arrogance take over. Michael wasn't used to Smith's subtle challenges, pushing back on Uber's claim that it was providing a public good, or that drivers who complained about pay just didn't understand the math. Michael didn't notice that Smith was tapping notes onto his phone as the conversation moved into more controversial areas.

"It's just bullshit," Michael said, referencing the waves of negative press coverage. "The way we've been singled out like we have." The worst, he noted, was Sarah Lacy.

Lacy wasn't a universally beloved figure. She and her partner, Paul Carr, would pick fights with other journalists as frequently as they skewered the people they covered. Michael knew this. Probably too well.

"What if we gave them a taste of their own medicine?" Michael continued. "What if we spent, I don't know, a million dollars hiring a few journalists and top oppo people," he said, referring to "opposition researchers"—those who get paid to dig up information on other people for a living. "They could look into *your* personal lives, *your* families. Help *us* actually fight back against the press," Michael said. He was convinced there was dirt to be found investigating Lacy's marriage and her relationship with her business partner, Carr.

Michael wasn't finished. "Ask 100 women whether they'd feel safer in an Uber or in a taxi," Michael went on, referring to a recent story in which Lacy posted that she would stop using Uber, fearing for her safety because of the drivers. "If any women decided to delete the Uber app, like she did? And then they go on to take a taxi ride and, god forbid, are assaulted? She should be held personally responsible for that," he said.

Smith couldn't believe his ears. Why was an executive at the most reviled tech company in Silicon Valley dumping this in his lap? Did he know who he was talking to?

Importantly, the dinner was supposed to be off the record. That could explain some of Michael's bravado. But Michael Wolff had forgotten to relay this detail to his friend Smith when Wolff forwarded him the invitation to the dinner party.*

Smith wanted to give Michael a chance to save himself. If Uber were actually to go through with this, he asked, the story would no longer be about Lacy, it would be about Uber. What if, someone at the table

---

* Wolff, who would later write an infamous, best-selling book about the Trump administration, had put "Travis Zalanick" [sic] in the subject line, showing the *BuzzFeed* editor just how much his friend Wolff knew about the ride-hailing chief's reputation or reasons for hosting the dinner. Wolff would later say he assumed Smith had known the dinner was off the record, an egregious error on Wolff's part.

suggested, it were to get out that Uber was behind the plan to dig up dirt on Lacy?

"That won't be a problem," Michael said. "Nobody would know it was us."

Smith, still taking notes, waited politely and finished the rest of his dinner as the conversation floated across the low-lit back room of the Waverly.

Then he got up, thanked his hosts, and went home to start typing on his MacBook.

NAIRI HOURDAJIAN WOKE UP on Sunday morning believing Uber's charm offensive had gone well. The comms chief knew it hadn't been perfect—Kalanick had gone off script when talking to reporters at the Gramercy Park meeting, Hourdajian felt; he came across as a bit too self-deprecating and pity-seeking. But she was proud of herself; she managed to keep Josh Mohrer, Uber's brash New York general manager, away from the Friday dinner even after he asked to be included. But she'd held firm. No, that weekend had gone as well as could be expected. Things were looking up; perhaps she could convince the world that her boss was not, in fact, an asshole. Or at bare minimum, at least *some* reporters might start believing her. The team packed up and prepared to fly back to San Francisco, having accomplished what they set out to do.

Thirty-six hours later, at 8:57 p.m. Monday evening, Ben Smith's story went live on *BuzzFeed News*' website.

In it, Smith outlined the details of the multiday charm offensive, the attempts at ingratiating themselves with reporters on Friday afternoon at the Gramercy, the fancy dinner and star-studded guest list at the Waverly Inn. The story revealed the details of Uber planning to start an oppositional research squad, with express intent of "slut-shaming" a prominent critic of the company.

"Uber's dirt-diggers, Michael said, could expose Lacy," the article said. "They could, in particular, prove a particular and very specific claim about her personal life."

The backlash was swift and furious. The *New York Times*, the *Wall Street Journal*, and a host of other publications immediately seized on the comments. Morning shows on NBC, ABC, and CBS fanned the flames, underscoring the comments as proof that Michael, Kalanick, Uber—the officers and leadership of the company—were indeed the greedy, slimeball degenerates of popular imagination.

What made the piece so damning was that it rang true. Kalanick *did* want to win at whatever cost. He *did* like the idea of hiring oppo researchers to attack his opponents. And not only did he want to win, he wanted to rub his opponents' noses in it, too.

After all, he and Michael had come up with idea long before the dinner, and spoke about it privately to one another at length. Close friends knew the two hated the sensationalist, uninformed media, whose only goal was to chip away at Uber's hard-fought progress. What the pair didn't know was that they couldn't fight the media the way they fought their corporate opponents. The aggression they used to bulldoze cities wouldn't shame reporters into backing off the story. For all of Kalanick's talents, he still couldn't see that this wasn't a turf war, it was a popularity contest. Now, that blind spot was becoming a liability.

That Monday evening, as Hourdajian's colleagues panicked at Uber's headquarters in San Francisco, she could only shake her head and wince.

Uber's charm offensive had failed.

Chapter 14

CULTURE WARS

IF SILICON VALLEY WAS DEFINED BY "THE CRAZY ONES, THE MISFITS, the rebels and the troublemakers," a rising countercultural force of hackers and techno-revolutionaries described in Apple's "Think Different" advertising campaign, then the post-recession era of the Valley was shaped by a different force: the rise of the MBA grads.

Before the 2008 crash, business school bought one a job as a junior investment banker at Goldman Sachs, or a six-figure consulting salary at McKinsey. But the times were changing. As the finance and consulting industries lost their luster after the financial crisis, business school graduates began to sense new opportunity out west.

The weather was better in Silicon Valley. Meals, laundry, and gym memberships were covered by the company. They wouldn't be required to do grunt work for older traders during the first years at J.P. Morgan. And best of all, techies weren't (yet) hated by the 99 percent, and didn't have Occupy Wall Street camped out in front of their offices. By 2015, some 16 percent of MBA grads went into the technology sector, the third most common destination. And of the more than 150 "unicorns" in the Valley by then, nearly a quarter of them were founded by business school graduates. Even Lyft's co-founder, John Zimmer, was an intern at Lehman Brothers before he turned techie.

More than most other tech companies, Uber prized the almighty Masters of Business Administration, a degree that signaled business acumen and, often, an alpha male mindset. Not every MBA grad was an asshole, by any means. It just seemed that many of the ones who *were* assholes tended to feel at home joining Uber.

At Uber, being cutthroat and competitive was considered an asset, not a liability. According to Uber company value number two, "In a meritocracy, the best idea always wins and the fiercest truth-seekers rise to the top." Fighting, Kalanick believed, was good; having a "champion's mindset" (value number four) was what surrounded him with "winners." And he only wanted to work with winners.

Once Uber stuck a bunch of alpha male MBAs together in a workplace, the "champion's mindset" became something else entirely. "Kill or be killed" was the unofficial motto at Uber, where if you weren't watching your back you might be betrayed by a colleague looking to get ahead. Success, many believed, only came at the expense of others. The will to power was the only way to rise into Kalanick's favor.

Josh Mohrer was the model Uber employee. As general manager for New York City, he was responsible for one of Uber's biggest money-making franchises worldwide. A math major undergrad with an MBA from New York University, Mohrer defined the Uber ideal. Stout and squat, Mohrer had a face like a boxer, thick-chinned and ready to take a punch. He had a boyish grin and a thinning hairline, which somehow made him look both older and younger than he actually was. Barely into his thirties, Mohrer leaned hard on his people—browbeating them when he needed to—never accepting excuses. And he loved to fight—important in a city with one of the strongest transportation unions in the world.

Mohrer would pit his employees against one another to see who could impress him or deliver better—a tactic espoused by Kalanick himself. Subtle intimidation of his underlings sometimes meant flicking at their flaws, like inspecting the receding hairline of an employee as they tried to discuss a project with their boss. He described the short-comings of individual employees in front of the entire office, praising winners and shaming losers.

Mohrer thought he was empowering his staff, and felt like his high expectations were a good management strategy. But around the office, according to two employees, he seemed like a shorter version of Biff Tannen, the high school bully antagonist from *Back to the Future*.

Winston, Mohrer's small, puffy white miniature poodle, barked or nipped at some staff, and never shut up. Mohrer loved handing the dog off to company executives when they came through New York, snapping photos of them holding Winston (with varying degrees of affection) and posting them to Winston's personal Twitter account, @WinnTheDog. One day, after he left Uber, Mohrer tweeted a photo of Winston taking a dump next to a Citi Bike, the city's blue rental bicycles—owned by Lyft.

Some women at the Chelsea office felt alienated by management. To some staff, Mohrer appeared more comfortable with his "bros," other alpha-male types who shared his frat-like mentality, and the office culture reflected as much.

But Mohrer always hit his numbers, no matter what. And that was what mattered at Uber. His business success kept Mohrer's position secure at the company for years.

Maintaining that high performance also made for intense pressure around the office. Employees across all cities worked late into the evenings. Some never took weekends off to enjoy time with their families. It wasn't uncommon for bosses to call employees in the middle of the night, or for staff to be asked to join a conference call at two o'clock in the morning from New York if you were trying to talk to offices in Southeast Asia or Australia. Though employees were fed for free at work, Mohrer followed Travis's lead and delayed dinner until 8:15 p.m.

One employee, viewed by some as a particular offender, was a favorite of Kalanick's, which meant he could get away with bad behavior towards his underlings. In a tense gathering with other Uber colleagues, this man called another employee a "faggot," something he never answered for despite employee complaints. Kalanick's protection meant you did not face consequences.

Other managers would threaten to demote workers if they didn't perform well enough. One manager in Rio would scream or throw coffee mugs at subordinates when he was upset. Workers were threatened by managers with baseball bats if they didn't meet targets. Once, this manager berated an employee about his performance so intensely, he made the man cry in front of the entire office. That same manager later dated one of his direct reports, causing discomfort among staff when he started favoring her in the workplace. Yet because Rio de Janeiro was one of Uber's top performing markets, the numerous HR complaints about that manager went unresolved. To leadership, nothing mattered—as long as you made your numbers.

Not that HR was a particularly robust department at Uber. Like compliance, it felt like an afterthought. Ryan Graves, the company's head of operations, was also in charge of human resources; Renee Atwood, Uber's head of "people and places," was supposed to be reporting issues to Graves as they came up. But Graves wasn't focused on the day-to-day minutiae of running a complicated HR operation. And Atwood appeared to be in over her head. HR could hardly keep up with employee complaints much less address or resolve them.

Even during recruiting, prospective employees were treated poorly. The company had designed an algorithm that determined the lowest possible salary a candidate might accept before making an offer to them, a ruthlessly efficient technique that saved Uber millions of dollars in equity grants.

Kalanick found other ways to save money, too. In more advanced markets where Uber was popular and required fewer subsidies, executives at the company sought ways to increase profit margins. Uber's margins were fixed for the most part; they took an approximately 20 to 25 percent cut of every ride while giving the driver the remainder of the fare.

Until 2014, that is, when one executive had the brilliant idea of introducing the "Safe Rides Fee," a new charge that added $1 to the cost of each trip. At the time Uber billed it as necessary for passengers: "This Safe Rides Fee supports our continued efforts to ensure the safest possible platform for Uber riders and drivers, including an industry-leading background check process, regular motor vehicle checks, driver safety education, development of safety features in the app, and insurance," went the company's blog post. If riders noticed the fee, they rarely complained. Many assumed it would just make their rides safer somehow.

The reality was much less noble. As Uber's insurance costs grew exponentially, the "Safe Rides Fee" was devised to add $1 of pure margin to each trip, according to employees who worked on the addition. That meant for each trip taken in the United States, Uber took in an extra dollar in cash. The drivers, of course, got no share of the extra buck. That number added up to hundreds of millions of dollars over years of operation, a sizeable new line of income. After the money was collected it was never earmarked specifically for improving safety. "Driver safety education" consisted of little more than a short, online video course. In-app safety features weren't a priority until years later. "We boosted our margins saying our rides were safer," one former employee said. "It was obscene."

Not that the overall tone of Uber was that of a professionally run organization *anyway*. Employees, sometimes fresh out of college, would occasionally post immature things to the company blog. One employee coined the term "rides of glory" to describe the Uber trip a customer takes home the morning after a one-night-stand. "In times of yore, you would have woken up in a panic, scrambling in the dark trying to find your fur coat or velvet smoking jacket or whatever it is you cool kids wear," the post said, authored by Bradley Voytek, one of Uber's data scientists. "Then that long walk home in the premorning dawn." Voytek, a cognitive neuroscientist by trade, joined Uber because he loved the insight that such an enormous data set gave him into human behavior. Watching trips across cities being carried out in real time was like having his own personal human ant farm.

"But that was then," Voytek continued, noting the volume of people across multiple cities who were Uber-ing home from a stranger's house the morning after. "The world has changed, and gone are the days of the Walk of Shame," he joked. "We live in Uber's world now."

Beyond immature employees and bullying managers, the real war was between departments for the most valuable prize inside the company: incentives.

"Incentives" was the name for the free money Uber doled out to users and drivers. Uber lost money on incentives, but it didn't matter; for one, if Uber kickstarted the demand flywheel hard enough, they'd train people to keep using Uber even after the company stopped providing freebies. Moreover, Kalanick knew he could always, *always* find more money.

By 2015, Uber was globally spending more than $2 billion annually incentivizing drivers and riders, a staggering burn rate for even the most well-capitalized startup. It became clear inside of Uber that whoever controlled the money held the power, so different parts of the organization started competing for a bigger share of the piggy bank. Incentives offered the fastest route to growth. And growth was rewarded with bonuses, promotions, and praise from the top. There was the growth division led by Ed Baker, an ex-Facebook VP who was known for bringing millions of new users to the social network. Other executives from the product, operations, and finance divisions joined the fray.

The CEO loved it. Kalanick's approach to management was to let each department fight for control until a winner emerged. It was the fairest way, in his mind, to determine the most talented staff.

What Kalanick didn't see—or chose to ignore—was all the politicking that went on behind his back. Everyone knew you couldn't *really* challenge TK—the staff's nickname for Kalanick—if you wanted to stay on his good side.

If you were bold enough to challenge Kalanick, you had better back up your arguments with cold, hard data. Kalanick wouldn't listen to anything else. For years, general managers begged Kalanick to let them build a tipping function into the app so that riders could toss a few extra dollars to drivers at the end of a ride. It was a simple gesture that would earn the company significant goodwill with their driver base; besides, Lyft offered it. Yet Kalanick remained staunchly against tipping. Kalanick felt Uber worked so well because of the frictionless payment experience. A passenger could exit the vehicle without even thinking of money; adding a tipping function would require the rider to open the app again, needlessly, in Kalanick's mind. He never understood—or never cared—how much tipping could improve Uber drivers' livelihoods.

Sometimes, people pushed back. Kalanick once went head to head with Aaron Schildkrout, a tall, wiry product leader who would later rise to become the head of Uber's driver product division. Schildkrout was sharp, a naturally argumentative hipster type. He almost always dressed in black, wore thick-rimmed dark eyeglasses and rarely combed his mop of dark brown tousled hair. Discussions with Schildkrout often ended up in the realm of the philosophical; he studied social theory at Harvard and the University of Chicago, and loved to think about *why* humans did things, not just what they did. Before Uber, Schildkrout had been a CEO of his own company, a dating startup, and he learned valuable lessons over his few years as a leader. One of the most important was knowing when a CEO needed challenging.

In one meeting, Kalanick had made a product decision, one of dozens he made throughout the day. But Schildkrout disagreed. He started rattling off a list of things that could go wrong with Kalanick's plan, and how he'd do it differently. The two went back and forth until they were shouting at each other across the room, a table full of silent employees sitting awkwardly between them. Schildkrout was wearing a grey pullover sweatshirt when he walked in the room. By the time the two had stopped fighting, the front of Schildkrout's sweatshirt was soaked through. But he had won the argument, and Kalanick respected him for it.

For young, promising engineers, winning Kalanick's favor was worth it. He was a great motivator, something between a wartime general and a self-help guru. Kalanick always positioned Uber's fight as "us against them." If Kalanick thought you were a true believer, someone who stayed "super pumped" for the cause, he noticed. Maybe he'd reward you with some attention, a quick "jam sesh" together in Uber's lobby, taking a lap and firing off ideas. Or a shout-out in an employee all-hands meeting. Whatever it was, employees loved being in his favor, and fought to stay there.

One of the highest honors for those in Kalanick's inner circle was admittance to clandestine, 10 p.m. strategy meetings at Uber's headquarters. In these evening sessions, Kalanick and a hand-picked crew would concoct new ways to spend the mountains of venture capital he had raised to battle competitors. Kalanick loved giving projects code names. He dubbed his late-night strategy meetings the "North American Championship Series," or NACS, a nod to Uber's competition with Lyft.

The luckiest employees got to work on "Black Gold," the code name for Uber's Asia strategy meetings. The name was special: "Black Gold" was a reference to political corruption, the "black" deeds carried out by gangs—the infamous Triads of organized crime in Taiwan.

For Uber, it meant playing dirty, because they were facing off against a Chinese competitor with a hell of a lot of money.

Chapter 15

EMPIRE BUILDING

FOR DECADES, WESTERN TECHNOLOGY EXECUTIVES HAVE DREAMED OF successfully launching an American software business in mainland China. Very few have succeeded.

When Travis Kalanick looked at the country he saw a near-perfect market for startups. Home to nearly 1.4 billion people, China presented an untapped ocean of potential Uber users. Nearly one-third of that population were millennials: young, urban, upwardly mobile with growing disposable income, ardent students of technology and the sciences, and almost *always* connected online.

As in America, this Chinese generation had grown up with ubiquitous access to the internet. Nearly 97 percent of Chinese internet users ages fourteen through forty-seven owned some sort of smartphone. Westerners had experienced the mass migration from desktop computer to smartphone. But China's millennials skipped the desktop, advancing directly to their phones. Like Kalanick, the Chinese believed in technology, embracing it much faster than Westerners. Kalanick needed them to embrace Uber.

This was far easier said than done. Larry Page and Sergey Brin, Mark Zuckerberg, Jeff Bezos, Dick Costolo, Evan Spiegel—almost all of the most influential Silicon Valley leaders of the past twenty years have made overtures to China to operate inside the coveted country. And almost all of them have failed. Every company in Silicon Valley had—*has*—their own version of the China problem.

Kalanick projected confidence that he knew how to crack the market. The Chinese were still reliant on taxis, Kalanick's most hated foe.

He was convinced once they saw Uber's better service they'd adopt Uber en masse. Moreover, he had another secret weapon: billions of dollars in free driver and rider subsidies, more than enough, he believed, to ignite Chinese consumer demand. China was going to be Uber's toughest battle yet. Privately, he harbored doubts he would be able to beat the Chinese on their own turf. Even so, Kalanick relished the coming fight.

He feared the Chinese government would be hostile. The Communist Party took pride in promoting and ensuring the success of *Chinese* companies on Chinese soil. Under Xi Jinping, the government had invested hundreds of millions in state-backed venture funds, which seeded a wave of startups, giving China the fastest growing economic sector in history. It had created so-called "special economic zones" in cities like Shenzhen, fostering Chinese innovation and startup incubation. The West still maintained global tech dominance, but of the top twenty technology companies in the world measured by market cap, nine of them were Chinese.

Government control of the internet meant the Party could play kingmaker, choosing to regulate selectively based on what it felt was beneficial to the state. It was against their nature for the Chinese government to warm up to a foreign invader, especially one as brash and hard-charging as Kalanick. Xi read the news, and surely knew Kalanick's reputation. Nevertheless Kalanick was confident he would prevail.

The only thing that worried Kalanick more than China's Big Brother–esque government was another "brother," this one a startup: Didi Chuxing. Roughly translated, it meant "Brother Travel." Colloquially, it was "honk honk taxi": "DiDi" mimicked the sound of a car's horn in Chinese.

While the name sounded playful, the company and its leadership were anything but. Didi Chuxing was the preeminent ride-hailing startup in China, built on years of analysis of how China's billion-plus citizens travel the country's congested streets. Cheng Wei, Didi Chuxing's CEO, was in his late twenties when he started the company, having only held a handful of jobs in sales prior. His bet on building a taxi-hailing business in 2012 ballooned into a multi-billion-dollar ride-hailing giant in just three years, backed by heavy-hitting venture investments from Tencent and Alibaba, two of China's biggest and most popular technology companies.

Didi Chuxing had everything Uber needed to succeed: scale, recognition, and, most of all, support from the government. It also had incredibly deep pockets, having raised billions in capital from Chinese investors over just a few years of operation. The top brass there had mettle, too. Before becoming Didi Chuxing, Didi Dache had been locked in a spending war with a major competitor, Kuaidi Dache. In 2015, the two companies came to a truce and merged, but not before burning millions of Yuan worth of free rides offered to users. By the time the two companies merged, app-based ride-hailing was woven into the fabric of everyday life.

Kalanick wasn't fazed. He had overcome corrupt politicians and taxi unions in every major city in the United States. He had fought Lyft and outsmarted its leadership. He had charged into cities worldwide, outspent his opponents, outflanked governments, and won more customers with a better product. Barreling ahead had worked before, and it could work again.

"I get excited about doing things people think can't be done," Kalanick later said.

TRAVIS KALANICK WAS EXCITED to accomplish the impossible. But Thuan Pham had to deal with the day-to-day reality of it. And by 2015, Pham was already swimming in problems.

Uber was growing faster than anyone could have anticipated. Pham had installed teams in several dozen cities across China—Chengdu, Beijing, Wuhan, and many others—putting up an ample fight against the aggressive DiDi. That made Travis happy. And when Pham's boss was happy, Pham was happy.

As chief technology officer of Uber, Pham was responsible for the entire engineering corps inside of Uber, a sprawling organization of hundreds of brilliant young hackers. Pham's team looked up to him; a disciplined executive with dark hair, bronzed skin, and square, gunmetal-colored glasses that cut a sharp contrast to his smile. To many of his engineers, he was a rare breed of CTO: empathetic to his staff and even emotional when dealing with tough company issues. His workers repaid him with their loyalty. Most of all, they respected his work ethic, especially his ability to respond to subordinates' emails at all hours of the day. Even when he took a vacation, Pham had his laptop open at the airport, answering emails on the runway up until the moment the flight attendant told him it was time to stow his electronic devices.

But today, Pham wasn't happy. Even as the number of trips people took in China were skyrocketing, so were Uber's incentives.

Everyone inside of Uber knew it was going to take an enormous number of free rides to gain significant market share in China. To head off concerns about the burn rate, Kalanick wrote a letter to investors warning them Uber may be spending heavily to gain a stronger foothold in China.

As CTO, Pham saw something the investors couldn't. Uber was spending $40 million to $50 million on subsidies in China every single week, an enormous sum just to convince riders and drivers to use Uber over DiDi.

When press started sniffing around, Kalanick sent Emil Michael to fend off nosy reporters, telling the press that Uber's operations in China were "more efficient" than many believed. If press had any inkling of what the numbers were—something Michael refused to discuss in interviews—their collective jaw would hit the floor.

In a letter Kalanick wrote to investors, that then leaked to the press, he noted that in just nine months, the number of trips taken in Chengdu and Hangzhou were more than four hundred times the number of trips taken in New York, one of Uber's largest cities, when those markets were the same age. "This kind of growth is remarkable and unprecedented," Kalanick wrote in the letter. "To put it frankly, China represents one of the largest untapped opportunities for Uber, potentially larger than the US."

What Kalanick left out was the fact that in many cities, more than half of those trips were fake, a complete waste of money brought in by investors.

Pham, who was responsible for dealing with the fraud, wasn't a stranger to tough situations. Born in Vietnam in 1967, Pham was thrown into conflict as young as twelve years old, as his mother piled him and his brother into a rickety wooden fishing boat, sailing out amid the rough waters of the South China Sea to escape the violence of the '79 Sino-Vietnamese War. Pham weathered deadly storms, was robbed by Thai pirates, and felt lucky to land at a refugee camp in Indonesia, even though he was soon shuttled off to an island for other Southeast Asian immigrants that lacked basic sanitation. After finally making it to the United States, Pham shared a small, roach-infested apartment in Maryland with another family while his mother worked multiple jobs to support them.

Pham studied hard in school and grew to love IBM PCs, which he discovered at a young age. An entry-level computing job at Hewlett-Packard led to the startup world and, eventually, a big break with a high-paying, high-level job at Uber as CTO. Like Kalanick, Pham worked hard, and never flinched from the pressure, intensity, or long hours as the company exploded.

The China challenge, though, was unprecedented. Pham's job was to turn the China strategy, an economic nightmare, into something that actually made financial sense.

Pham spun up a crisis team, poaching top security and fraud detection talent from local Bay Area competitors to form a fifty-person fraud squad at Uber's HQ in San Francisco. He ordered local managers in China to review new sign-ups more rigorously. They introduced identity verification features and other tactics to screen drivers and riders over time.

China wasn't the only porous market. Scams were endemic in every market worldwide. In New York in 2014, insiders noted that nearly 20 percent of the company's gross revenue went to fraudulent rides. Thieves skimmed the same amount in London. From Washington DC to Los Angeles, Uber was bleeding money by the millions in its most important markets.

Pham's fraud specialists soon proved invaluable—and not just in China. In Brooklyn, the team watched as credit card thieves used stolen card numbers to run drug trafficking and prostitution rings using Uber vehicles. The ruse was simple: the dealers would buy stolen credit card numbers from the Dark Web, then plug those numbers into the app to charge Uber trips to the stolen accounts. Over hundreds of trips per week they delivered drugs and call girls throughout New York City— all paid by Uber incentives, or through chargebacks from credit card companies after the original card owners reported the fraud.

After monitoring the criminals for months, Uber eventually partnered with the New York Police Department to help take the scammers down in a complicated sting operation. Over the course of a single Uber ride, the police would obtain a report from a credit card company, call the driver of the vehicle and tell them to pull over, then arrest the rider on a number of charges, including credit card fraud, possession of narcotics, prostitution, and so on. Though they would never brag about it publicly, the fraud team helped the NYPD take out the entire operation.

Pham's crew then built machine-learning models on past criminal behavior, training Uber's systems to spot patterns of fraud as it occurred. After the team found its stride, fraud in markets like New York dropped to the low single-digit numbers; Pham was proud of his team, and so was Kalanick.

Fighting Chinese fraud, however, required another level of vigilance. Chinese scammers engaged Uber's engineers in an aggressive arms race in which the anti-fraud team fought to overcome the fraudsters' increasingly clever methods. In the United States, fraud was simpler; criminals usually either went for joyrides or used Uber to make illegal deliveries, all using stolen credit card numbers. But in China, drivers and riders colluded to scam Uber out of billions in incentives, divvying the rewards.

Most scammers found each other over text-based Chinese internet forums, a simple, anonymous way to match people who wanted to make a quick buck. They developed their own codified language; drivers seeking a fake ride would ask for "an injection," a reference to the small, red digital pin that signaled a user's location inside the Uber app. A "nurse," or scammer, could respond in kind to give a "shot" to the original poster by creating a new fake account and going on a fake ride with the driver. The two parties would then split the bonus incentive payment from Uber. Repeated over and over across dozens of cities, small driver bonuses mushroomed into millions in squandered cash.

Kalanick couldn't stop the incentives because he had to keep pace with DiDi, who were just as willing to burn money to attract ridership. To juice growth, Kalanick had made the new user sign-up process as simple as possible. Joining Uber only required a name, email address, phone number, and credit card number, all of which were easily replicable. Fraudsters simply entered fake names and emails. Then they used apps like "Burner" or "TextNow" to create thousands of fake telephone numbers to be matched with stolen credit card numbers. But requiring Chinese users to add other, more precise, forms of identification would add more friction to the process. And, as Kalanick's data scientists found in their research, adding friction slowed growth. For Kalanick, putting a dent in growth was not an option.

Kalanick's solution was to grow and rely upon the anti-fraud team. But scammers grew more shrewd over time. Eventually, hustlers found that searching forums for riders was inefficient and time-consuming, so they ended up creating "riders" themselves. Some drivers would purchase caseloads of cheap cell phones, creating multiple driver and rider accounts for each different phone. A scammer would request rides from his "passenger" phones, and use his "driver" phones to accept those rides. He would then drive around the streets of Chengdu with dozens of phones spread across the front and back seats of his car, racking up fares for each of the "trips" he was completing for his fake customers.

The fraud team eventually discovered the trickery. Little blips flickered along the complicated topography of the Chinese city maps on monitors lining the control rooms back at Uber headquarters in San Francisco. Each little blip represented a scammer's vehicle, which was followed closely on the map by a trail of a dozen fake "passengers" who were taking the trip. It was as if the fraud team were watching dozens of digital centipedes skittering across their computer screens, each one getting fat on Uber incentives.

Some scammers created giant makeshift circuit boards filled with hundreds of slots to insert SIM cards, the small microchips that allow mobile phones to communicate with a cellular network. Each SIM card in the circuit board acted as a new number that could automatically respond to a verification text for a newly created account, which the scammers then used to rack up more fake rides and bonuses. After the SIMs had been used, a scam artist replaced all the SIMs on the board with a fresh set of numbers and started the process all over again. Have hundreds or even thousands of "drivers" repeat that dozens of times per day, seven days a week, and it amounted to serious Uber losses.

Pham's anti-fraud team was good. But there was only so much they could do; even as Uber saw success in cities like New York and San Francisco, Kalanick kept throwing more into the money pit that was China. All Pham and his engineers could do was try to stanch the bleeding.

FRAUDSTERS WEREN'T THE ONLY problems with China. When it came to business, Uber shared something in common with the Chinese; they were both willing to play dirty. Ethics went out the window in China as DiDi and Uber fought for advantage.

DiDi's city managers would pay local taxi operators to protest Uber's peer-to-peer car services. They'd send fake texts to Uber's drivers, claiming that Uber had shut down in China and that drivers should switch over to driving for DiDi. One of DiDi's preferred tactics was to send new recruits over to Uber to join as engineers. As soon as they were hired they acted as moles, feeding proprietary Uber information back to DiDi and carrying out corporate sabotage on some of Uber's internal systems.

While DiDi was busy infiltrating Uber's ranks, the Chinese startup also received help from some of its largest, most powerful allies. Just as Google backed Uber in the United States, Tencent, one of the three largest technology companies in China, was one of DiDi's biggest investors.

Tencent would occasionally block Uber's account from WeChat, China's most popular social network and messaging app, a serious blow to the Western company. WeChat is the Facebook of China.

Blocking Uber on WeChat meant Tencent had scrubbed the company from the most important Chinese social media environment. Worse, being blocked cut Uber off from WeChat's "wallet," a feature that allows people to buy goods and services without cash or a credit card.

At first, Uber hadn't understood how popular mobile wallets were in China. They launched only accepting credit cards, a form of payment that the Chinese just didn't use; it took time for Uber to begin to accept mobile payments from WeChat and Alipay—some would say too much time. They eventually fought their way into various Chinese mobile wallets, only to be intermittently blocked from the biggest one by Tencent.

Some of Uber's China problems were self-inflicted. For one, Uber still relied on Google Maps to guide drivers from pickup to drop-off. That proved to be a terrible choice; while Google has mapped much of the developed world with unmatched precision, China remains one of Google's blind spots. Uber's Google-powered navigation software often confused drivers with awful directions, and irritated riders who were frustrated when their driver took a slower route.

Uber had serious issues beyond China as well. All across Asia Kalanick was fighting taxi operators, governments, and well-funded competitors like Ola in India and Grab in Southeast Asia, two cutthroat ride-hailing startups that were willing to play just as dirty as DiDi.

Kalanick sent a twenty-four-year-old employee—Akshay BD—to be a front-line community manager in Bangalore. BD was scrappy, chasing cab drivers down the street to get them to sign up for driving for Uber. He had the hustle Kalanick wanted in an Uber general manager, especially one trying to drum up demand in one of the world's largest markets.

But BD wasn't prepared for what Ola and taxi companies would do to push back. In Mumbai, local taxi operators muscled up at Uber's offices and tried to intimidate employees. Violence was not uncommon; in Bangalore, whenever BD took a ride home from work, he refused to let Uber drivers take him directly to his house; he knew competitors might follow him. In Hyderabad, one Uber driver committed suicide after he wasn't able to make his car loan payments on time. An angry mob of drivers—some who drove for Uber, others employed by taxi organizations all too happy to stoke anger—showed up outside of Uber's offices in early 2017 with the dead body of the thirty-four-year-old driver, M Kondaiah, dumping the corpse on the company's front doorstep. If Uber's wages for drivers in India weren't so low, the group claimed, Kondaiah would still be alive today.

Security incidents usually increased around planned price cuts. From Uber headquarters in San Francisco, Kalanick would occasionally cut the fares for Uber rides in dozens of markets at a time. The effect rippled outward from the Bay Area to the rest of the world, affecting the livelihoods of millions of drivers in an instant. Kalanick did it to spur rider growth; it sometimes incited violence.

One incident involved an Indian man who arrived at an Uber outpost in hysterics, upset that Uber had yet again slashed prices. The man took out a canister, doused his body in gasoline and then brandished a lighter, threatening to set himself ablaze unless Uber raised its rates again. Security guards tackled the man, wrestled him to the ground, and stripped the lighter from his hands. This was not the only occurrence; a rash of suicides by self-immolation would soon follow.

One of the worst driver-related incidents in India occurred in December of 2014. A twenty-six-year-old finance worker had called an Uber after a work dinner to take her to her home in Gurgaon, a city outside of New Delhi. Shortly after her Uber arrived she dozed off in the back of the car. That was when her driver, Shiv Kumar Yadav, noticed she was asleep and diverted from the intended route.

Yadav switched off his cell phone, making the two untraceable to police or Uber headquarters. He found a secluded area, parked the car, climbed in the back seat and raped the young woman. Afterward, Yadav bullied her to stay silent; he said he would murder her if she told the police. Yadav drove her home. The woman called the police at 1:25 a.m. on Saturday. She had snapped a photo of Yadav's license plate as he drove away from her apartment. The police arrested him the following day.

The story went viral almost immediately. The public—both in India as well as abroad—immediately blamed the incident on the lax security at Uber. Occurring just weeks after Kalanick's public blow up with the journalist Sarah Lacy and the botched "charm offensive," the rape accusation fueled the idea that Uber was misogynistic, a company that didn't care about women, and offered a service that wasn't safe. American press skewered Uber for the incident, which reinforced every negative stereotype that people held about Uber.

The Indian response was much more severe. Indian officials, sensing public outrage, immediately shut down all ride-hailing services in New Delhi, pending an investigation. General managers in Indian cities like Bangalore shuttered their headquarters and moved into hotels, an attempt to escape nonstop protests and threats at work. For six weeks, Uber employees in India even brought their parents and families into hotels with them; taxi officials were beating up Uber employees in the street.

Southeast Asia was another debacle. Grab, the predominant ride-hailing company in the region, was a tenacious competitor. Uber would spend nearly $1 billion fighting Grab. The result was an astonishing loss of nearly 50 percent of their market. After just four years, Uber held just 25 percent of the Southeast Asian market. Years later, Uber would have to sell off its Southeast Asia business for a 27.5 percent stake in Grab.

All of this—the losses, the corporate subterfuge, the nonstop bare-knuckle street fighting, the literal violence—had an effect on Kalanick's psyche. Kalanick was already a tense, competitive person. But China and Southeast Asia only served to grow his persecution complex. He began to feel that he was always being sabotaged or that friends or employees were trying to deceive him and harm the company. After the war for China, Kalanick's cynicism would spread to other parts of the business; it would never subside.

THOSE WATCHING UBER back in the United States noted the trouble overseas, but most observers saw the company doing no wrong. Kalanick was living large.

With a blank checkbook at his disposal and no investors or board members to hold him accountable, Kalanick began to build a series of Uber offices that symbolized Uber's success.

In Pittsburgh, where Uber was focused on engineering self-driving cars, Kalanick hired architects and industrial designers to build a futuristic-looking office from the ground up. The office was a model in extravagance, home to a few hundred employees. The firm placed two dozen different types of chairs, scattered across the building's enormous office, for no other reason than they understood Kalanick loved different types of chairs. The office, when all was said and done, cost upwards of $40 million to create, or roughly $200,000 for each of the two hundred or so employees who worked in the satellite operation. Uber Santa Monica was home to a lavish beachfront property, also costing tens of millions of dollars.

But the crown jewel was San Francisco. After outgrowing a handful of other offices, Kalanick leased several floors at 1455 Market Street— a bunker-like space in the middle of downtown—and soon rebuilt in high Uber style. Holes were knocked between two of the concrete floors to build a transparent glass staircase connecting the two levels. The multi-million-dollar staircase led to his favorite of the many Uber spaces, this one designed to reflect Kalanick's taste. He dubbed the aesthetic "Blade Runner meets Paris," a slew of black granite and see-through glass conference rooms, to be inhabited round the clock by engineers hunched over their silver MacBooks.

Managers spent hours strategizing in the most clandestine place in the building: the "War Room." Custom designed with boutique architects and furniture designers, the War Room was a large conference room placed dead center of Uber's primary office floor, a box encased in glass held for important strategy meetings. Digital clocks on the wall displayed the hour in San Francisco, New York, London, Dubai, Singapore—all on twenty-four-hour time—as if company leaders were in the White House Situation Room.

If the occasion was secretive enough, one could flip a switch that changed all of the glass to a frosted, translucent look, a way to hide company secrets from outsiders—or from other parts of the organization. Kalanick's new offices grew more and more opulent over time. But Kalanick never worried about money, as he could always raise more.

Chapter 16

THE APPLE PROBLEM

AS UBER BLED CASH BY THE BUCKETFUL IN CHINA, KALANICK WAS breathing down his engineers' necks to solve the problem. It was a recurring theme at Uber: something went wrong, the boss wanted it taken care of, and he didn't much care how you got it done. Just *get it done*.

When Kalanick's CTO, Thuan Pham, began to spin up an anti-fraud strike team, he was given tremendous latitude. Uber fraud engineers would have to be thoughtful, fast on their feet, and ready to improvise. Kalanick said he would protect the team from internal politics, and pledged to give its members whatever funds or support they needed.

One of the recruits was Quentin,* a sharp, thirty-year-old product manager who had won awards as a grad student at MIT and after college worked on search products at Google. Colleagues described Quentin as clever, kind to his co-workers, soft-spoken—diametrically opposite the alpha male "Uber bro" employee archetype. Quentin would not be doing keg stands with Uber's operations managers out in the field.

One of Quentin's defining qualities, co-workers said, was his nervous energy, and the caution with which he approached the world and interacted with others. Even his body language, they said, was defensive; in conversations he kept his body slightly turned away. And he gave long, hard looks at people, as if he was sizing them up. An apt personality, they believed, for a job assessing risk and security.

At the beginning of 2014, Uber employed around 500 people. By

* I've changed my source's name to protect their anonymity.

October of that year, Uber had more than tripled in size, and was adding new employees by the day. Under Quentin, Uber's risk, account
security, fraud and abuse prevention teams grew to more than 150
people. Everyone worked hard at Uber, but Quentin's team worked
harder than most. He and a few close colleagues helped to orchestrate some of the drug busts in New York, limited widespread fraud
in China, and helped repair other areas in which Uber was bleeding
money and facing liabilities. He was valuable.

When he started in March 2014, Quentin's team faced a very specific
headache. Two years earlier, Apple released a version of its iOS mobile
software that killed outside access to the unique identification number
of every iPhone, the so-called IMEI number, or "international mobile
equipment identity" number.

The update was a hallmark of Tim Cook's Apple. Unlike its rivals
Google, Facebook, and Amazon, Apple's business didn't rely on
hoovering up personal data from its customers. Facebook and Google
were advertising companies, and as such relied on discovering every
digital detail of its customers' lives in order to target them with ads.
Uber's method of identifying fraudsters made use of digital surveillance
techniques common to Silicon Valley's largest companies.

That practice ran against some of Apple's long-espoused principles,
specifically an individual's right to privacy. Steve Jobs had valued consumer privacy, but his successor, Tim Cook, was a fanatic. He believed
Apple's users should have complete control of their private digital lives.
And if an Apple customer decided to wipe their iPhone clean of data, no
one else—individuals, family members, companies, law enforcement—
should be able to find a trace of that data on the device afterwards.
Wiping an iPhone was final; the data was gone.

The unanticipated iOS software update was very bad news for Uber.
Chinese fraudsters loved to use stolen iPhones to create fake accounts
and sign up for the service. If Uber's security team discovered one of
the accounts was fake and blocked it, all a fraudster had to do was
erase the iPhone of its data and create another new account, which took
only a few minutes and was endlessly repeatable. To counter the tactic,
Uber had spent months building a database of IMEI numbers, which

helped the company keep track of which iPhones had already been used to create new Uber accounts. Before the 2012 iOS update, if Uber saw someone using the same devices to create new accounts over and over again, they knew they had found a scam artist and could quickly ban them from the network. After 2012, however, Uber lost access to the serial numbers and went back to square one.

But then, in 2014, Quentin's team found a way around it. After Apple's iOS software release, about a half dozen companies sprang up overnight that claimed they could detect the sacred IMEI. Quentin tested a few of them before landing on InAuth, Inc., a small firm based in Boston. With just a smattering of code inside Uber's mobile app, InAuth could track down the device identification number of the iPhone used to install the app, a technique known as "fingerprinting" in the security and fraud industry. Once a phone was "fingerprinted," it was much easier for Uber to tell if it was being used for fraud. Just a few months after starting at Uber, Quentin signed a contract with InAuth.

It worked *perfectly*. Before Uber had started using InAuth, fraud in China and other major cities cost Uber tens of millions of dollars *every week*—and occasionally even more than that. After Uber built a new version of its app with the InAuth code installed, Quentin watched the fraud numbers fall off a cliff. When a scammer tried to create a new account on a device Uber had fingerprinted, Uber's anti-fraud systems would kick in and the account would be banned automatically. Finally, after years of being ripped off, Uber had found a way to fight back.

There was one problem: InAuth's service blatantly violated Apple's rules regarding user privacy. So everything between Uber and InAuth had to be kept secret. if Apple found out, both Uber and InAuth could be in serious trouble, and could even get Uber's app banned from the iPhone.

At some point in his or her career, every mobile software engineer in Silicon Valley has come up against the vague, byzantine rules of the App Store. Each year, Apple would update its mobile software. A simple tweak in Apple's software practices could make or break an entire startup business plan. To build mobile software, especially for Apple, was to be in a state of constant anxiety and frustration. When developers submitted new apps to the App Store they would wait for a response like pilgrims at the Oracle of Delphi. Sometimes Apple would answer helpfully. Other times Apple would say nothing at all.

Quentin and his team skirted the privacy rules because they felt like they didn't have a choice. They needed to deal with the enormous fraud problem, and Apple wasn't giving Uber any other options. If Uber and InAuth could keep a low profile, perhaps the fraud team could escape detection.

They would have no such luck. In mid-November 2014, *BuzzFeed* ran its story about the infamous dinner where Emil Michael suggested doing oppositional research on journalists. Most public attention at the time focused on Michael.

But during the "charm offensive" Josh Mohrer, Uber's brash and cocky general manager in Manhattan, had made a grave mistake. In an interview that week he let slip a mention of an early version of "Heaven," a tool that provided a "God View" of riders on trips in real time. The reporter had taken an Uber to meet with Mohrer that afternoon. Mohrer bragged that he had tracked her the whole way. The comment would not go unnoticed.

Eight days after the first story broke, Quentin's team was hit with a bombshell. As scrutiny intensified in the wake of Uber's recent scandals, an enterprising young hacker in Arizona named Joe Giron had decoded Uber's Android application and found the list of data access permissions Uber's app requested upon installation. The litany went far beyond what most Uber users expected: phone book, camera access, text message conversation logs, access to Wi-Fi connections. These were permissions that were suspect for *any* app to request, much less a taxi service. Why would a ride-hailing app need access to their customers' text messages or camera? It was seen as a broad overreach into users' privacy. Not only was Uber willing to go after journalists, but the company also wanted to know everything about you and your phone.

The blog post blew up. After circulating across security forums and other internet sites, it landed on *Hacker News*, a message forum widely read by engineers and the Silicon Valley elite.

What those readers *didn't know* was that the armchair hacker had stumbled upon the secret InAuth code library, written inside of the

Uber app as part of their secret deal. In order to fingerprint devices, InAuth required *far more* data than the average smartphone app, which meant asking for all sort of extended permissions. InAuth created device profiles based on this data to triangulate the users' IMEI numbers. It was a clever technique, and companies besides Uber paid millions to use it. But the practice upset consumers when they discovered how much information they had unknowingly given Uber.

Back at Uber HQ, the fraud team members were freaking out. The public wasn't supposed to know Uber had a deal with InAuth, much less read the code they had licensed. Should they address the issue with the public? And what if Apple started snooping around? Uber had recently submitted their newest iOS app build. What were they supposed to tell Apple if they found out Uber was breaking the rules?

At first, nothing happened. But a few weeks later they got their answer: the App Store declined Uber's latest software update. Quentin's team had been caught.

AS THE MAN IN CHARGE of the App Store, Eddy Cue had seen the best—and worst—of the startup world.

Eddy Cue reported directly to Apple CEO Tim Cook, and no one else. He was the guy who saw rising startup stars before nearly anyone else in Silicon Valley, because their apps skyrocketed to the top of his charts. When they did, Eddy Cue made it a point to meet the founders. By 2014, the fifty-year-old senior vice president of Apple's internet software and services business had known about Travis Kalanick for about a year. Cue and Cook saw the potential in Uber early on, and absolutely *loved* how it used the iPhone's technology. Cue and Cook had a sit-down with Kalanick, after Uber raised millions from Google Ventures and TPG.

Both Cue and Cook walked away from the meeting struck by Kalanick's passion and talent, but they weren't charmed. As Kalanick and Emil Michael spoke at length about Uber's ambitions, Cue was struck by the founder's arrogance. Kalanick waved off issues like Uber's bad reputation in the press and the threat of regulation.

"I know what the hell I'm doing," Kalanick said to the Apple exec-utives, who between them had fifty years of experience at the high-est levels of the computing industry. "No one else knows what they're doing in ride-sharing. We have it figured out."

During the meeting, Cue thought challenging Kalanick a bit might bring out his self-effacing side. "Why do the Google investment at all?" Cue wondered aloud. "It feels a little bit like letting the fox in the henhouse. They've been into self-driving for years. We always figured something like what you guys are doing would be on their roadmap someday," Cue said.

Cook nodded, pointing out that potential threat may extend to Uber's board of directors as well. "Are you at all concerned about Drummond being in the room?" Cook asked, noting the board seat Kalanick gave to David Drummond, Google's chief legal officer and SVP of corporate development. Cook and Cue saw him as a proxy for Google CEO Larry Page.

"The board is irrelevant," Kalanick said, waving them off of the idea. "I hand pick all of these guys. They do what I tell them, and the way I've structured things, I do what I want."

Cue was taken aback. Many founders at least *performed* a sense of humility in public—a strategic modesty that Kalanick clearly lacked.

After the meeting, Cue and Cook remained in regular touch with Uber. iPhones were only as good as the apps that people wished to use on the devices, so Apple made it a priority to keep tabs on its top apps. The executives caught up every three to six months, almost always asking Kalanick and Michael to make the hour-long Uber ride south to Apple's headquarters in the sunny Cupertino suburbs.

And yet, Uber was never what Apple would call a perfect partner. The startup frequently frustrated App Store executives, those directors below Cue who were responsible for tracking top-performing partners.

Most of the problems came in Uber's software updates. Every time an App Store company wanted to update its software, they would have to send a new "build," or new software version of the app, to the App Store for approval. For Apple, handling Uber's new builds was a par-ticular pain. When Uber sent an update, Apple engineers would often catch them trying to sneak backdoor tricks into its code. One version of Uber's consumer app, for example, was able to convert itself from the app that riders download to the special app built only for drivers—reducing "friction" for new users—a small but meaningful breaking of Apple's rules. The new build didn't fly. Apple caught the misbehavior and gave the company a light scolding. Uber was required to have one app for riders and a separate one for drivers.

As the nits inside Uber's updates piled up over time, Cue's lieutenants closely monitored developments within Uber's app; the engineers studied Uber's code so rigorously they could tell when the startup was trying to pull another trick.

For a while, Cue was willing to give Kalanick's engineers the benefit of the doubt. Not all of Apple's rules were crystal clear, and Uber was a *very* popular app with iPhone customers. Hackers being hackers, the App Store moderators saw all sorts of little tricks and shortcuts inside the code of apps in the store, some worse than others. Uber's constant sleight of hand was a pain, but relying on the App Store team to police them was worth the resources.

But things went downhill fast at the end of 2014. App Store leaders had seen the *Hacker News* post where Uber's Android app had been decompiled and exposed for the data-sucking beast that it was. Sure enough, Uber's iOS app was asking for the same types of permissions as well. Uber's "fingerprinting" solution wasn't going to fly. As the holidays approached and engineers rushed to get their code approved before everyone took off for vacation, Apple began rejecting Uber's attempts to push the fingerprinting techniques inside the iOS app.

Back at Uber headquarters in San Francisco, the company's engineers were scrambling to overcome the constant App Store rejections. In typical Apple fashion, each denial came without a real explanation *why* Apple had turned Uber down. Uber employees knew it was probably about InAuth's code, but didn't want to tip their hand if Apple hadn't discovered it.

After a long brainstorming session between members of Uber's fraud and mobile teams, one frustrated mobile engineer stood up. The engineer, a previous Apple employee, knew how Uber could get around the

App Store problem. "I have an idea," he said, before walking out of the conference room and back to the laptop on his desk. "I can handle this."

IT WAS ONE THING for Uber's engineers to fudge the rules every now and then on a new build submission. Loads of developers submitting to the App Store did it.

But this new idea was as brazen as the Trojan Horse. The engineer's idea was to trick Apple by using a technique called "geofencing," using the GPS and IP address data from the phone to tell Uber where the user was located. A "geofence" acts much like it sounds; if the user is within a specific geographic radius, the app would perform a certain way. In Uber's case, if the Uber app was used within the Bay Area or near Apple's Cupertino headquarters, it wouldn't run the InAuth "library" of code, which asked for the personal data needed to fingerprint phones.

What that Uber engineer assumed—incorrectly, as it turned out—was that all of Apple's App Store code reviewers were located in Cupertino and the San Francisco Bay Area. Eventually, an Apple reviewer who wasn't based in California stumbled upon the InAuth code library. Uber's ruse was up.

Cue was apoplectic. Fudging your way around Apple's rules was one thing. But active subterfuge—intentionally hiding an app's behavior from Apple administrators—was a cardinal sin. Uber was actively deceiving Apple in an elaborate and sophisticated way.

Seething, he sat back in his office chair at Apple's headquarters, pulled out his iPhone, and dialed a number.

Kalanick answered. He was cheerful. The Uber CEO knew he always needed to stay on Cue's good side.

Cue wasn't having it. "We need to talk. We have a real problem." Cue went into some of the specifics of what Uber was doing with its apps, and made it clear he was pissed off.

"You need to come down here and sort this out with us," Cue said. "I'll have my staff get this set up. Goodbye." Then Cue hung up. He hadn't even waited for Kalanick to say goodbye.

Kalanick was freaking out. He worried Apple might do something drastic.

He called a meeting, roping Quentin and a few of his team members into a meeting room at Uber headquarters. As Kalanick shut the door, he started asking questions, all of which amounted to: "What the fuck happened?"

Quentin's team knew, at least generally, what had gone down. He brought in the mobile engineer—who by then was scared out of his mind—and had him explain the technique he used to fool Apple.

As usual, Kalanick paced around the room as the gravity of his team's actions sank in. In his defense, Kalanick had never told the engineer to lie or cheat Apple. After all, the people working on that team were layers below Kalanick. He expected his leaders to handle their staff appropriately.

What Kalanick *did* tell his teams was consistent: "We need to win, no matter what. Do whatever it takes." That message, across every team, up and down every part of the organization, was at the core of each employee's understanding of Uber. Win, at all costs.

The fraud team started preparing its explanation for what happened—and its apology to Eddy Cue.

THOUGH IT IS ONE OF the Valley's most secretive and opaque corporations, everything about Apple's Cupertino campus works hard to convey openness and transparency.

Stark white office buildings rise above lush, well-manicured lawns at 1 Infinite Loop. The main entrance echoes the aesthetic of Apple's retail stores: sheets of glass, solid white walls and a half-domed roof, shielding the building from the hot California sun.

As the group from Uber walked into the building they were ushered to a private conference room. They had prepared a careful presentation for their hosts.

Cue strode into the room, followed by a few of his lieutenants from the App Store. Flanking Cue was Phil Schiller, Apple's senior vice president of marketing. Since 1997, Schiller had worked for Apple reporting directly to Steve Jobs. Under Jobs, Schiller promoted the revamped iMac in 1998, an egg-shaped blast of color that came in bright orange, lime green, deep turquoise, and other colors. He promoted the iPod in all of its various iterations, helping to create a record-breaking hit. The two Apple execs, both in their early fifties, had a combined net worth in the hundreds of millions.

Cue hammered Kalanick from the start. "We want you to walk us through exactly what happened here, from the beginning to how we ended up in this room today."

Kalanick stammered, shaken, but started from the beginning. He walked Cue and Schiller through the massive fraud across the platform, through the ingenious solutions scammers had and the problems Apple's iOS updates had created for halting fraud. Emil Michael, the point person for dealing with Cue and the Apple blowup, had prepped Kalanick well.

Kalanick was trying on a new face for this meeting, one of conciliatory regret. He knew he could get away with telling the government and city authorities to kiss his ass. However, on rare occasions, he could sense he needed to humble himself. It almost never happened. But here, at Apple HQ, in front of its top brass—he kissed the ring.

"We want to hear you commit to us," Cue said to Kalanick, as the group wrapped up the long, tense meeting. "We want to know you will never, ever do this again. Make this promise, or you're gone, you're out."

Cue meant business. He had brought the matter to his boss, Tim Cook, and both of them considered this a serious infraction. No one, no matter how successful the app or company, could lie to Apple and get away with it. For Cook, there was no greater sin than breaching the privacy of his users. Cook would later fight the FBI in public, refusing to unlock the smartphone of a mass murderer in San Bernardino, and would slam Facebook at public events for the company's intrusive privacy practices. He had no problem supporting Cue on this decision: if Uber didn't cut it out, Cook and Cue would ban Uber from the App Store.

Kalanick knew they were serious. If word of this showdown got out to the public it would trigger a major scandal. Worse, he knew what an App Store lockout could mean for Uber. His startup was now valued in the tens of billions of dollars, and iOS downloads accounted for a majority of Uber's business. Taking Uber off every iPhone in the world would kill his company. Kalanick assured the Apple executives that this would never happen again.

Cue could accept that. But Uber was on probation. They left the meeting with a few stipulations, mostly about how Uber engineers would now be required to submit supporting documentation every time they pushed a new software build to the App Store.

And if Kalanick's team tried to pull a stunt like this again, Cue wouldn't be as understanding. Uber would be gone.

WEEKS LATER, Kalanick headed back down to Apple for a regularly scheduled catch-up with Cook and Cue. The first meeting with Cue, Schiller and the top App Store leaders was rough. But this was the one Kalanick was really dreading.

Kalanick tried to play it cool. As he walked back through the front door of Apple's campus, he wore his favorite pair of Nikes—Darwins were a deep, bright red with matching red laces and a mesh outer coating[*]—and striped hot pink and blue socks to give himself an extra pop of color. He looked good on the outside.

On the inside, he was nervous. This was the first time he had seen Cook in person since the blowup between Apple and Uber. He didn't know how the CEO was going to react.

After the meeting began, Cook, in his calm southern drawl, raised the issue. He wanted to make sure the problem was behind them.

Kalanick shifted in his chair. He had been expecting this, but was still uncomfortable to hear it. He explained it was true—more deferential than ever—but as he assured Cue, it wouldn't happen again.

Cook nodded. He let the tense moment pass, and the group went on to discuss the rest of the agenda. But in his subtle way, Cook was draw-

---

[*] Kalanick particularly loved how Darwins gripped the cement floors of Uber HQ when he paced his laps. He wore the shoes to most public events, including his interview with *Vanity Fair* editor Graydon Carter in 2016.

ing a line in the sand. If Uber ever, *ever* tried to deceive Apple again, it would be the end for Uber on his company's platform.

Kalanick Ubered north, away from the Apple campus, and later met up with a friend. As he debriefed the friend on the afternoon's events he confessed he was shaken. But only momentarily. The showdown had sent adrenaline surging through him. He had withstood an upset Tim Cook—Tim fucking Cook!—Kalanick said, and his company wasn't obliterated.

Uber had survived. As his friend watched, Kalanick's fear melted away and was replaced by a renewed sense of confidence—even swagger. If Uber could take on Apple, it could take on *anyone*.

Chapter 17

"THE BEST DEFENSE . . ."

THE SHOWDOWN WITH APPLE WAS A BIG PROBLEM. BUT EVEN AS THAT crisis was unfolding, Travis Kalanick had an even *bigger* problem to deal with. To solve it, his CTO, Thuan Pham, had hired a guy named Joe Sullivan. What Joe Sullivan saw was a security nightmare.

As chief security officer at Facebook, Sullivan was used to chaos. He had seen it all over his six years at the social network. Sullivan was responsible for protecting Facebook's users from identity theft, drug sales, gun sales, kiddie porn distribution. While Mark Zuckerberg was down the hall discovering new frontiers of the internet to conquer, Sullivan was tracking down digital thieves—the kind of men who, for example, blackmailed women after stealing nude photos from their phones.

But when Sullivan got the email from Thuan Pham, Uber's chief technical officer, asking for help, Sullivan was intrigued. He had read about the ride-hailing company—no one could escape the headlines about the embattled unicorn. Uber sounded like a *hot mess*. Tracking riders, digging up dirt on journalists, slurping up user data—at least, that was its reputation.

Rider tracking in particular was a wild invasion of privacy. Kalanick saw it as a neat party trick—literally. When Uber first launched its service in Chicago in 2011, the company invited a small group of high-profile Chicagoans to a private party at the Elysian Hotel. There, he debuted "Heaven." The guests watched as a giant screen showed hundreds of Uber riders zooming across a map of Chicago in real time. Kalanick and his partner, Ryan Graves, grinned; the crowd was stunned.

While Uber had "Heaven," Kalanick also held court over "Hell."

That was the nickname of one of Uber's most highly guarded and extremely valuable internal programs; "Hell" was devised to monitor the locations of all Uber drivers who also drove for Lyft. Uber employees at headquarters would create fake Lyft accounts, which tracked nearby vehicles—up to eight per fake account. Information about those vehicles was then sent back to Uber and stored in a database. "Hell" created a way for Uber to monitor the real-time positions of Lyft drivers. And because many of those drivers worked for Uber as well, Uber could monitor the rates Lyft was offering for drivers and outbid them, thereby swaying drivers to work more regularly for Uber. "Hell," as Sullivan saw it, was sneaky. It was also highly unethical and would be a public relations nightmare if it ever leaked.

"Heaven and Hell" were just the beginning. Those programs fell under the umbrella of "competitive intelligence"—a friendlier phrase than corporate surveillance—which was shortened to an even more genial acronym, COIN. Everyone in the Valley had a version of COIN, in one way or another. The most widely used form involved scraping competitor data from websites, apps, and other publicly available repositories. "Scraping" was computer-speak for automating the collection of information through written programs and coded scripts. Uber's most useful tool scraped information on pricing changes within the Lyft app, allowing Uber to systematically undercut its competitor.

Uber also purchased receipts from companies like Slice Intelligence. These data-brokerage firms bought reams of anonymized purchasing data from credit card companies and retailers, sliced up the results, analyzed them by sector, and packaged them for resale to other companies. Aggregate data for trip receipts from Lyft, for instance, allowed Uber to confirm its competitor's prices. Combine that data with Uber's scraped location and pricing data and the company could create a remarkably complete picture of Lyft's business. Sullivan knew it wasn't sporting. But it worked.

Besides surveillance, there were severe safety issues. The India rape scandal was just the tip of the iceberg. Unbeknownst to outsiders, Uber operations teams dealt with thousands of misconduct cases every year, including increasing instances of sexual assault. As the service grew, millions and ultimately billions of rides were taken. The power of large numbers meant that assaults and sex crimes were probably inevitable. But Uber had so lowered the bar to become a driver that people who might have been prevented from driving in the official taxi industry could easily join Uber. The problem became so significant that later, the company would create its own taxonomy of twenty-one different classifications of sexual misconduct and assault in order to properly organize the sheer number of annual incidents reported.

It would have been a public relations nightmare if the public knew that hundreds of drivers had been accused of sexually assaulting customers. When a new rape accusation or lawsuit was leveled against the company or a driver, some Uber employees would remind others that drivers are always "innocent until proven guilty." Kalanick himself would repeat the phrase often, especially to the security and legal teams. Technically, it was true, and Uber had certainly seen its share of false claims and scams. But perhaps more than the assaulted riders, or the accused drivers, Kalanick felt it was *Uber* that was being persecuted. Outsiders were always scheming against Uber; enemies wanted to see his company fail. Uber was the real victim, he felt. "Innocent until proven guilty," Kalanick reminded his employees. On occasion, when a sexual assault victim decided not to pursue litigation or if the evidence in a police report was not conclusive enough to prosecute, a round of cheers would ring out across the fifth floor of Uber HQ.

Beyond privacy and safety issues, Uber had another big problem. When Sullivan heard about it he almost didn't believe it. According to executives at the company, Uber had been the victim of a massive hack earlier in 2014, a serious breach of the company's data that compromised the names and license numbers of more than 50,000 Uber drivers. Uber had kept the hack secret. It didn't know how to tell the public, much less if it even wanted to do so. Kalanick didn't know the law, and had no interest in making these calls. Though he certainly didn't want to spur a public backlash, he always thought it was up to the legal and security teams to figure out what the solution was—and most importantly, to make it go away. Sullivan knew it wasn't that simple; Uber was required by California law to notify authorities of a data breach.

The breach had happened in May, and Uber discovered the effects of it in September. When Sullivan was arranging to join Uber, it was December—and the company hadn't said a word.

During the recruiting process, Kalanick asked Sullivan to give a presentation to Uber's executives on what Sullivan's vision for security at Uber would be, if he got the job. Sullivan said he wanted to make security an integral part of Uber's marketing strategy. Consumers, he believed, should think of Uber as far safer than taking a taxi. "Security should be a brand differentiator for us, not a minimum viable component," he said.

Sullivan considered his options. He had been offered the job of chief security officer, overseeing a ragtag security team. Some thirty employees scattered across different groups inside of Uber. If Sullivan was going to help Uber—a sprawling, global operation—he'd have to bulk up the team. He'd also have to report directly to the CEO—a request the company accepted.

Uber needed Sullivan far more than Sullivan needed Uber. But Joe was ready for a challenge. And by then, he had bought the sales pitch and taken a liking to the smooth-talking CEO, Travis Kalanick.

SULLIVAN DIDN'T COME FROM the tech world. The oldest of seven children, he "rebelled" against his hippie parents—his father a sculptor and painter, his mother a schoolteacher and writer—by going to law school. While young tech entrepreneurs were building software with wide-eyed optimism, Sullivan spent his twenties as a federal prosecutor, confronting the worst of what humanity had to offer. Robert S. Mueller, a decorated war hero who would later go on to investigate President Donald J. Trump, handpicked Sullivan to work in the computer hacking and intellectual property cybercrime unit, a prestige position in the Northern District US Attorney's Office in San Francisco. Sullivan had studied cyberlaw at the University of Miami, where he earned his JD, and threw himself into challenging cases involving trade secrets and corporate espionage during the late '90s boom. By the time the bubble had burst in 2000 he had made a name for himself.

Sullivan stood tall at around 6'2", yet his posture was always slightly hunched, hands tucked into his pockets. His brown, bushy eyebrows and neatly combed chestnut hair gave him a non-threatening look. After years in government suits, he compromised with dadcore jeans and button-downs, and eventually moved to a more tech-friendly jeans and T-shirt combo. His high cheekbones, broad forehead, and wide-set eyes made his default expression a kind of restful stoicism, even in the face of complex information security problems.

He spoke quickly and clinically, his dispassionate attitude forged over his years as a lawyer. The most emotion you'd see was a raised eyebrow, or perhaps a knowing smirk when telling war stories from his days as a prosecutor. Laughter never came in more than a chuckle, like the joke was a secret he kept to himself.

Sullivan didn't exude the natural charisma of a flashy trial lawyer, but people liked him. He was geeky without being entirely antisocial, he was willing to work hard, and he went after the bad guys. Everyone who knew Joe said he was solid—an all-around dependable guy.

After trying his fill of cybercrime cases on a government salary, Sullivan got the itch to go in-house. In 2002, Sullivan landed a job at eBay, then a tech powerhouse with growing revenue, bright prospects and millions of daily auctions from buyers and sellers online.

It was also rife with fraud. As a senior director of trust and safety, Sullivan spent most of his time hunting down scammers who used the platform to con web novices out of thousands of dollars. As millions of people came online for the first time, they weren't ready for the fraudsters, hawking fake listings for valuable Beanie Babies and collectible baseball cards that had never existed in the first place.

Most scams were as simple as a seller completing a sale and then never mailing the merchandise to the buyer. But some were more intricate. One scheme involved a con artist offering to pay an honest seller outside of eBay, then sending a bounced check. If the merchant complained, they would have little recourse since the purchase wasn't completed on eBay itself. The worst scam was often the simplest: a seller mailing a customer nothing but an empty box. Tens of thousands of these frauds occurred on eBay every year, and were becoming only more prevalent as the site grew in popularity.

At eBay, Sullivan's job was part detective, part digital police officer. It was just like going after the thieves and scammers he encountered as a prosecutor. Only this time, it was better. In court, he had to put together a meticulous case to take down a single defendant. Maybe a few at a time if they went after a syndicate. At eBay, his teams of anti-fraud experts caught hundreds of scammers every day, booting them from the platform. He created entire systems designed to defang the bad guys. And when a big, organized crime syndicate came along and tried its hand at eBay scams, Sullivan and his teams were there to stop them.

Sullivan's favorite story involved the Romanians. Romania was a nexus of fraud. Until 2003, Romania didn't have a single cybercrime law on the books. Combine that laxity with a number of organized criminal outfits and a generation of savvy programmers and you had a pirate's cove of malefactors. The scams would usually involve offering high-priced electronics for a deep discount, which would fetch immediate bidders across eBay. After someone sent $2,000 USD for a big-screen TV, for example, the Romanians would disappear. The fraudsters worked out of internet cafes in Bucharest and accepted only Western Union wire transfers, making it difficult for police to locate them. And since the syndicates were run by the Romanian or Russian mafia, local law enforcement never pursued cases, fearing their own safety.

Sullivan wasn't afraid. After he and his colleagues took down one of the biggest Romanian eBay fraud rings, eBay flew him to Bucharest to testify in court—at his own request. As Sullivan took the stand, he was flanked by two beefy local police guards. Each of them held an AK-47 and wore a jet-black balaclava—a woolen mask that fully covered the face—for fear they would be identified and later killed by the local mafia after the trial. Sullivan, donning his old uniform of suit and tie, delivered hours of testimony that helped put the fraudsters behind bars. He didn't wear a mask.

After eBay and a two-year stint at eBay's sister company, PayPal, Sullivan was presented with an even more intriguing challenge. By the end of 2008, a young, buzzy startup had come calling. Facebook—then closing in on 150 million users—had an opening on its legal team. Sullivan leapt at the chance; Facebook's growth was explosive, and Mark Zuckerberg's ambitions were boundless. He wanted to bring the entire world online and plug it into his social network. That kind of opportunity was a no-brainer for Sullivan. He took the job.

If eBay gave Sullivan a chance to operate like a Navy SEAL, working on security at Facebook was like commanding his own private army. Facebook was a daily destination for scammers and fraudsters, just like eBay. But it also harbored pedophiles, stalkers, vengeful ex-boyfriends, blackmailers—you name it. In the six-and-a-half years Sullivan spent at Facebook, the company rose to become the world's largest repository of personal information, and he was the man charged with watching over all of it. After just a year, he was promoted to chief security officer.

Sullivan's group actively pursued so-called "bad actors," those intending to do harm on the internet. They weaponized lawsuits against spammers and scammers who flooded Facebook with garbage posts. They played cat-and-mouse games with cyberbullies, and fingered rings of Russian cybercriminals, turning them over to the FBI.

His approach was different than other security types in Silicon Valley.

"A lot of companies stop at playing defense," Sullivan once said in an interview. "We spend a lot of time trying to figure out who's sitting on the other side of cybercrime."

Sullivan's tactics were best exemplified one weekend during his time at Facebook, when he got a frantic call from a friend, a female co-worker from Facebook. She had been browsing Match.com one evening, looking for a date, when things started heating up with a construction worker from San Jose. As the flirting went on, she sent the man a topless photo. The stranger's next message alarmed her: the man told her he had researched her background, and knew she worked at a famous Silicon Valley company. If she didn't wire him $10,000 cash, he threatened to email the topless photo to her entire company.

Sullivan knew what to do. He and a colleague took control of her Match.com account, and attempted to lure the blackmailer into revealing his identity. The best way, Sullivan knew, was to push the scammer toward a payment system. For digital detectives like Sullivan, online payments often provided the best chance at finding clues to an attacker's identity. Certain banks, for instance, would block attempts at money transfers to specific areas, which narrowed down the list of potential countries where the scammer could be located. Sullivan would also add incorrect details when making a payment, an intentional maneuver that made the transactions fail to go through. After the payments failed enough times, the attacker would give additional details about his account location, which helped Sullivan narrow the location details further.

Backtracking the blackmailer's steps through the payment system led Sullivan to a former Google intern, now located in Nigeria. After finding his address in Lagos, Sullivan hired a local lawyer to confront the guy at a coffee shop in Nigeria. The intern immediately confessed to the scam and handed over his computer and email account information.

After they gained access, they discovered the scammer's activities had gone far beyond Sullivan's female friend; the intern was part of an enormous, ongoing Match.com scam. He had been extorting dozens of Silicon Valley female employees out of money for months, dangling the threat of sending their nude photos out to their companies if they didn't pay up. Not only did Sullivan save his friend's reputation, he was able to notify the other women being extorted that they had finally caught the blackmailer, ending months of anguish.

Whether it was hackers in Romania running massive fraud schemes or blackmailers bilking innocent women, Joe Sullivan was good at finding people on the internet, and keeping people safe. It was the reason he had been recruited to Uber. And it was why Sullivan ultimately said yes to the job. He looked at Uber and saw a rat's nest of problems: widespread fraud, competitors across four continents, hackers laying siege to the company's valuable stockpile of personal information. Plus, Uber offered him the chance to be more than an internet cop; the very nature of Uber's service meant dealing with things that can go wrong in the physical world, with millions of Uber riders in actual cars every single day.

Months before Sullivan joined Uber, he helped the company clean up the mess around the breach of its systems; Uber reported the breach, as is legally required of companies, in February 2015—nine months after the hack had happened. It would not be the company's last data breach; another attacker would crack Uber's systems in 2016. It would, however, be the last time Sullivan and Kalanick would come forward voluntarily to admit Uber had been hacked. The decision to keep quiet would prove more costly than either man could have imagined.

But by the time Sullivan arrived at Uber in April 2015, he realized he had a much bigger problem on his hands than fraud or thievery.

He needed to keep Uber's drivers from getting murdered.

NOT TWO WEEKS into Sullivan's new job, he got an urgent call on his cell phone. One of Uber's drivers had been killed in Guadalajara, and operations managers on the ground suspected the local taxi companies were responsible.

For months, Uber Mexico had been under attack by the local taxi cartels. The violence had started slowly at first; a physical altercation here, vandalization there. But things soon escalated. Much like their kin in American cities, Mexican taxi operators had spent thousands of dollars on licenses, permits, training classes, and other state-mandated items just to pick up passengers in Mexico. But now the unions watched helplessly as Uber siphoned off business. As the cabbies grew more desperate, beatings, ransackings and robberies of Uber drivers grew common. Many were assaulted to intimidate others from joining Uber.

"We are not going to leave them alone," Esteban Meza de la Cruz, a taxi driver and union leader who represented about 13,000 drivers, said at the time. "We are tracking them and hunting them down."

By the time Sullivan had arrived, violence had spread from busted lips and bruised heads. People were dying, and it was happening all over the world. Law enforcement offered little help. The death of a taxi driver wasn't exactly a top priority for Guadalajaran police. Sullivan's calls went unanswered. Frustrated, Sullivan started calling old friends from the intelligence community. One former FBI contact shed light on the situation: "Guadalajara is cartel country," Sullivan's friend at the Bureau told him. "We don't send people there."

Countries like Brazil were even worse. Kalanick had tapped Ed Baker, a former Facebook growth executive, to grow South America. He encouraged city managers in São Paolo or Rio de Janeiro to sign up as many riders and drivers as possible. To limit "friction" in the sign-up experience, Uber allowed riders to sign up without requiring them to provide identity beyond an email—easily faked—or a phone number. Further, Brazil was largely a cash-based economy where credit cards weren't in common use, so there was no payment or identity data to gather on the individual riders.

For thieves and angry taxi cartels, it was the perfect crime. A person could sign up for Uber anonymously with a faked email, then play a version of "Uber roulette": They'd hail Ubers, then cause mayhem. Cars were stolen and burned, drivers assaulted, robbed, and occasionally murdered. The company stuck with Baker's low-friction system, even as violence increased.

Osvaldo Luis Modolo Filho, a fifty-two-year-old driver, was murdered by a teenage couple who hailed a ride using a fake name and chose to pay in cash. After stabbing Modolo repeatedly with a pair of blue-handled kitchen knives, the couple took off in Modolo's black SUV, leaving him in the middle of the street.

Brazil was in upheaval when Uber arrived in 2015. Unemployment was at an all-time high, and violent crime and murder rates across Brazil were skyrocketing. While the lack of jobs meant many more Brazilians were willing to drive for Uber, the cash bankroll of each day's earnings made them a tempting target for thieves. At least sixteen drivers were murdered in Brazil before Kalanick's product team improved identity verification and security in the app.

Kalanick and the other executives at Uber were not indifferent to the danger drivers faced in emerging markets. But they had major blind spots because of their fixation on growth, and their casual application of financial incentives often enflamed existing socio-cultural problems. Kalanick believed that there were things inherent in the Uber software that made it safer than a regular taxi, namely that the rides were recorded and trackable by GPS. He further hoped Uber could fix the problem of driver safety through more tech solutions.

But Sullivan saw all this and knew he needed to act fast. He would build a world-class security organization, divided into branches to handle threats from financial fraud, to digital espionage to physical security. He requested hundreds of staffers—security engineers to handle Uber's systems, ex-CIA and NSA types on contract to handle on-the-ground operations and field investigations, and many others. Kalanick agreed, and gave Joe Sullivan a blank check.

But Kalanick had one very important requirement: Uber wouldn't only play defense.

Chapter 18

CLASH OF THE SELF-DRIVING CARS

TRAVIS KALANICK WAS FUMING IN THE GRAND BALLROOM OF THE TER-
ranea Resort—a seaside haven for the rich off the coast of Rancho
Palos Verdes, California. It was the opening night of the 2014 annual
Code Conference, a confab for the tech elite. On stage, Sergey Brin was
in the middle of a historic speech, but Kalanick was on his iPhone firing
off messages to David Drummond. Brin—who was ostensibly Kalanick's partner and investor—had just unveiled something that could
threaten Uber's existence: a fully autonomous self-driving car.

"The reason I'm excited for this self-driving car project is the ability
for it to change the world around you," Brin told the audience. The
technologists, venture capitalists, and journalists were buzzing with
excitement. The Google co-founder showed up that night as the key-
note speaker in a white T-shirt, black pants, and a worn pair of Crocs.
Brin preferred comfort over style.

As the video played, the audience saw an egg-shaped, stark white
two-seater vehicle doing laps around a parking lot. It was ugly and
small. The front of the vehicle looked like a smiley face, as if Humpty
Dumpty had turned into a golf cart. *Blade Runner* this was not.

None of that mattered. The car didn't need a steering wheel, so it
could be any shape. In the Dumpty-mobile sat two people. Neither did
anything to drive the car as it zipped effortlessly around a Mountain
View parking lot. As far as Kalanick was concerned, Google's egg-
shaped, self-driving monstrosity was a work of art.

Google, long considered an ally and a partner, seemed to be turning
on him. Google's little car would destroy Uber, and would do it *smil-*

ing. If Google had a ride-sharing service that didn't need drivers, they could charge almost nothing, steal all of Uber's customers, and destroy its business.

Brin was being interviewed on stage by the journalist Kara Swisher, who ran the Code Conference. She asked him point blank if Google had plans to ever create a ride-hailing service, like Uber. Kalanick might have hoped to hear Brin deny it. He didn't.

"I think some of these kinds of business questions—how will the service be operated, will we operate it ourselves, will we work with partners—are things that we'll sort out when it's closer to being widely deployed," Brin told Swisher, noncommittal. "I think that these initial test vehicles, we'll probably just operate a service ourselves because it's going to be a very specialized thing. But longer term, it's not clear."

Kalanick was irate. Uber was becoming a major force—both as a tech company and a transportation machine—but it didn't have a fully autonomous vehicle. It wasn't even researching one.

Kalanick still believed and acted as if Uber was the underdog wherever it went—a frame of mind that wouldn't change for his entire tenure as CEO. In the beginning, it was Uber against the greedy, unethical taxi companies who had the sleazy local politicians in their pockets. Later, it was Uber against Lyft, the well-funded startup whose warm and fuzzy branding—those pink mustaches—was just cover for its ruthless executives. And now, it looked like it was going to be Uber against Google, the global corporate technology giant.

His anger turned slowly to fear. Google's search advertising business practically minted money. That gave Google the freedom to pursue wild projects, even if they lost money or, in some cases, were patently absurd.* Google's self-driving car research—years in the making by that point—would be a major cost center for almost any other company in Silicon Valley. For Google, it was a rounding error.

---

* See: Google Glass, the thousand-dollar face computer that flopped magnificently after Google realized it had created a legion of "Glassholes"—people who used the technology to take photos of unsuspecting others. The project didn't last very long, but ate up hundreds of millions of dollars before it was shuttered.

As Kalanick would later tell friends, it was after the 2014 Code Conference that he started sweating. As Brin left the stage, Kalanick kept sending frantic texts and emails.

He needed to talk to David Drummond.

DRUMMOND, AS IT TURNED OUT, was expecting Kalanick's messages.

Google's Gulfstream V left the tarmac at San Francisco International Airport the Tuesday after the Memorial Day weekend bound for Los Angeles. Google executives heading to the Code Conference had mulled over how they would tell Kalanick about Brin's onstage demonstration, which would occur that evening. They decided it made the most sense that Drummond, who sat on Uber's board, should break the news to Kalanick.

Drummond usually knew how to handle these situations: with empathy. Tall and well-built, he could pass for a fit ex-linebacker. But his hazel eyes and toothy grin made him look harmless, like the smart and affable corporate lawyer he was. Those qualities, in addition to being one of the few African-American executives who had reached the top rungs of the Silicon Valley ladder, made Drummond stand out among his peers. But for all his skill and confidence, Drummond was conflict averse, which kept him from telling Kalanick about Google's plans until the eleventh hour.

Drummond already knew it was a touchy subject. Kalanick had spies all over the Valley. Uber's "competitive intelligence" operation—that is, the sprawling, systematic COIN program led by Joe Sullivan and his lieutenant, Mat Henley—grew larger by the day. Kalanick often heard whispers of Google's self-driving car project, or occasionally, an errant rumor that Google was starting a self-driving taxi service. Every time Kalanick would hear a rumor like this, he'd fire off an email to Drummond.

"We get stuff like this more than I would like," Kalanick once wrote to Drummond, forwarding intel about a Google self-driving car service. "A meeting with Larry [Page] could calm this down if it's not true but he has been avoiding any meeting with me since last fall. Without any dialogue we get pushed into the assumption that Google is competing in the short term and has probably been planning to do so for quite a bit longer than has been let on," Kalanick continued. It went that way for months. Something would pop up, Drummond would smooth things over, and everything would go back to normal until the next rumor came along.

On the day of the Code Conference, Drummond finally called Kalanick and gave him a heads-up about the demo. People familiar with the call would later describe it as tense; Kalanick was understandably upset. He felt betrayed by his own backers.

After Brin's session came to a close, Drummond asked Kalanick to take a walk with him around the Terranea. Kalanick was melting down, but Drummond laid on the platitudes even thicker than usual, according to someone familiar with the conversation. As a partnerships and biz dev guy, Drummond knew how to do the kind of handling and soothing that his bosses, Page and Brin, never bothered to do.

Kalanick tried to cool off, wanting to believe Drummond. The executive was on Uber's board of directors. His company had invested hundreds of millions in Uber's future. For Kalanick, there was hope, indeed, that Drummond was actually telling the truth.

But later that evening, the idea of Kalanick "remaining calm" flew out the window. Every year on opening night at the Code Conference, the organizers threw a large seaside dinner for the attendees. The most powerful chief executives, however, joined a private dinner elsewhere on the hotel's campus. That year, Kalanick had scored an invite and brought his girlfriend Gabi Holzwarth, a charming musician and dancer. The two of them were introduced by chance through Shervin Pishevar, an early Uber investor and friend of Kalanick.

Holzwarth, then twenty-four, was a classically trained violinist who grew up studying music. She often played in public as a street performer in San Francisco and Palo Alto, where she was raised. When Pishevar ran into Holzwarth performing music in front of a candy store, he hired her to perform at a fundraiser for Cory Booker, held at Pishevar's house, where she and Kalanick first met.

Holzwarth possessed a fierce spirit and pursued the arts from a young age. Kalanick loved that fiery spirit, warm personality, resilient attitude, and the way she could speak to pretty much anyone. As Kalanick's star rose, so did the couple's profile. They attended dozens of high-level parties together—the *Time* 100 Gala, the *Vanity Fair* Oscar Party, the Met Gala—he in his tuxedo, she in designer ball gowns.

At the Code Conference's private dinner, Kalanick and Holzwarth were seated with the heavy hitters; powerful CEOs of the Valley's biggest companies. He should have enjoyed his meal—this was the ultimate validation of Kalanick's success and influence. Instead, he spent most of the evening watching Sergey Brin chat up his girlfriend.

Holzwarth was polite, good at conversing with even the most awkward engineer. But Brin, who was going through a messy, public divorce after having an affair with one of his employees, ignored Kalanick, oblivious to the optics. Before the meal ended, Kalanick snapped an iPhone photo of Brin's cozy-looking chat with Holzwarth and texted it to Drummond. He later told Drummond he saw Brin place his hand on Holzwarth's leg, and that he believed that Brin's behavior was a liability for Google.

Despite Drummond's talents, there was no smoothing over what happened that night. After the dinner, Brin asked Holzwarth to hang out with him and talk by the pool later that evening. Kalanick stewed. Google was going out of its way to screw him. And now he watched as the man who was going to kill his company tried to steal his girlfriend.

WHEN IT CAME TO self-driving cars, Kalanick was further behind than he could truly appreciate. Larry Page—a transportation obsessive—had invested more than a billion dollars and tens of thousands of employee hours into the problem by the time he and Sergey felt comfortable enough to show their egg-shaped car to the world. No one was more determined to bring robot cars to life than Larry Page.

No one except, perhaps, for Anthony Levandowski. The lanky, cantankerous engineer was still working on "Project Chauffeur," the company's pet name for autonomous vehicle research. But his position was growing tenuous.

For one, Levandowski was a poor leader. He would get in constant fights with colleagues over the speed—or lack thereof—at which Google was willing to work on the self-driving project. But Page *liked* the fact that Levandowski didn't always play by the rules. Men like Levandowski, Page believed, would bring Google to the next phase of autonomous vehicle research.

Levandowski had a certain effect on Page, too. While often divisive, Levandowski could be charming. The two men would dine together occasionally—that in itself a peculiar activity for Page—imagining a future driven by robotic cars. Whatever flaws Levandowski had, Page needed him in the Googleplex.

But by 2015, Page's personal attention and millions of dollars in bonuses wasn't enough to keep the golden boy of autonomy happy. Levandowski was tired of his risk-averse colleagues. He was tired of being told "No." Google never seemed comfortable with the project, and dragged its feet, he thought. Levandowski believed they could do better. That *he* could do better.

And so he started pitching a few of his trusted Google colleagues on a new idea: long-haul trucking, a space whose last great innovation was No-Doz. He practiced his pitch on co-workers at dinners he hosted off-campus: imagine a world where self-driven trucks moved goods constantly from city to city. A world in which the sleep-deprived truck driver was no longer a threat. Trucking was an enormous industry, employing 7.4 million Americans and creating $738.9 billion in revenue each year. Trucks drive 5.6 percent of all vehicle miles in the United States, according to Department of Transportation data, and are responsible for some 10 percent of highway fatalities. Automating it would be worth billions. It helped that self-driving trucks didn't directly challenge Google. At least, that was what Levandowski told his colleagues. They would call it Ottomotto—Otto, for short.

By 2016, Levandowski had left Google, taking with him a small cadre of colleagues, including a close partner named Lior Ron, who worked for years on Google's popular Maps software. Levandowski made his feelings clear in a final email he sent to Page: "I want to be in the driver seat, not the passenger seat, and right now [it] feels like I'm in the trunk."

Less than six months later, in the summer of the 2016 presidential election, Otto was up and running. Levandowski and Ron, who became Otto's co-founder, had already expanded the startup to forty-one employees. They had logged more than 10,000 miles on the road with their experimental equipment in three Volvo trucks. With them came fifteen former Googlers, more than half of them autonomous vehicle engineering specialists, a rare and valuable breed in Silicon Valley.

In a rare move, Otto took no venture capital at all. The entire group of Xooglers—the preferred noun for "ex-Googlers"—were rich, and could afford to fund the project themselves. Levandowski was richest of all, having made millions off the sale of his companies to Google years earlier.

But Otto's secret weapon wasn't the well-lined pockets of its founding team, nor the foresight to charge into an open field. It was that Levandowski was finally free of Google's corporate and legal bureaucracy. Now he could do things *his* way. Back at Google, he was scolded for breaking rules and bending regulations. At Otto, Levandowski had no such restrictions.

When the startup was ready to film a demo of its self-driving hardware kit—which could be fitted, off-the-shelf, to existing big-rig trucks—Levandowski called the lobbyist who convinced Nevada regulators to write a new law for Google self-driving cars, and had him request a permit for Otto to film on a stretch of highway in the state. After the Nevada Department of Motor Vehicles declined his request, Levandowski ignored them and filmed it anyway. The sweeping, aerial views of a stark-white eighteen-wheeler dotted with Otto's black signage stood out marvelously against the warm tones of the Mojave Desert. A regulator grumbled that Levandowski's move was illegal. He never faced any actual consequences.

For Levandowski, it was worth it: everyone who saw the launch video loved it. If he had played by the rules, as he had mostly done at Google, he'd still be waiting for approval. Inside Otto, engineers printed out orange-colored stickers and pasted them around the San Francisco headquarters with a message they knew Levandowski would love: "Safety Third."

THEIR MEETING FELT almost predestined.

Travis Kalanick and Anthony Levandowski were first introduced to one another in 2015 at the suggestion of Sebastian Thrun, a former Google executive and bigwig in the world of self-driving cars. Soon after, as Levandowski was preparing to leave Google and start anew, he began meeting with the Uber CEO in secret.

The two men clicked immediately. Levandowski, a born futurist and affable, six-foot-seven showman, roused something inside of Kalanick. The men, both in their forties, imagined a future filled with self-driving vehicles, with Levandowski's engineering talent fueled by Kalanick's enormous ride-hailing network. In Levandowski, Kalanick felt he had found a "brother from another mother," he'd later say.

That first meeting developed into a series of clandestine conversations. Levandowski went to work for Google during the day in Mountain View, and would then return to San Francisco in the evening to meet Kalanick and talk about their future partnership. To keep from attracting attention, they would arrive separately at San Francisco's Ferry Building, a beloved city landmark. After each picked up a bag of takeout, they'd walk north and westward, up the pier and toward the Golden Gate Bridge, where they'd begin discussing their self-driving dreams.

Kalanick knew almost nothing about autonomous tech, but Levandowski filled him in on technical details. A self-driving car needed an enormous amount of equipment just to understand the terrain it traveled, much less navigate it safely. The chassis was fitted with lasers, 360-degree cameras, an array of sensors, radar beacons. Lidar, short for "light detection and ranging," helped the car's software absorb terabytes of data about the landscape.

Anthony's laser "is the sauce," Kalanick once wrote on a whiteboard in a meeting. They would spend hours making up code names and communicating in secret slang. If the autonomous vehicles could drive themselves, Kalanick mused, they could create a "Super Duper" version of Uber, or "Uber Super Duper." Instead of taking 30 percent of driver earnings—the company's current business model—Uber would instead take the entire fare. That meant billions upon billions more in revenue. The abbreviation, USD, lent itself to a cool internal codename: "$," a simple dollar sign.

They talked like teenagers obsessed with a science project. Kalanick would arrive home buzzing with excitement after a meeting with Levandowski, waving his phone in his girlfriend's face. "Look how far we went this time!" Kalanick said, noting the steps his iPhone pedometer had tracked across the city.

Levandowski would eventually create a separate company, Otto, right after leaving Google, as if he were interested in pursuing his own trucking startup. Then he would take venture investment meetings up and down Sand Hill Road—the famed home of top-tier Silicon Valley venture capitalists, including Andreessen Horowitz and Kleiner Perkins—to drum up money for the endeavor. But he would shrug those meetings off, opting instead to raise no outside capital. (Since they were operating independently mostly for appearance's sake, it made no sense to fork over equity to outside investors.) Then came the coup: Uber would acquire Otto for millions, a grand proclamation of Uber's intentions to pursue self-driving technology.

Kalanick, seeking to defend himself against Google's self-driving division, had begun staffing up. He had opened an entire center in Pittsburgh dedicated to self-driving car research—the Advanced Technologies Group—in a joint program with researchers at Carnegie Mellon University. The Uber Advanced Technologies Center itself served mostly as cover to raid CMU's robotics department. Matt Sweeney, an early employee and lieutenant to Kalanick, led some forty engineers out the door, over Pittsburgh's many bridges, and into the arms of Uber's new research team. The university was furious.

But buying Otto would mean something different. Acquiring Anthony Levandowski's startup, effectively poaching Google's autonomous research unit, would signal Kalanick's dedication. The price tag sealed the deal: a cool $680 million, or 1 percent of the value of Uber's entire operation at the time. Plus, Levandowski and his team were entitled to 20 percent profit-sharing of any self-driving trucking business they created. It was a monster deal. And a kick in Google's teeth.

In return, Kalanick would get all of Otto's data, a road map for the direction the company was headed, complete control over its intellectual property and patents, and "a pound of flesh"—Kalanickspeak for Levandowski's dedication to Uber and the cause.

On August 18, 2016, Kalanick and Levandowski unveiled the acquisition. The press covered it as the coup it was; with one of Larry Page's protégés at the helm, Uber was suddenly ready to challenge Google in the race for self-driving cars.

"The golden time is over," Kalanick said in a meeting with a top engineering manager, discussing the deal. "It is war time."

Forty miles south at Google's campus, executives woke up to news of the acquisition.

They were furious.

Chapter 19

SMOOTH SAILING

THINGS WERE GOING WELL FOR TRAVIS KALANICK.

It wasn't that long ago that he had met with the co-founders of Lyft—Logan Green and John Zimmer—to discuss the possibility of a merger. Though Uber was soundly beating Lyft in the war for customers, the pink, mustachioed company still managed to raise money every six months or so to keep itself afloat. Executives at Uber figured it would be cheaper to purchase Lyft outright instead of continuing the ongoing price war.

Kalanick had invited John Zimmer, Lyft's president, to his apartment high up in the Castro hills, along with his lieutenant, Emil Michael. Over cartons of Chinese food, the two sides presented what each thought would be a fair deal. But they had vastly different ideas of what fairness entailed. Lyft's founders wanted a 10 percent stake in Uber for selling their company.

Kalanick and Michael wanted something closer to 8 percent. As the sides worked toward a compromise—apparently one that did not include the number 9—a venture partner who had joined the discussion asked for much more—a 17 percent stake. The talks basically ended there.

Kalanick didn't really want to buy Lyft, anyway. He wasn't into Zimmer. Something about the Lyft president's personality irked Kalanick. The Uber CEO didn't want to work alongside Zimmer. He wanted to professionally humiliate him. Lyft was going to run out of money soon enough.

In retrospect, Kalanick felt lucky for having passed on Lyft. Between Michael and himself, Uber had perfected the art of the fundraising.

Now, they had convinced the public investment arm of Saudi Arabia to invest $3.5 billion in Uber, privately valuing the company at $62.5 billion—an unprecedented figure for any private technology company.

The deal with the Saudis, announced in June 2016, allowed Kalanick to cement his power at the top of Uber. In a move that would anticipate later events, Kalanick directed his dealmaking stewards to draw up paperwork that gave him, and him alone, the power to appoint three additional members to Uber's board of directors. The move made some directors nervous—especially Bill Gurley. If approved, it would give Kalanick the power to stack the board against any challenge.

But those same directors also saw Kalanick show up with $3.5 billion in new investment. This was more than a cash infusion. The Saudi investment was a war chest. At the moment, Uber was competing against DiDi in China, against Grab and Go-Jek in Southeast Asia, against Ola in India, and against Lyft in the United States. These were costly, painful wars—with battles on multiple fronts on multiple continents against well-funded adversaries. The Saudi capital gave Kalanick the firepower to overwhelm all his enemies at once.

So, after some deliberation, Uber's board of directors signed off on the deal.

What improved Kalanick's mood even more was the state of play at Lyft. By the end of 2016, Lyft was struggling, bleeding capital in a subsidy war with Uber, but without the security of Kalanick's financial backing. Kalanick took pleasure in hurting Green and Zimmer, and showed them no mercy. Joe Sullivan, Uber's security chief, monitored Lyft's websites, open-source repositories, and data, seeking a knockout blow.

Kalanick was presented with a delicious new secret weapon by a group of engineers on "Workation." A "Workation" was an annual Uber tradition: instead of spending two weeks in December relaxing, employees would volunteer to spend two weeks working on any kind of project they wanted. Over the course of one December Workation, a group of employees built a prototype Uber driver app that repurposed certain parts of a driver's smartphone—specifically, the accelerometer and gyroscope—to detect the sound of notifications that came from the Lyft app. If Uber knew that a driver worked for Lyft, Uber could market itself differently to the driver—likely with cash bonuses—to entice them away from the pink moustache.

In a meeting, the engineers presented the project to managers, lawyers, and Kalanick himself. The executives around the table were both excited and nervous. This was a powerful new weapon in the war against Lyft. But detecting sounds in a driver's car without permission might cross an ethical line. After the presentation ended, Kalanick sat in silence. No one spoke.

"Okay," Kalanick barked, breaking the tension. "I think this should be a thing," nodding with approval. He stood up and looked the engineers in the eye: "I don't want the FTC calling me about this, either," he said. Kalanick thanked everyone for coming, turned toward the door and promptly dismissed the meeting. The feature was ultimately never implemented.

In Silicon Valley, customer privacy had long taken a backseat to companies' desire to collect data. But Uber took the neglect one step further. Kalanick treated user privacy as an afterthought. At one point, Kalanick changed Uber's settings so the app could track people even after they had ended their ride. Customers protested and demanded tighter privacy settings, but Kalanick wouldn't acquiesce for years; he wanted to gain insight into user behavior by seeing where people went after getting dropped off.

Uber would outsmart Lyft at nearly every turn. Green and Zimmer were competitive and ambitious, but Kalanick always faster, and more willing to use questionable tactics. Kalanick didn't just attack Lyft's userbase, he went after their best personnel. Travis VanderZanden was an entrepreneur who sold his startup, Cherry—the "Uber for carwashes"—to Lyft in 2013. VanderZanden was a "hustler," which Kalanick admired. In just a year, VanderZanden had risen to be Lyft's chief operating officer, one of the highest positions in the company, before he double-crossed his partners to join Uber in 2014.

This was classic Kalanick. When one of his underlings would deliver him good news (which was usually some form of bad news for a competitor), he always grinned—his charming, boyish grin—then rubbed his hands together and, if he was sitting, rose to pace and think of Uber's next move. He hated his former mentor, Michael Ovitz, for screwing him over during the Scour startup years. But he had also learned from Ovitz. When the superagent had reigned over Creative Artists Agency, and dominated Hollywood for twenty years, Sun Tzu's *The Art of War* had been his bible. Now it became Kalanick's, too:

So in war, the way is to avoid what is strong and to strike at what is weak.

Lyft was weak; Uber was strong. Both companies were willing to make war, but Uber was faster, better capitalized, and more ruthless than Lyft. Whereas Green and Zimmer acted the part of nice guys, Travis Kalanick would do *anything* to win. As Lyft's coffers emptied and the founders struggled to raise money, it looked like Uber was going to prevail.

"HOLY SHIT. Am I in over my head here?" Jeff Jones wondered.

Perched at an empty standing desk in the foyer of Uber's headquarters, Jones and an assistant watched the names of drivers dancing down his Facebook page, all blasting him with expletives and angry questions. For a man used to the mild temperaments of people from Minnesota, this was uncomfortable. "These people are *furious*!" he said, looking around for sympathy. Other than his assistant and a handful of employees sitting on black leather sofas nearby, engrossed in their MacBooks, Jones was alone.

Jeff Jones was a career executive. Workers had been upset with him before. But it wasn't an everyday occurrence. People *liked* Jones. Even at almost fifty and mostly silver-haired, he still looked like a Boy Scout. His face was bright and fresh—chipper, even—and he would set the tone of a meeting by flashing a big, open smile. A year spent at Fork Union Military Academy playing baseball contributed to his personal discipline and upright posture. That and a natural pep and charisma had helped him navigate corporate America. First came Gap Inc. and Coca-Cola, before making his name as a marketing whiz at Target.

During his tenure at Target, customers had mostly loved the big, red bullseye; *Targét*, they called it, with a mock French accent. But as chief marketing officer, Jones had also shepherded Target through one of the worst periods in its history, after a 2013 data breach left the personal information and financial data of tens of millions of Target customers vulnerable to hackers. He knew what it felt like for people to be mad at him.

Even still, there was angry, and then there was *Uber-driver angry*. Travis Kalanick knew his business was flying high, and people were taking more Uber trips than ever. But he also knew he had a driver problem, which had begun to affect his bottom line. Uber's driver "churn"—the length of time it takes for a person to start driving for Uber and then stop and not return—was abominably high.

Everyone in the company knew why, and soon, Jones would know, too: driving for Uber was miserable. The company jerked them around with rapidly fluctuating hourly rates and terrible communication with headquarters. After Uber launched its carpooling product in New York, the office sent a survey to its drivers to see how things had gone. As a roomful of Uber employees examined the results, one manager expressed disgust with the spelling and grammatical errors the drivers included in their responses. "God, I can't believe these people's votes count the same as ours," he quipped to his subordinates.

Drivers, as a result, felt they were disposable to Uber. And in truth, they were. In internal presentations, product managers would stress that "satisfaction ratings" among drivers—already low—had plummeted in early 2016. Roughly a quarter of Uber's drivers churned out every three months. People hated driving for Uber so much, the company had to recruit new drivers from the widest labor pools possible. That included the obvious, like Lyft and taxi drivers, and the not-so-obvious, like minimum wage–earning workers at McDonald's, Wal-Mart, even entry-level employees at Jeff Jones's alma mater, *Targét*.

Jones was first enticed when, like Anthony Levandowski before him, he met Kalanick at a TED conference. After Kalanick left the stage, the two struck up a conversation about how to improve Uber's abysmal reputation. Everyone loved the product itself, but hated the brand.

And Jones was a brand guy. It didn't take long for Kalanick to lure him over to Uber. Jones's title was "President of Ridesharing," a portfolio as vague as it was wide.

In practice, Jones took over most of the marketing duties of Ryan Graves, SVP of operations. Graves was an "OG" at Uber, there from the start, but he was no marketing guru. The company's reputation was in the toilet; it needed a professional. So Graves was pushed aside, and given a consolation prize of "focusing on some of Uber's experiments," like food and package delivery services.

For Jones, his job was twofold: spin up marketing, and fix the driver problem. Graves had neglected this project. He had never built a proper, functioning human resources apparatus for his employees, nor did he create an effective way of fielding complaints from Uber's millions of freelance "driver-partners."

Now just a few weeks into his new job, Jones found himself in front of a laptop at Uber's headquarters, faced with hundreds of pissed-off Uber drivers. His plan was to begin improving driver relations by introducing himself with a question-and-answer session conducted over Facebook. Drivers seized the opportunity to express their frustration.

"What are you going to do about YOUR DRIVERS when driverless cars come on the road?" "Will you be giving drivers stock options once there are driverless cars on the roads?" "Has Uber forgotten that drivers built their company?" "Why should drivers be put out of work when they were the ones made Uber successful?" Drivers pelted Jones with questions and accusations. Jones was catching *years* of pent-up aggression from Uber's driver force. He had only managed to answer twelve questions in the thirty minutes he scheduled—clearly not enough time, he realized, to deal with years of anger and baggage. After his assistant jumped on the thread to announce that Jones had to run, the thread erupted.

"You made it crystal clear (if there was any doubt) that Uber does NOT care about it's [sic] drivers. From the bottom of our hearts, ::middle finger::," one wrote.

Jones shook his head at the MacBook screen. What had he gotten himself into?

WHILE JONES WAS GETTING the digital bird, Kalanick was embodying his new lifestyle—that of the billionaire* playboy.

Back in the Scour days, Kalanick had lived with his parents. During the early years of Uber, he had preferred to wrap himself in the warm comfort of an Excel spreadsheet than to stuff dollar bills into G-strings at the Gold Club. (One night he and some friends had actually gone to the Gold Club and Kalanick pulled out his laptop and started working.) Now that Uber had become a unicorn, Kalanick leveled up, largely with help from one man: Shervin Pishevar, a friend of Kalanick's and an early investor in Uber. Pishevar helped Kalanick bring out the true baller within.

Pishevar, a stocky, slick-haired VC, was the kind of Silicon Valley investor whose friendships matched his rivalries. Pishevar might shower an entrepreneur with compliments one day, then battle them over a term sheet the next. Most of all, Pishevar loved being in the presence of power, and had a keen sense for when such opportunities might present themselves.

One such opportunity presented itself, over time, with his new friend Kalanick; he eventually convinced Kalanick to let Pishevar's firm, Menlo Ventures, invest in Uber. One of Pishevar's partners, Shawn Carolan, did much of the work to make the deal happen. But Pishevar managed to take most of the public-facing credit for it; at one point, Pishevar shaved the word "UBER" into the hair on the back of his head, an attempt to prove his devotion to Kalanick's company.

In later years, Pishevar would be accused of sexual misconduct by multiple women. One alleged incident involved Austin Geidt, one of Kalanick's earliest hires and longest tenured employees. At Uber's "Roaring '20s" theme holiday party in 2014—at which Pishevar showed up with a live pony on a leash—Pishevar was said to have groped Geidt by sliding his hand up her leg and under her dress. Pishevar disputed her account; another person who was with him that eve-

---

* Billionaire on paper, that is. Kalanick was still living off money he made from the sale of Red Swoosh. He didn't sell a single share of Uber stock during the entire time he ran the company.

ning claimed Pishevar "wouldn't have been able to touch Geidt because he was holding the pony's leash in one hand and a drink in the other."

Kalanick was a rockstar now, Pishevar said, and encouraged Travis to embrace the lifestyle. Once when Kalanick flew to Los Angeles from Panama, Pishevar sent his assistant to meet Kalanick at the airport. In the back of the car was a suit for Kalanick to change into. They would Uber to parties in Beverly Hills, mingling with celebrities like Sophia Bush and Edward Norton. Leonardo DiCaprio was a frequent guest in their social circle.

Friends close to Kalanick called it "aspirational baller syndrome." Long before Uber, Kalanick had always wanted to be the badass who hops in limousines, dates the hottest girl, and graces the right parties. Now, he got to live his dream. Kalanick was making up for years of longing. The cost of admission: opportunities for small but significant amounts of Uber equity, available primarily to members of this new celebrity circle.* (Some of it was strategic, too; in the startup world, celebrities often promoted up-and-coming apps in exchange for equity or cash.)

Parties in remote, exotic locations particularly appealed to Travis and Emil Michael, who would later become his new wingman. Kalanick's girlfriend, Gabi Holzwarth, helped him organize a gathering of friends and stars on the Spanish island of Ibiza. To Kalanick and his crew, the idea of jet-setting, fame, and fun was instantly appealing.

After some time around celebrities, Kalanick mused with Michael about how they needed a "big star" to join Uber's board of directors. A hot startup, they believed, needed a heavy hitter to turn heads in Hollywood.

Oprah Winfrey was the prize. Kalanick met Winfrey in Ibiza and became fixated on the idea of her becoming a board member. Everyone in Silicon Valley wanted Oprah on their board. She was a self-made, entrepreneurial black woman with millions of adoring followers and

---

* One founder, Oren Michels, cut Kalanick a check for $5,000 early on in Uber's history. By the end of 2017, that $5,000 had multiplied in value 3,300 times, worth somewhere close to $20 million. Mr. Michels made more off his $5,000 Uber investment than he did when Intel purchased his entire startup in 2013.

a global empire. Many sought access via Gayle King, the *CBS This Morning* co-host, and Winfrey's longtime friend. But few had made any progress. Pishevar tried to throw a dinner for King to soften her up. Kalanick dispatched Gabi Holzwarth to sweet-talk her. They pulled out all the stops, but King didn't bite, and Oprah was never really interested.

Kalanick did better with Shawn Carter, also known as the hip-hop mogul Jay-Z. Carter was an early investor in Uber, as was his wife, Beyoncé Knowles. Carter and Knowles had the foresight to know Uber was going to be big. During one venture capital round, Carter once wired more money to Uber's bank account than he was supposed to, an attempt to increase his equity stake. Kalanick and Michael, together during the moment, were thrilled at the idea of rebuffing the "Big Pimpin'" star Jay-Z. They let Carter down gently, and wired some of the money back, saying they already had too many interested investors.

Parties at strip clubs became regular occurrences, often expensed on the company's corporate account. A few execs would usually bill the evening as client entertainment, or business development—and one or two other executives would sign off on it, as was the case for an incident in South Korea in 2014, something that would come back to haunt the company later. They had a pet phrase to describe expensing strip clubs to the corporate card: "Tits on Travis."

THE TONE OF UBER'S CULTURE was being set from the top. Kalanick knew what he wanted in his employees—who were mostly white, male, and in their twenties—and made his hiring decisions based on that instinct. The result was a workforce that largely reflected Kalanick himself.

Every global office was unique. Kalanick wanted to empower his workers—"let builders build," according to the Uber company value—and urged employees to be responsible for their own fiefdoms. Yet still, because Uber had hired thousands of Kalanick clones, many satellite offices had flickers of similarity.

Southeast Asia, for instance, was a hotbed of partying for Uber operations employees and managers. Cocaine and booze were common, as was harassment—and even worse.

One female employee in Uber's Malaysian office was heading home from work one evening in 2015, when she noticed a group of men following her. It was a local gang, she realized, and began frantically texting people for help. One of those people was her boss, the local Uber general manager. She said that she needed help, and that she was scared she was going to be raped.

As her ride home continued, her manager responded: "Don't worry, Uber has great health care," he texted. "We will pay for your medical bills."

The Thailand office at the time was perhaps even worse, a toxic workplace where drug use and visits from sex workers were not unheard of. No one from Uber kept the behavior in check.

One particularly raucous evening, a bunch of Uber Thailand employees were up late drinking and snorting coke, a semiregular occurrence at that office. One female Uber employee with the group had decided she didn't want to do drugs with her colleagues, and tried to abstain. Before she could leave, her manager grabbed the woman and shook her, bruising her. Then he grabbed the back of her head and shoved her face-first into the pile of cocaine on the table, forcing her to snort the drugs in front of them.

The New York office was largely defined by its machismo, sexism, and aggression. São Paolo saw angry managers throwing coffee cups across the room or screaming at employees when they weren't happy with results. It wasn't unheard of for managers to sleep with subordinates.

These dark events rarely led to consequences for management, and other employees—if they knew about the wrongdoing—either ignored the problems or squashed their concerns. But for many, the drawbacks did not outweigh the excitement. Even if they had to white-knuckle it through bad times, there was a pervasive feeling that Uber, the world's preeminent ride-hailing service, would soon become a global behemoth on the order of Google, Amazon, or Apple. Uber had billions in the bank, was poaching top talent from companies across the Valley, and had its sights set on conquering international markets. When employees' restricted shares vested, they would earn an absurdly sweet payday.

Kalanick's fortieth birthday was a bash he wouldn't forget—a multi-yacht party in the Aegean sea featuring top shelf booze and a group of models flown in for good measure. By the end of 2016, life was good for Travis Kalanick. He was rich and powerful, and his empire was growing further by the day.

AS 2017 BEGAN, a young woman was just beginning her job at Stripe, a payments startup based in San Francisco. It had been two months since she had left her job at Uber, and she hadn't told anyone in much detail why she departed such a hot startup. Every time she reflected on her year at Uber she felt disgust, sadness, anger. Working for Uber wasn't anything like she thought it would be.

Friends and family wouldn't stop asking her why she left, but she had never found the words. By February, however, she had managed to take stock of her experience, and began to describe her time at Uber in a post on her personal blog, susanjfowler.com. The entry clocked in at more than three thousand words, the length of a magazine article. She was nervous when she scanned the title: "Reflecting On One Very, Very Strange Year At Uber." Would anyone actually read this? Would anyone actually care?

"It's a strange, fascinating, and slightly horrifying story that deserves to be told while it is still fresh in my mind," Susan Fowler wrote in the introductory paragraph of her blog post.

"So here we go."

PART IV

Chapter 20

THREE MONTHS PRIOR

THREE AND A HALF MONTHS BEFORE SUSAN FOWLER HIT "PUBLISH" on her blog post, the tech world had been thrown for a loop.

Since the dot-com crash of the early 2000s, and certainly throughout the smartphone era, the press had largely flattered the American technology sector. Headlines in the *Wall Street Journal*, the *New York Times*, and other major publications admired the progress achieved by tech's boy geniuses. Mark Zuckerberg was a visionary whose social network connected friends and family worldwide. Twitter had enabled democracy to flourish in the Middle East. Google wunderkinds had built beautiful maps that made life easier, and given everyone a free email account. Elon Musk's ambitions were transcendent: he would save the world with electric Teslas and conquer the stars with SpaceX.

Though many had written about the negative aspects of tech, the American press and public often overlooked Facebook's towering monopoly on social media, Amazon's takeover of internet infrastructure, the disappearance of privacy enabled by Google's advertising technology, the noxious, racist trolls enabled by Twitter, and the outlandish and harmful theories fed to users by YouTube's automated algorithms—the earth is flat, vaccinations cause autism, 9/11 was an inside job. That generous view of technology would curdle on the night of November 8, 2016, when Donald Trump unexpectedly won the US presidential election.

But while the election cast a pall over tech in general, the night also served as the turning point for Uber in particular. The company's troubles did not stem from the election, of course, nor did they cause the result, but they were soon caught up in the chaos that followed. The maelstrom marked the beginning of one of the worst twelve-month periods in American corporate history.

WHEN THE TECH WORKERS AWOKE on the morning after the election, their mental image of themselves—as bastions of youth and democratic idealism, helping to create a more efficient, healthier, more connected country—had been shattered.

Donald John Trump was the president of the United States. The thrice-married real estate mogul who spent the last decade taking birther-conspiracy potshots at Barack Obama on Twitter was now the commander in chief. Silicon Valley had donated millions to the Clinton campaign; techies were eyeing jobs in the Clinton administration.

Now the public was pointing fingers. Facebook, Google, Twitter, Reddit, and Instagram had won Trump the election. Cambridge Analytica had manipulated social media—Facebook embedded its own employees in the Trump campaign. Tech had gone from the youth-led leveling force that had brought Obama to the White House to a nefarious, psychological propaganda machine. The public suddenly realized the scope and targeting power of Google's and Facebook's advertising engines. Members of Congress, sensing unrest, began singling out the tech companies. So did the media.

"The most obvious way in which Facebook enabled a Trump victory has been its inability (or refusal) to address the problem of hoax or fake news," a report from *New York* magazine claimed. The headline expressed the creeping sentiment that "Donald Trump Won Because of Facebook." That doubt and worry began spilling over into the minds of tech workers, too. Even inside of Facebook, the most zealous of the true believers began to question the world-shaping power of the platform they had built.

Twitter, too, came in for condemnation. They had given a platform to a billionaire troll, which he leveraged into maximum, round-the-clock exposure. Trump had banked more than $2 billion in "earned media," that is, free attention—far surpassing that of any other candidate. Now, each tweet was a presidential proclamation.

Where once the public and media had adored Big Tech—Facebook and Twitter gave people a voice, while Uber and Lyft gave anyone a ride—now the public devoured stories of state-sponsored hackers using vast databases of personal information to influence the election. Suddenly, nefarious forces in Silicon Valley had led the country off a cliff, and Big Tech was profiting from the strife.

TRAVIS KALANICK HAD SPENT the past two years steeling Uber for a Clinton presidency.

He spun up teams of lobbyists in every market that mattered. He wanted them ready to deal with an incoming administration that was a friend to unions and an enemy to companies that relied on contract workers. Clinton hadn't come after Big Tech quite yet; she was closely tied to major donors in the Valley, including Facebook's Sheryl Sandberg, John Doerr of Kleiner Perkins, and Marc Benioff of Salesforce. But if there *was* a company a Clinton presidency might come after, it could be the most hated startup in the country: Uber.

But Trump's upset victory caught everyone at Uber off guard. Most of the rank and file, a largely Democratic- and Libertarian-leaning force, were tearing their hair out at the thought of a Trump presidency. (Even many of the Republicans of Uber's ranks found the idea ludicrous.) Thuan Pham, Uber's chief technical officer, wrote an internal letter blasting Trump's election as "a huge step backward," calling the new president an "ignoramus" and comparing his win to the ascendancy of ruthless dictators like Chairman Mao Zedong in China.

But as Trump's victory became inevitable on election night, Travis Kalanick was beginning to see the silver lining. A Republican administration was less likely to come after Uber, especially if he positioned his company as one of the largest job-creating startups in history. Anyone who owned a car could be put to work, and Kalanick could take credit for that. Perhaps the next four years wouldn't be so bad.

Besides, he had enough headaches to deal with. After two years and billions of dollars in losses and fraud, Kalanick's investors demanded he abandon China. No American tech company had been able to crack the country, and Uber wasn't going to be the first. Despite Kalanick's efforts, the Chinese government had chosen to support DiDi, a Chinese company, and remained hostile to Uber.

Kalanick was loath to concede. He had hoped to twist the knife; the Strategic Services Group attempted to photograph Jean Liu, the president of DiDi, when the *New York Times* pushed out the news of Uber's $3.5-billion funding round from the Saudis.

But while Kalanick may have had the stomach for a further battle, his backers did not.

Bill Gurley, the "Chicken Little" of Uber's board, was chafing at the burn rate in China. Another of Kalanick's antagonists on the board, David Bonderman, was starting to make noise, too. Kalanick brought on Bonderman, a giant of private equity at TPG Capital, back during a funding round in 2013. But now Bonderman was criticizing the way Kalanick was funding the losing fight in China.

Some of Uber's institutional shareholders held calls with DiDi's biggest investors about settling the conflict. Kalanick was pissed, yet unsurprised; investors, he had maintained, will always screw you in the end. On August 1, Uber conceded the fight; DiDi would take over Uber's business, and Uber would suspend operations in China.

For investors, it was a win. No more enormous cash drain, no squandering the profits from booming markets. And to sweeten the deal, Uber received a 17.7 percent equity stake in DiDi, something that would grow in value and could prove immensely lucrative when DiDi decided to go public. Emil Michael negotiated the deal hard, and considered it one of his crowning achievements at the company. But for Kalanick, the defeat was more bitter than sweet. He would not outdo Page, or Dorsey, or even Zuckerberg and become the first American tech CEO to conquer the Chinese market.

He had something else on his mind. With Trump as the victor, a business-friendly Republican administration might end the vilification of tech and cut labor and transportation regulations. But they would need to act fast. President-elect Trump had already began putting together a handful of policy councils stacked with some of tech's biggest leaders, and Travis wanted to be on it. He and his team pulled strings to make sure he had a seat at the table. One month after the election, top tech CEOs were called to attend a technology summit with Trump during the transition. Kalanick was stuck in India and missed the photo opportunity, but a direct line to Trump was something Kalanick was happy to have.

His employees disagreed. Grumbles traveled the hallways of 1455 Market Street, as many Uber employees wondered why their boss needed to embrace a man they considered xenophobic, ignorant, and racist. At internal all-hands meetings, they urged their boss to reconsider and step away from the council.

Kalanick defended his decision, figuring it was better to have a seat at the table than not have one at all. He could manage a little frustration in the ranks.

Chapter 21

#DELETEUBER

AS TRAVIS FOUGHT HIS WAY ONTO THE TRUMP BUSINESS ADVISORY council, a Chicago tech worker named Dan O'Sullivan still believed Donald Trump was full of shit.

The president spent his entire first week arguing with the press over the size of his inauguration crowd. ("The biggest ever inauguration audience!" Trump's press office announced, an obviously false statement.) Trump was a buffoon, O'Sullivan thought, an idiot foisted upon the office by an electorate poisoned by Fox News. By the time he left office, O'Sullivan prayed, Trump would be thwarted by his advisors and accomplish little of what he promised on the stump in 2016.

The Long Island–born son of a nurse and an Irish telephone lineman, Dan O'Sullivan grew up worlds away from Trump's gold-plated tower in Manhattan. He was proud of his blue-collar background. His great-great-uncle, Mike Quill, co-founded the Transport Workers Union in New York City back in 1934. Quill's ties to the Communist Party earned him the nickname "Red Mike." On the night of his sister's birth, O'Sullivan's father was out on strike with fellow linemen in the Communication Workers Union.

After kicking around schools in Long Island and Maine, Dan O'Sullivan landed in Chicago, a place he liked though knew little about. At six-foot-three and pushing 220 pounds, O'Sullivan looked like a different kind of lineman—more Chicago Bear than Bell Atlantic like his father. He picked up a Chicago accent quickly, cutting short his "U's" and "A's." His nasally vowels gave many the mistaken impression he was a native Chicagoan.

O'Sullivan dreamed of being a writer, and started freelancing political pieces for *Gawker*, *Jacobin*, and other left-leaning outlets. To pay
the bills, he landed in a call center at a tech company, a lower-level peon
answering angry customer support questions. The work was depress-
ing, but he spent his off-hours pursuing his passion, hustling for opportunities to write.

More vivid than his dreary call center job was O'Sullivan's digital
life on Twitter. He mostly used it to follow political accounts and news
and to connect with other writers. He started chatting with other left-
ists and joking around with people who began as anonymous avatars
in his Twitter feed, then slowly grew to become his online friends. Even
as Dan despaired at Trump's popularity and success, at least he could
make fun of Trump's buffoonery with his friends on Twitter.

O'Sullivan cherished his digital anonymity. He was opinionated and
crass on Twitter, and knew his obscenities towards Trump might not
please his employer. And if he had to find a new job, some of the esoteric, vulgar in-jokes he shared with Twitter friends wouldn't thrill a
recruiter.

Still, Twitter was worth it. He chose a handle for himself, a pun his
online friends could remember him by: @Bro_Pair.

THE ORDER CAME as night fell on Friday, January 27, a week after
Trump took the oath of office. Effective immediately, Trump was closing the nation's borders. Singling out predominantly Muslim countries,
he barred refugees from places like Syria, which was in the midst of a
violent civil war that was driving thousands to seek asylum from potential slaughter.

"We don't want them here," Trump said, referring to so-called "rad-
ical Islamic terrorists"—his name for Muslims—during the signing ceremony. "We want to ensure that we are not admitting into our country
the very threats our soldiers are fighting overseas. We only want to
admit those into our country who will support our country, and love
deeply our people."

Trump had presaged such a proposal at the end of 2015 on the campaign trail, in which he called for a complete restriction of all Muslims from entering the United States as a response to bloody terrorist attacks in San Bernardino, California and Paris, France. Christians and other religious practitioners, he said, should be granted immigration priority over Muslims seeking asylum. The Muslim ban played extremely well at rallies. Trump's base loved it. At the time, of course, politicians from both parties condemned the idea as inhumane and unconstitutional. But the outrage at the time passed almost as quickly as it arrived.

Now it was 2017, Donald Trump was the president of the United States, and he was following through with a campaign promise. Among ardent Trump opponents like Dan O'Sullivan, the Muslim Ban brought forth all of the rage that had simmered since November 9. The announcement confirmed that Trump would be every bit as monstrous as they had imagined.

That energy wasn't squandered. Millions of people across the country rushed to airports and other places where immigrants seeking asylum might be turned away by the TSA, ICE, or other federal agencies. Thousands of lawyers arrived clad in neon yellow hats and T-shirts to offer pro bono legal advice to immigrants stuck in limbo. Throngs of protesters flooded baggage claim areas and TSA security lines with chants of outrage against Trump, carrying hastily written cardboard signs and posters with pro-immigrant messages.

As the protests continued through Friday night and into Saturday morning, the Muslim community of taxi drivers in New York banded together to strike at the airport, in part to show solidarity, and also to give America a glimpse of the country without Muslim workers. "NO PICKUPS @ JFK Airport 6 PM to 7 PM today," the New York Taxi Workers Alliance posted to its Twitter account shortly after 2:00 p.m. Saturday afternoon. "Drivers stand in solidarity with thousands protesting inhumane & unconstitutional #MuslimBan."

As taxi workers organized, employees in Uber's New York office watched and began to worry. People were traveling to airports in droves, often using Uber to get there. JFK was slammed, its terminals were drawing one of the largest crowds in the country that weekend. If passengers kept Ubering to JFK in large numbers, Uber's "surge pricing" would kick in. That meant people would be charged multiples of the base fare—two, three, four times as much or even greater—just to go and protest. Managers in New York and San Francisco could predict the negative headlines if surge pricing kicked in: big bad Uber fleecing honest citizens during a humanitarian protest.

Uber didn't need that headache now. A manager in San Francisco gave New York the all-clear to turn off surge pricing for Uber trips to JFK. Later that evening, @Uber_NYC sent a tweet: "Surge pricing has been turned off at #JFK Airport. This may result in longer wait times," the tweet read. "Please be patient."

The tweet would end up costing Uber millions.

O'SULLIVAN COULDN'T BELIEVE what he was seeing.

Election night had broken him. He wrote a final piece for the leftist magazine *Jacobin* on the Trump victory—a half-delirious meditation on Trumpism and the forces it took to bring America to propel such a man to victory—and subsequently swore off political writing for good. He wandered the empty streets of Chicago in a stupor after the race was called, sensing a deep depression coming on, one that would carry into 2017 and add another ten pounds to his frame.

The swearing-in ceremony in January was painful to watch. He winced as the group of tycoons and robber barons surrounded Trump at the Capitol, celebrating the triumph of evil over good. The travel ban carried out less than a week later seemed sadistic to him. The cruel execution of the announcement perfectly symbolized Stephen Miller and Steve Bannon—two of Trump's most xenophobic, nationalistic advisors—and their desire to inflict pain on immigrants.

But O'Sullivan felt a glimmer of hope as the news reported crowds of people gathering at the airport to protest Trump's unjust ban. Thousands of other people like him, fed up with fear and anger, were fighting the administration through protest, one of the most American acts there is. And as @Bro_Pair, he scanned his Twitter account and monitored chatter from reporters, newspapers, and his digital friends who, too, were speaking out against the president. As Saturday wore on,

@Bro_Pair noticed a tweet from the New York Taxi Workers Alliance scroll through his Twitter feed, noting their strike on the JFK airport. He appreciated the solidarity.

A few minutes later, he noticed another tweet—this one from Uber, claiming it was shutting off surge pricing at JFK.

Up until that point, O'Sullivan had never really liked Uber. He had passively followed its various controversies; everyone in tech did. To the leftist O'Sullivan, Travis Kalanick was an avatar of Silicon Valley's capitalist id, concerned only with user and revenue growth, not the lives of everyday workers like himself. He used Uber occasionally—it was, after all, a great product and very convenient—but always felt guilty afterwards.

But at that moment, seeing Uber's tweet pass through his feed, he saw it as an act of subversion—a betrayal of solidarity. O'Sullivan and others interpreted Uber's tweet as a company trying to profit off the backs of striking cab workers, a cash grab during a vulnerable public moment. Even beyond the immediate circumstances, the tweet reminded him of his larger ideological grievances towards Uber, and the core of how its business operates. The contract-based labor model that eschewed directly employing drivers. The campaigns against drivers who wanted to unionize. To him, this faceless, monolithic tech company would never defend its Muslim cab drivers. O'Sullivan couldn't pinpoint whether it was his deep, familial ties to organized labor, the frustration he felt towards his shitty call center tech job, or the deep-seated need to fight back against Trump. He just snapped: he had had it with Uber.

Sitting alone in his cold apartment in the dead of a Chicago winter, he started typing a response to Uber's tweet, still fuming with anger. "congrats to @Uber_NYC on breaking a strike to profit off of refugees being consigned to Hell," @Bro_Pair tweeted, "eat shit and die." He quickly followed up with an idea for a hashtag, something people could add to their angry tweets about the company: "#deleteUber."

"Don't like @Uber's exploitative anti-labor policies & Trump collaboration, now profiting off xenophobia? #deleteUber," he tweeted. O'Sullivan dug into Uber's support pages on its website to figure out how to actually delete his Uber account, a feat that was surprisingly difficult and required filling out a form and sending it to engineers at the company. O'Sullivan started tweeting out screenshots and links to the online account deletion form, making it simpler for others to find it and delete their own accounts.

The hashtag began to resonate. Others tweeted angrily at Uber, joining @Bro_Pair. People started adding #deleteUber to the end of their tweets. As seething Americans sought an outlet for their helpless rage, the idea that Uber was not just subverting the protest but actively trying to profit from it was maddening. Hundreds of people started replying and retweeting @Bro_Pair's tweet, catching the attention of other angry onlookers. Hundreds turned to thousands, which turned to tens of thousands of people chanting, digitally: #deleteUber.

To O'Sullivan's amazement, people started tweeting their screenshots of their account deletions back to him. "You're fascist colluding scabs," one user's screenshot said. "Taking advantage of the taxi strike in NYC is a disgusting example of predatory capitalism and collusion with an overtly fascist administration," another user wrote, tweeting back at @Bro_Pair. Another person added: "Catch a rideshare to hell."

O'Sullivan was dumbstruck. *Celebrities* were tweeting him screenshots of themselves deleting Uber. The press started calling him for interviews. He had tapped into a rage shared by more people than he had realized. Most immediately, those who retweeted him expressed anger towards the Trump administration and its discriminatory actions. But deleting Uber went beyond that; it became something people could do, an action they could broadcast as part of their protest, a repudiation of tech culture, of fake news, of Silicon Valley—the industry that many believed duped Americans into electing Trump in the first place. To #deleteUber wasn't just to remove a ride-hailing app from one's phone. It was also to give a giant middle finger to greed, to "bro culture," to Big Tech—to everything the app stood for.

As O'Sullivan logged out of the @Bro_Pair account on Twitter and turned off his computer later that night, he felt a twinge of happiness for the first time in months. #deleteUber was trending across Twitter around the entire world. The press was covering the fallout, and Uber was scrambling to try and contain the damage.

"Okay I have to go to bed," @Bro_Pair tweeted. "But this has been the only good thing I've seen come from hashtags ever. thank you all, keep it going."

He signed his tweet with a hashtag: "#deleteUber."

ALL HELL BROKE LOOSE at 1455 Market Street.

As the #deleteUber hashtag gained traction, engineers had account deletion requests flood in by the thousands from across the world. Up until that point, the company had received few deletion requests. Everyone loved the product, and those who didn't merely erased the app from their phone without deleting their account. There was no automated mechanism in place to handle such requests. By the time @Bro_Pair's protest spurred a mass revolt, Kalanick was forced to assign an engineer the task of implementing a system to process the flood of account deletions.

Uber's public relations team scrambled to try and convince reporters that Uber wasn't breaking a strike but actually trying to *help* protesters get to the JFK protests by eliminating surge pricing. Kalanick had attempted a mealy-mouthed apology that weekend, noting that he planned to raise Uber's issues with the travel ban the following week with President Trump in person. He was days away from the first meeting of Trump's policy council of executives. But the statement had the opposite effect, instead reminding people that Kalanick was actively working with the administration. Outsiders saw Kalanick's position as a tacit endorsement of Trump. Eventually, his own employees began to see it that way, too.

"I understand that many people internally and externally may not agree with that decision, and that's OK," Kalanick said to employees in an email. "It's the magic of living in America that people are free to disagree."

His thinking on keeping his seat on the council didn't last long. In the span of a week, more than 500,000 people deleted their Uber accounts entirely, not counting the incalculable others who simply deleted the app from their phones. Uber's all-important ridership growth curves—

for years always hockey-sticking up and to the right—started turning downward. Kalanick began to sweat.

Lyft, at that point running out of money and on the verge of surrender, benefitted enormously from the backlash. People began to ditch Uber and switch over to Lyft. (Protest felt good, but people still needed to be able to call a car sometimes.) Lyft's executives then pulled a well-executed PR stunt, publicly donating $1 million to the American Civil Liberties Union over four years, making themselves look like white knights while Uber was groveling before Trump.

The resultant surge in ridership brought Lyft back from the brink of failure. At last showing positive signs of growth, Lyft soon attracted investment from Kohlberg Kravis Roberts, the private equity firm, buoying the ride-hailing company with more than a half-billion dollars in additional capital.

Lyft's fundraising sank Kalanick's spirits. He had spent the entire summer trying, and failing, to defeat his largest competitor in China. And now, just as the new year began, his chance to kill his strongest American opponent had slipped away as well. He was *so close* to rubbing John Zimmer's nose in defeat. No longer.

Less than a week later, at the Tuesday all-hands meeting, multiple employees confronted Kalanick for keeping his position on Trump's advisory council. Two different engineers asked him what it would take for him to step down from the position, a question he repeatedly dodged. But by Thursday, with ridership losses mounting and employees fast losing faith in their leader, Kalanick acceded.

With less than twenty-four hours before he was scheduled to be at his first advisory council meeting at the White House, a call was arranged between Kalanick and President Trump so he could tell him he was withdrawing from his position.

The call was brief and awkward; Kalanick apologized and gave a pitiful explanation. Trump grumbled through it. The two men had never met before, but Kalanick ended the call knowing that he had annoyed the president of the United States.

Later that day, he wrote a conciliatory email to staff, noting he had left the council, though for many both inside and outside of Uber, the concession felt too little, too late. It didn't stop the downturn of Uber's growth numbers, either, as ill will toward the company continued to damage the brand and overall ridership. But for the moment, Kalanick had neutralized the immediate threat and knocked Uber's name out of negative headlines.

For the moment.

Chapter 22

"ONE VERY, VERY STRANGE YEAR AT UBER . . ."

IN NOVEMBER 2015, JUST UNDER A YEAR BEFORE THE PRESIDENTIAL election that brought Trump to power, Uber onboarded a new engineer. The twenty-four-year-old philosophy and physics major was one of dozens of engineers hired that January, and joined a cohort of hundreds of new employees, of whom less than 40 percent were women. This new engineer joined a department that was overwhelmingly male; some 85 percent of Uber engineers were men, according to a later study. Raised in a small town in Arizona, she was an unlikely candidate for an engineering job at Uber. But for Susan Fowler, it was a dream come true. Like sailing "over the moon," she would later tell a reporter.

Fowler had worked at a pair of startups directly out of college, but an engineering gig at one of the Valley's hottest companies was a personal coup. Fowler hadn't followed the typical pathway to a big time engineering job. She had no MIT degree, no intense undergraduate focus on computer science, no serious engineering internships. But she was driven.

The second child among seven, Fowler grew up in Yarnell, Arizona, a rural town made briefly famous because of a deadly 2013 wildfire, but otherwise unknown to outsiders. She was homeschooled along with her siblings. Most of her knowledge came through exploring the library, devouring Plutarch, Epictetus, or Seneca. (She loved the Stoics.) The family was not flush with cash. Her father was an evangelical preacher who sold pay telephones on the side. Her first jobs, as a stable hand

and a part-time nanny, helped supplement the family income. God was present in the Fowler household, and young Susan was open to exploring other branches of philosophy—but she preferred to do it on her own at the local library.

At sixteen, she was suddenly inspired to go to college, though with no real help from her family. Susan searched frantically for information on how to apply. She had no idea what an application looked like, or that she needed letters of recommendation—much less how to get them when she hadn't gone to high school. But going to college was her dream. Luck and a stellar entrance essay landed Fowler a full scholarship to Arizona State, where she finished lower-level classes before transferring to the University of Pennsylvania. Fowler, whose book learning came mostly from the Yarnell Public Library, had made it to the Ivy League.

Many engineers in Silicon Valley fit a stereotype: white twentysomething males, skinny, awkward and unsocialized, good with numbers but less adept with other people. Fowler was the opposite. She was warm, friendly to strangers, at ease with conversation. Mousey and slight, she had the accent of a Southwestern preacher's daughter, a lilting voice full of long vowels and "y'alls." With shoulder-length hair and deep brown eyes, Fowler was a natural beauty, fair-featured, with chestnut bangs that fell just over her eyes. She made people feel that she was excited to see them. It was hard not to return her wide smile when Susan Fowler gave you a warm "Hello!"

Her outward sweetness belied a fire within; Fowler accomplished things she put her mind to. Whether it was writing her way into college or breaking into the bro-y world of startups, she always pushed ahead, no matter what difficulties she saw on the path.

It wasn't all easy; she fumbled through a first semester at Penn, and advisors—skeptical of her education from the "Home School of Susan Fowler," attempted to steer her away from studying physics.

Fowler wasn't having it. She called on the president of Penn, Amy Gutmann, and left a message with her office. Her dream, Fowler said, was to study physics at an Ivy League university. And Gutmann said in a commencement speech that Penn would help students fulfill their dreams. Gutmann acquiesced. The president told Fowler she was absolutely right, and encouraged Fowler to press on. After the rough start, Fowler regained her footing, eventually graduating in 2014 with a degree in physics and philosophy.

And now, just a few years after leaving Penn, Susan Fowler was a site reliability engineer at Uber, the glittering unicorn of Silicon Valley. Uber represented an entirely new challenge: how to succeed in one of the most aggressive, most masculine, and most high-profile companies in Silicon Valley.

The same month Uber hired her, Fowler met the love of her life. Chad Rigetti had "Michael Fassbender–worthy" good looks and an enthusiasm for quantum computing theory. Fowler was attracted to him almost immediately. At the end of their first date—dinner and a movie—Fowler reached for her iPhone to call for an Uber home.

"No, no, no," Rigetti said. "I don't use Uber."

Fowler was confused. She was, after all, an Uber employee.

Rigetti said all the negativity around the company bothered him; as an entrepreneur running his own startup, he didn't like Uber and chose not to support it. Rigetti swore off using the app as a result.

It was an omen Fowler would remember later.

AFTER TWO WEEKS of introductory training, Susan Fowler began work with her new team in December 2015. That same day she received a string of chat messages from her manager.

Fowler was still riding the new-hire high. She got to pick the team she wanted to work with, a pleasant surprise. Site reliability engineers, SREs, played a crucial role at Uber. They kept the platform up and running—hence the job title. At companies like Facebook or Twitter, SREs worked to keep the service online 24/7 so that people could post status updates or tweets whenever they wanted. For Uber, SREs were focused on keeping the hundreds of thousands of drivers working for the service connected at all times. SREs were told even a few minutes of downtime could threaten Uber's very existence; if riders got frustrated they would choose another service. The work of keeping Uber online thrilled Fowler.

Some of Uber's worst crises fell upon the shoulders of harried SREs. Halloween night in 2014 was a date scarred into the minds of Uber employees: The company's supply and demand system went down that evening, wildly overcharging people on one of the busiest Uber nights of the year. The next morning, angry riders woke up to Uber bills as high as $360 in their inboxes.

And then, on the first day with her important new team, her manager started hitting on her. Apropos of nothing, he told her that he was in an open relationship. While his girlfriend was having no trouble finding new sexual partners, he was struggling to do the same. He said he was trying to "stay out of trouble at work," but that he "couldn't help but get into trouble" since he spent all his time at work anyway.

Fowler was taken aback by her manager's insinuations. She knew the Valley was a treacherous place for women engineers—it seemed every department across every tech company had a skeezy man or two looking to bag a colleague—but getting propositioned over the company's uChat system on her *first real day of work* was a new low. It wasn't exactly someone she could shrug off, either; she reported directly to him.

Nor was Uber some rinky-dink startup. By the beginning of 2016 it was a full-fledged private corporation, with offices in dozens of countries. She had faith that a company of Uber's size would do the right thing if she called out her new manager's behavior. As her superior prattled on about his wish list of sexual conquests, Fowler took screenshots of the conversation and reported him to the human resources department. Uber was a big corporation; HR would know what to do. She expected him to be out the door by the end of the week (if not the end of the day).

WHAT FOWLER DIDN'T KNOW was that becoming a "big corporation"—like so many others in the Valley—was Travis Kalanick's nightmare. In his mind, Uber needed to stay scrappy, to "do more with less" and "always be hustlin'." Growing into a boring, faceless megacorp meant employees would become complacent, lazy, inefficient. Nothing would be lamer than for Uber to turn into Cisco, a bloated behemoth where midlevel executives still tucked in their polo shirts.

But avoiding the "big company" feel also meant avoiding bureaucracy, like a proper human resources department. All Kalanick cared about was recruiting. He saw HR as a tool to onboard swaths of new talent and quickly dismiss the inevitable bad hires, rather than as a way to retain and manage Uber's standing workforce. Managerial coaching and training were almost completely ignored. A handful of people looked after the working lives of thousands of full-time employees.* To Kalanick, the phrase "HR" meant behavior codes, sensitivity training, sexual harassment policies, misconduct reporting procedures, formal reviews—all things that make a hard-charging young man roll his eyes. Nevertheless, the company was more than doubling in size every year; by early 2016, it employed more than six thousand people, not counting drivers. Kalanick may not have wanted to instill systems that would give Uber a "big company" feel, but he could deny it no longer: Uber *was* a big company.

Beyond complaints and workplace problems, employees felt HR hadn't created systems to properly evaluate workers. Performance reviews were little more than a list of three positive and three negative attributes about a worker—the "T3 B3" process, devised by Kalanick—followed by a largely arbitrary number score. Those scores fluctuated wildly, often depending on how close a given employee was with the manager or department head who was doing the grading. And the backdrop to the entire grading system was Uber's fourteen cultural values: a worker might receive poor marks for a lack of "hustle." (Uber's cultural value wasn't *sometimes be hustlin'.*" It was *always*.) Managers made their evaluations in private and came back with a score, with little explanation of how they arrived at it. Positive or negative, the

---

* The recruits never stopped flooding in. By the end of Fowler's "very strange year" in 2016, Uber's workforce would swell to nearly ten thousand.

score was your score. And one's year-end bonuses, salary increases, and overall career trajectory inside Uber hinged upon that score.

Over time, scoring and advancing through the organization required politicking, cozying up to the right leaders, and, above all else, delivering products or ideas that led to growth. Your quality as an employee, or as a person, didn't really matter. At the end of the day, growth—trips, users, drivers, revenue—won all arguments.

Often, the emphasis on growth created unintended side effects, or "negative externalities," in management-speak. Managers would pursue growth even if it led to staggering inefficiency in *other* parts of the business. For example: In Uber's earliest days, the company sent free iPhone 4 devices to all new drivers. In order to get drivers on the road as quickly as possible, managers started sending out iPhone 4s as soon as someone signed up. But some eager managers began mailing phones out *before* drivers passed their background checks or completed other paperwork. Growth of new drivers exploded, which made the managers in charge look better. But so did a rash of iPhone thefts and fraudulent sign ups, costing the company dearly in what amounted to free iPhone giveaways to scammers.

Uber's ill-fated Xchange leasing program was another example. At one point in Uber's history, someone had the idea that there might be thousands of potential drivers who didn't have enough collateral or credit history to secure a car loan. But Uber could overlook that and lease the cars anyway, requiring only that the lessee work off their obligation immediately by driving for Uber. So Uber began leasing to high-risk individuals with poor or nonexistent credit ratings. It worked—sort of. Growth went through the roof as people who were never eligible for loans before were suddenly being given car leases. Thousands of new drivers came onto the platform, and the managers in charge were given hefty rewards for the idea. It was the ride-hailing equivalent of a subprime mortgage.

And just like 2008, the negative consequences came soon after. Uber noticed that the rate of safety incidents spiked after the company began the Xchange leasing program. They later figured out that many of the Xchange leasing drivers—those with poor or nonexistent credit histories—were the ones responsible for these incidents, which ranged from speeding tickets to sexual assault. The managers had created a moral hazard, indirectly causing pain for thousands, and potentially triggering a public relations and legal nightmare.

Further, car dealerships were pushing these marginal drivers into more expensive leasing options, thereby lowering drivers' opportunity to profit from their work. And after driving the cars around the clock, drivers were returning the vehicles in far worse condition than when they began the lease. Despite all the driver growth, Uber soon found it was losing upwards of $9,000 per vehicle on each Xchange leasing deal, *far* above the initial estimated losses of $500 per car. Never mind that the company was giving people subprime loans that they couldn't pay back while ruining their credit—all for a gig-economy job that returned less and less each year as the company garnished drivers' wages.

Still, despite the waste and ill effects caused by imbalanced incentives, Kalanick never ceased rewarding growth. Growth was what made the difference between an average employee and a high performer who delivered results. High performers were untouchable.

That was another Uber value: The Champion's Mindset.

FOWLER DIDN'T GET the response she expected.

The HR representative told her since it was her manager's first sexual harassment offense, he'd receive a stern reprimand. Further, since he was a "high performer," he likely wouldn't be fired for what was "probably just an innocent mistake on his part." Fowler was told she had a choice: stay on her current team under him and almost certainly receive a bad performance review when it came time for her evaluation, or find another team she wished to work with and switch.

As she saw it, it wasn't much of a choice. HR didn't seem to care about her experience or the fact that her manager might harass other women beyond her. And she was scared he would give her a bad review when it came time to evaluate her performance. So, she left her team, spending the next few weeks looking inside the company for another good fit.

Fowler was worried. In less than a month at work, she had been harassed by her boss, exposed herself to potential retaliation by reporting him, and now had to find a new role. She began having second thoughts about her dream job. But within a few weeks, she landed with another team of site reliability engineers, settled into her position, and was doing the work she wanted to do when she first arrived. She even managed to write a book for a technical publisher based on the work she did for her new team.

But as time passed, she started meeting other women across Uber whose experiences at Uber echoed her own. Her former manager had behaved inappropriately with other female colleagues, she discovered, which clashed with what she heard from HR, who said it was an isolated incident. Now, she began to understand that he had a history of bad behavior with women at Uber, but his high-performance reputation had kept him safe from dismissal.[*]

The more she dug into HR and gathered data from colleagues, the worse the company seemed. Uber's employee performance system had created an alpha, kill-or-be-killed environment. In a later meeting, Fowler recalled a director boasting about withholding information from one executive to curry favor with another (and it worked). Backstabbing was not only endorsed, but encouraged.

"Projects were abandoned left and right," Fowler would later say. "Nobody knew what our organizational priorities would be one day to the next, and very little ever got done." There was a constant fear that an employee's team would be dissolved or absorbed into another warring faction of the company, or that this month's leader would implement some massive reorganization, only to abandon it when a new person rose to replace them. "It was an organization in complete, unrelenting chaos," Fowler believed.

Things were toughest for women. Fowler recalled that when she had joined her specific department it consisted of 25 percent women, a low

[*] Not long after Fowler found a new role in the company, her manager came on to yet another Uber employee, who again reported him to management. He was terminated and left the company in April 2016.

rate by most corporate standards but a *stellar* ratio for a dude-centric place like Uber. Kalanick, after all, had been quoted in *GQ* nicknaming his company "Boober" for all the women it brought him.

It was the leather jackets that truly stuck out in Fowler's mind. Earlier in the year, all of the site reliability engineers were promised leather jackets as a gift from the company, a nice team-building perk to reward employees. Uber had taken all of their measurements and would buy them for the group later in the year. Weeks later, the six remaining women in Fowler's division, including Fowler, received an email. The director told the group of women that they wouldn't be getting leather jackets after all; Uber got a group discount on the 120 *men's* jackets they were able to find. But since there were so few women in the organization, they weren't able to find a bulk rate. That lack of a deal, the director said, made it untenable to justify placing a jacket order for the *six* women in the organization.

Fowler, shocked at the decision, pushed back. It just wasn't fair. The director's reply was blunt. "If we women really wanted equality, then we should realize we were getting equality by not getting the leather jackets," she was told. In the director's mind, making special accommodations for women demeaned them, undermining the meritocracy. The director would do the same thing if the roles were reversed and men were the ones to miss out on the jackets; it didn't occur to him that, in male-dominated Silicon Valley, that scenario would never occur.

After a back and forth with HR and top executives about the jackets and general issues with how Uber treated women, Fowler had had enough. Disgusted with Uber, she negotiated a job offer from another tech company. A few months after the jacket incident, she left Uber for good.

IT WAS RAINING on that Sunday morning in early 2017, just two months after Fowler had left Uber, when she decided to go public. Uber was just coming off a disastrous whirlwind of press in the wake of Kalanick's decision to stay on President Trump's advisory council, and then, under pressure from employees, to decline the position.

Fowler had typed up some three thousand words on her time at Uber and pasted them into her personal WordPress blog. The incident with her manager, her nightmare battles with the HR department, the leather jacket situation—all of it went into the post. She had no idea what would happen after she hit publish, if anything were to happen at all.

Susan Fowler gave one last look at the words on the screen. "Reflecting On One Very, Very Strange Year At Uber," the title of the post read. She took a breath.

And then she hit publish.

Chapter 23

. . . THE HARDER
THEY FALL

TRAVIS KALANICK WOKE UP TO AN IPHONE ON NUCLEAR MELTDOWN.

Within hours, the link to Susan Fowler's blog post had been shared internally, across private messages and chat rooms hundreds of times. Uber employees were buzzing with ire, excitement, confusion. It was raining in San Francisco that Sunday morning, but Kalanick was in Los Angeles. Groggy, he began returning the flood of calls that had come in from Uber's top executives about Fowler's whistleblowing memo.

Fowler had never ranked high enough in the company to cross his radar. And yet this one woman—a single engineer in Uber's sprawling workforce—was rattling the entire organization. Press calls began flooding into the public relations department, seeking comment on Fowler's post. Fowler herself had gone dark, leaving reporters' calls unanswered, saying nothing beyond what she had written on her blog.

Of all the scandals Uber had suffered to date, this Fowler memo struck the company the hardest. Chat rooms were in chaos. Email chains to leadership from angry employees were filled with demands and more allegations. Fowler's memo was just the beginning. Her post had burst open a dam, through which now flowed a river of pent-up employee complaints, years in the making. Worse, for Travis, employees began airing some of their bad Uber experiences in public, on Twitter.

"This is outrageous and awful. My experience with Uber HR was similarly callous & unsupportive," tweeted Chris Messina, another Uber employee who had recently left the company. "In Susan's case, it was reprehensible."

The frustration unleashed by the Fowler's explosion hadn't come out of nowhere; the rank-and-file resentment towards Kalanick had been building since his initial refusal to step down from Trump's business advisory council. Tech employees had changed since the election. Before November 2016, workers felt the hard-charging, visionary founders were on the right side of history. In the age of Trump, however, the idea of one's CEO as an oppressive and embarrassing despot became intolerable. By the time Kalanick stepped down from the council, worker attitudes toward him had devolved. Perhaps their boss was just as bad as the president.

Over the course of the past few months, Uber employees had begun failing the Bay Area cocktail party test. For employees, Uber had become a scarlet letter. Where once wearing Uber black had been a point of pride—like Facebook blue—now, admitting you worked at 1455 Market Street immediately short-circuited a conversation and drew strange looks. Implicit was the question, "How can you work for *Uber*?"

It didn't feel good. And people started quitting. Whereas over the course of 2014 through 2016, Uber was hiring *thousands* of Google employees away, now Google began rehiring Uber's conscience-stricken workforce in droves; Airbnb, Facebook, even Lyft started to pick off Uber employees. Uber needed to fix its morale problem. Fowler's blog post had only made it worse.

Kalanick snapped into action. He hopped a flight back up to San Francisco and arrived at Uber's Market Street headquarters early Monday morning, ready to deal with the Fowler situation.

During the meeting, one board member raised the idea of an internal investigation conducted by an outside entity. Kalanick needed a flashy heavy hitter, like Covington & Burling, the DC-based law firm, to show Uber was taking the matter seriously. Covington had hired Eric Holder, Barack Obama's attorney general. Kalanick knew Holder already; the former AG had done some work for Uber in the past. Holder had integrity. Picking him and his partner, Tammy Albarrán, to lead the investigation might provide good optics.

Others were more cautious. Rachel Whetstone, Kalanick's senior vice president of communications and public policy, was nervous. She was an operator, a longtime communications and policy executive who had worked for Google for nearly a decade, rising to the top of the comms food chain, before coming to Uber. Thin and anxious, with wisps of long, strawberry blonde hair and a posh British accent, Whetstone came from the cutthroat world of Conservative British politics before diving headlong into the tech sector. She was a natural strategist, had a knack for seeing around corners, figuring out where the press was going to strike next, and bracing for impact. She earned her seat at the table by discussing long-term policy decisions with executives as a peer, not a subordinate. After kicking upstairs David Plouffe—who was better suited to schmoozing politicians and composing speeches than running a daily press shop—Kalanick promoted Whetstone to take his place.

Whetstone's and Kalanick's relationship had grown strained over the course of the past few months. Kalanick believed Whetstone and her deputy, Jill Hazelbaker—another Google alumna and former political operative—were doing a terrible job shaping Uber's image, evidenced by the company's consistently bad coverage. The comms team, on the other hand, believed they were doing their best to defend the company with what they were given: an unlikeable, inflexible CEO and a raucous workplace staffed with thousands of men shaped in Kalanick's image. By the time Fowler's post hit the web, Kalanick had begun questioning Whetstone's strategies aloud in front of other executives.

During the Monday morning meeting with Kalanick and the rest of the leadership team, Whetstone offered advice Eric Schmidt, the former CEO of Google, had told her years ago: "Once you bring in outsiders, it's the fastest way to lose control." It was one thing for Uber to root around in its own garbage and discipline or fire employees. It was another to bring in some of the best lawyers in the country and tell them to have at it. Fresh, inquisitive eyes would surely unearth new, horrifying skeletons. Even so, it was Whetstone who first mentioned the possibility of bringing Holder on; if Uber was going to bring in outside investigators, she thought, it should be someone like him.

Travis needed no convincing. He had been upset after reading Fowler's post and wanted it handled immediately.

The way he had it handled, however, would lead to deeper issues than he could have forseen. Kalanick didn't understand what such an

investigation would be like—much less how thorough Holder's investi-
gators would turn out to be—but he instructed Emil Michael to contact
Holder and hire him on the spot. In a memo sent later that afternoon,
Kalanick tried to reassure his upset employees:

Team,
It's been a tough 24 hours. I know the company is hurting, and
understand everyone has been waiting for more information on
where things stand and what actions we are going to take.

First, Eric Holder, former US Attorney General under President
Obama, and Tammy Albarrán—both partners at the leading law
firm Covington & Burling—will conduct an independent review
into the specific issues relating to the work place environment
raised by Susan Fowler, as well as diversity and inclusion at Uber
more broadly . . .

Second, Arianna is flying out to join me and Liane [Hornsey,
Director of HR] at our all hands meeting tomorrow to discuss
what's happened and next steps . . .

Third, there have been many questions about the gender diver-
sity of Uber's technology teams. If you look across our engineer-
ing, product management, and scientist roles, 15.1% of employees
are women and this has not changed substantively in the last
year. As points of reference, Facebook is at 17%, Google at 18%
and Twitter is at 10%. Liane and I will be working to publish a
broader diversity report* for the company in the coming months.

---

* Kalanick's sudden openness to a diversity study—an internal tabulation and breakdown
of the gender and ethnicity breakdown of Uber's workforce—was met with bewilderment
by employees. For years, staffers had pushed Kalanick to publish a diversity report, which by
2017 was an increasingly common transparency tactic offered in the predominantly white,
predominantly male world of Silicon Valley. Joe Sullivan, Kalanick's own chief security officer, often pushed Kalanick the hardest. But Kalanick refused, over and over; diversity reports
went against the spirit of Uber's cultural values. Uber, after all, was a "meritocracy" in his
eyes. Uber only hired the "best," he believed, and was otherwise blind to gender and ethnic
differences. As with the Trump council decision, many saw the diversity report announcement
as too little, too late.

I believe in creating a workplace where a deep sense of justice underpins everything we do. . . . It is my number one priority that we come through this a better organization, where we live our values and fight for and support those who experience injustice.
Thanks,
Travis

Executives expected pushback, but the letter seemed to ease tension internally—at least until the all-hands meeting the next morning. For the moment, uChat cooled down, and employees went back to work.

Kalanick believed he was doing the right thing. And as others would later say, to his credit, he moved quickly and decisively to try and rectify what had happened to Fowler. He leaned heavily on another board member to repair Uber's image in the public eye, someone who would grow closer to Kalanick over the next six months than anyone else in his life.

That board member was Arianna Huffington.

KALANICK NEVER PLANNED to trust Arianna Huffington with his life and career; it just happened to turn out that way.

When Kalanick and lieutenant Emil Michael were dreaming up ideas for a perfect celebrity board member, Oprah Winfrey topped both of their lists. But when they failed to entice the megastar, Kalanick began thinking about another celebrity he had already known for years: Huffington.

The two first met at a technology conference in 2012, where Kalanick took Huffington aside during a break between speakers to show her how Uber worked. Back then, Uber was still a luxury service for the wealthy—Uber X wouldn't come for another few months—and Huffington was an ideal early adopter. "@travisk showing me his super cool app, Uber: everyone's private driver uber.com," Huffington tweeted, pushing a photo of the two together at the conference to Huffington's millions of followers. For Kalanick, it was a big moment; Huffington was a celebrity, the exact type of client he wanted carted around in Uber's black car service.

Huffington's star had risen long before anyone had imagined the idea of Uber, much less the iPhone that had made the startup possible. Born in Greece in 1950 to Konstantinos and Elli Stassinopoulos and raised in Athens, Arianna grew up close to her family and sister, Agapi, until her parents had marital troubles. Their father, an unfaithful journalist, separated from their mother when Arianna was young. The two girls stayed with their mother, a warm, intelligent woman who spoke four languages and who was supportive of her daughters. Their beginnings were humble, but her mother valued higher learning. "Your dowry is your education," Elli told the girls. Her mother moved them to London just so Arianna could take her entrance exams to Cambridge.

It paid off. Arianna was naturally intelligent, like her mother, who pushed the two girls to climb the social ladder. Arianna won a partial scholarship to Cambridge, the start of her journey into an elite social class. In school she excelled, studying economics at Cambridge and, later, comparative religion in India. Instead of gravitating toward partying and drugs as a teenager in the 1960s, Arianna preferred debate and civics. A fortuitous appearance in a televised debate on feminism at the end of her school tenure brought a publisher to her doorstep. That eventually led to the writing of her first book, *The Female Woman*, published in 1973, which took a much more conservative stance on women's issues—a reaction to the women's lib movement that would begin Arianna's long career as a public contrarian.

The first book led to many others—more than a dozen by the time she met Kalanick—and the development of her voice as a fearless writer with bold opinions. In 1981, she penned a biography of Maria Callas, a famed Greek soprano. In '88 she moved on to a book on Picasso. Both were bestsellers.[*]

In the eighties, she met Michael Huffington, a Republican banker

---

[*] Both books also came with their share of controversy. When her Callas biography came out, Arianna was accused of plagiarizing passages from another Callas biographer's past work. For her Picasso release, an art historian accused her of worse: "What she did was steal twenty years of my work," Lydia Gasman, the professor, told a journalist in 1994. Arianna, who has consistently denied all claims of plagiarism, settled the first case out of court; her second accuser never filed a lawsuit.

and politician. After a brief courtship and marriage in 1986, the contrarian author became Mrs. Arianna Huffington, the wife of a Republican House of Representatives member and, eventually, a prominent Republican herself. Huffington wrote the occasional piece for the *National Review*, aligned with Bob Dole and Newt Gingrich, played the conservative talking head on weekly radio shows and writing outfits, and in the nineties moved into regular guest spots on *Larry King Live* and Bill Maher's political talk show.

Huffington commanded every room she entered. At nearly six feet tall with a shock of copper-red hair, Arianna Huffington had striking looks, but more distinctive was her over-the-top accent, boisterous and full. "Daaah-ling," she'd call people even if they'd only just met, as if speaking to an old friend.

Huffington possessed enormous charisma. Both friends and enemies alike marveled at her skill: Want an introduction for work? Huffington knew everyone in New York, Los Angeles, and DC. Need a blurb quote for your book cover? Huffington could provide it—she had written fifteen books herself. And after she blurbed your jacket cover, she might host your book party and invite her celebrity friends.

She was a master of reinvention. In her earlier years she explored mysticism, later on came meditation. After years as a Republican, she did an about-face and fashioned herself a progressive, embracing eco-friendly policies and supporting John Kerry's presidential run.

Her progressive streak, coupled with John Kerry's loss to George W. Bush in 2004, eventually led to Huffington's first stab at a true online media destination, what the *New Yorker* called "a kind of liberal foil to the Drudge Report." With venture funding and an old tech executive partner, in 2005 she launched the *Huffington Post*. The site pioneered an early form of "citizen journalism"—in reality, freelancers farmed the web for others' articles to summarize, aggregate, and repost on the *Huffington Post*'s website. Mainstream journalists pilloried the idea. Huffington and her partners laughed all the way to the bank; she sold the *Huffington Post* to AOL in 2011 for $315 million, personally netting more than $20 million.

Huffington was impossible to pigeonhole. There was no cause, no point of view that others felt was consistent throughout her life. The one thing that remained true about Arianna Huffington was that she seemed unclassifiable to anyone but herself. Her only constant was change.

"Is it possible to come up with a unified theory of Arianna?" one writer said of Huffington in 2006, while reviewing her eleventh book. "What does she believe?"

At sixty-six, after being edged out of power within AOL, she moved into personal care and health, launching a lifestyle brand, Thrive Global, and promoting a new book.

"There are two schools of thought about Arianna," Mort Janklow, a former agent to Huffington for her Picasso book, told *Vanity Fair* in 1994. "One is that it's all deliberate and calculated and she's ruthless. The other is that she really convinces herself beforehand. She sells herself first."

Her political career had paved a pathway to digital media. Media begat her new venture into health and wellness. And as wellness continued, she looked westward and saw the transformative nature of Silicon Valley.

After their initial 2012 meeting, Huffington slowly grew closer to Kalanick. They would appear onstage at conferences together. Huffington invited Travis to a Christmas party at her home one year, and Travis brought his parents, Bonnie and Donald, as guests. By 2016, she was in.

Huffington's friendship came at a pivotal moment in Kalanick's personal life. The end of 2016 was difficult for him. He and Gabi Holzwarth, his girlfriend of two years, had recently split up. The only non-work relationships Kalanick had mostly consisted of his parents and Holzwarth. Now, Holzwarth was gone. The couple couldn't withstand Kalanick's grueling work schedule. Devoted to Uber, he spent nearly every waking hour at the office. Holzwarth took off to Europe with a friend for a few weeks to try and blow off steam; Kalanick stayed at work.

When Kalanick decided Huffington was the one for Uber's board, in early 2016, he came to her bungalow in Brentwood for a talk. Kalanick paced around the room, explaining his ideas to Huffington. One day there would be an Uber that moved more than just people—food, retail items, and packages, *everything*[*]—and his company would be the one laying the infrastructure to make it happen. He foresaw self-driving Ubers, fleets of them, navigating San Francisco. Someday, there would even be flying Ubers, carting people through the air, from city to city. Four hours later, he was still moving and talking. She was enamored with his vision, his passion for what he wanted Uber to be. And Kalanick felt great warmth from Arianna, an almost maternal sense of encouragement for his goals.

The two sealed the deal over omelets, which Huffington cooked in her kitchen and Travis ate while pacing around the room. Arianna Huffington would be Uber's newest board member. She would have Kalanick's back.

---

[*] "Everything" except one item of business: The mail. Kalanick was deeply against becoming a modern-day version of the United States Postal Service. An unattractive market, in his eyes, he could cede the regular mail to Amazon if Bezos wanted it.

Chapter 24

NO ONE STEALS FROM LARRY PAGE

AT THE END OF 2016, MONTHS BEFORE SUSAN FOWLER'S BLOG POST, Travis Kalanick had a different problem brewing: forty miles south of San Francisco, Larry Page was fuming.

Anthony Levandowski—his star pupil and golden child—had left the company in January 2016. On his way out the door, Levandowski collected $120 million in bonuses for his contributions to Google's self-driving-car project. After investing eight years, hundreds of millions of dollars, and the time and resources of dozens of employees in the project, Google's top brass felt like Levandowski was leaving the self-driving program in the lurch. Worse, Google employees were defecting en masse to his wayward protégé's new self-driving-car startup, taking valuable knowledge and experience with them.

For Page, it was personal. He had long ago checked out of the daily minutiae of the search engine business. In 2015 Google had changed its corporate structure, creating a parent holding company, Alphabet, with Page as CEO. Google's search business was still printing money—billions every single quarter—which gave Alphabet's other companies the ability to pursue diverse projects.

It also freed the reclusive Page from the public eye. He hated the scrutiny that followed the CEO of Google, and wanted more time to pursue his own projects. The idea of self-driving cars had been a private goal for a long time. And self-driving was just the beginning; Kitty Hawk, a side project backed by his personal bank account, was working on a first consumer-ready version of a flying car. Page wanted to make his childhood dreams of the future come true in his own lifetime.

Although Google was the first Big Tech company to devote substantial resources and money to self-driving-car research, executives admitted they were slow to move and test the cars more aggressively. Competitors like Apple and Tesla were gaining traction in the space. After Levandowski left, Page made changes to how the self-driving wing would operate. Formerly operating under the Google "X" wing of "Moonshots," Page spun self-driving research out into its own, separate company. It was called Waymo, derived from the idea that the work will create "a new way forward in mobility." Page tapped John Krafcik, a former president of Hyundai Motor America, to be Waymo's CEO. Waymo had a years-long head start on the competition. The new company planned to capitalize on that lead, before they were overtaken.

Levandowski revealed his new startup, Otto, in May of 2016, four months after leaving Google. Then, in August, just three months later, he sold the new company to Uber for over $600 million. Page was immediately alarmed; the company was already embroiled in arbitration with Levandowski, suing him months ago for allegedly using Google's confidential salary information to lure employees over to Otto. Immediately turning around a sale to Uber raised Page's hackles even further. He had some of his deputies begin forensic investigation on Levandowski's old Google workplace accounts and the circumstances of the engineer's departure. Something didn't smell right.

Page's hunch proved correct. After running forensics on Levandowski's Google-issued work laptop, investigators discovered that in the weeks before he left, Levandowski downloaded more than 14,000 confidential files related to the self-driving program from Google's servers directly to his personal laptop. Among the files were designs for Waymo's proprietary lidar[*] circuit boards, one of the crucial components necessary for most self-driving cars to function. After downloading the files, he copied and transferred all of the 9.7 gigabytes of Waymo data over to a personal, external hard drive. When he finished,

---

[*] Short for "Light Detection And Ranging," lidar is an important part of most prototype vehicles among the tech and automotive companies competing to successfully create fleets of autonomous vehicles.

Levandowski installed a new operating system, erasing the contents of his work laptop hard drive. "After Levandowski wiped this laptop," Waymo's attorneys would later say, "he only used it for a few minutes, and then inexplicably never used it again."

Other employees who followed Levandowski to Otto downloaded proprietary information as well, including "confidential supplier lists, manufacturing details and statements of work with highly technical information." And around the same time Levandowski left, his partner, Lior Ron, had searched Google for some incriminating phrases, including "how to secretly delete files mac" and "how to permanently delete google drive files from my computer."

The details were damning. But Page's investigators may not have pieced all of it together were it not for a mistake from one of Waymo's own lidar component suppliers. In February 2017, months after the Otto acquisition, the manufacturer accidentally included a Waymo employee on an email, an email which happened to include a schematic of a component from Uber's most recent lidar design. The Waymo engineer noticed something peculiar; Uber's lidar component looked like a carbon copy of Waymo's hardware.

Page had trusted Levandowski. For years the Google CEO had paid his protégé handsomely and put up with his insubordination, protecting Levandowski from managers who wanted to fire him. And now, Page's star pupil was betraying him.

On February 23, 2017, lawyers from the firm Quinn, Emanuel, Urquhart & Sullivan filed a lawsuit in the Northern District of California federal court on behalf of Waymo. The suit claimed that both Otto and Uber stole Waymo's intellectual property and trade secrets, infringed on multiple Waymo patents, and that the two companies had conspired and committed fraudulent, unlawful, and unfair business acts in concert with one another. The theft, Waymo claimed, was a rejuvenation of Uber's thus far unsuccessful efforts to build self-driving technology on its own.

"Otto and Uber have taken Waymo's intellectual property so that they could avoid incurring the risk, time, and expense of independently developing their own technology," Waymo's attorneys said in the filing. "Ultimately, this calculated theft reportedly netted Otto employees over half a billion dollars and allowed Uber to revive a stalled program, all at Waymo's expense."

It was a drastic measure by Page, who personally ordered the lawsuit. In Silicon Valley, companies copy their competitors' projects with a frequency that would alarm a fashion designer, an auto manufacturer, or those in other, less derivative industries. Facebook copied Snapchat's core feature, while hundreds of Instagram clones populated Apple's App Store at different points over the past six years. Lawsuits, however, were a different story; suing a former employee was an enormous risk. All sorts of embarrassing emails and documents might come out during the legal discovery process. Prospective employees might think twice about joining a company willing to sue them if they left.

With this lawsuit, Waymo was sending a message—to Levandowski and to the rest of Silicon Valley. *No one* steals from Larry Page and gets away with it.

IT WAS THE MIDDLE OF winter of 2017, and Jeff Jones, the man responsible for Uber's public perception, was trying to shake everyone in the top ranks of the company awake. Uber didn't have an image problem. Uber had a *Travis* problem.

As president of ridesharing and the only person on the executive leadership team with a history of marketing experience, Jones took it upon himself to study the root of the hatred of Uber's brand, something he hadn't anticipated before he joined. Jones knew people thought Travis was an asshole, but he wasn't prepared for *this*. Susan Fowler's blog post had made things exponentially worse.

The Waymo lawsuit—which landed just four days after Fowler's post—created an enormous new problem: Uber's new self-driving leader appeared to be a literal thief and potential criminal. And that wasn't even the worst of it. Three days later, one of Uber's marquee hires, Amit Singhal—the man responsible for perfecting Google's search algorithms—was forced to resign from Uber before he could even begin his new job. Kalanick had announced his hire just a month previously, thrilling Uber's employees. Instead, just days after Waymo's lawsuit dropped, the press uncovered the fact that Singhal was pushed out of Google for claims of sexual harassment, something that Google executives had covered up during his departure. (Singhal has consistently denied the allegations.) Kalanick didn't know about the claim when he hired him. For Uber, the timing could not have been worse.

But Jones wanted more data. When he first started at Uber, Jones told Kalanick he wanted to commission surveys into how people viewed Uber, and how those same people viewed *Kalanick*, separately, as well. The company didn't really have any data on such questions, and Jones wanted to see what they said.

Months later, the data came back. Jones called most of the executive leadership team to join him on a two-day leadership offsite retreat away from the office. He asked Kalanick not to attend—he wanted to go over the data with the executive leadership team alone, not in front of the big boss, and hoped Kalanick could respect that. Kalanick bristled at the request, but Jones was adamant, and ultimately Kalanick stood down.

In late February, the group—roughly a dozen executives from all of Uber's different divisions—gathered in downtown San Francisco's Le Méridien, a hotel off Battery Street in the Financial District, to go over the results of the survey, among other things. Jones had booked a meeting room for the discussion; he had a PowerPoint presentation prepared so that the rest of the executive leadership team could understand the data.

The results were clear: People enjoyed using Uber as a service. But when you brought up Travis Kalanick, customers recoiled. Kalanick's negative profile was actively making Uber's brand *worse*.

Later that day, Jones got a text from Kalanick. The CEO was coming over to join the meeting. Kalanick didn't like feeling left out while all his top lieutenants were discussing the future of his company. As Kalanick walked into the hotel meeting room filled with his executives, he saw charts, surveys, and studies taped to the walls. In the center of a room was a giant piece of paper with a sentence written on it. The group came up with what it believed Uber's image was to outsiders, written in bold, black ink: A bunch of young bro bullies that have achieved ridiculous success. It was a hard point to argue.

Nonetheless, Kalanick began to push back on Jones's findings immediately, rebutting the data he saw on the wall.

"Nuh-uh," Kalanick said. "I don't believe it, man. I don't see it."

His lieutenants were flabbergasted. Even in the midst of the most sustained set of crises in Uber's history, Kalanick couldn't see the literal writing on the wall. Aaron Schildkrout, who led Uber's driver product development, leapt to defend Jones and the data. Daniel Graf and Rachel Holt—two other well-respected leaders—joined him. Kalanick didn't love Jones at that point, but he respected Graf and Schildkrout, and Holt had been with him since the early days of Uber. And all three were supportive of the surveys. If anyone could get him to listen, it would be them.

The argument was interrupted. Rachel Whetstone, Uber's communications head, got a phone call, and stepped out of the room into the hallway to take it. Moments later, Whetstone signaled for Jill Hazelbaker, her second in command at policy and comms, to join her in the hallway. Something bad was happening, but none of the executives in the room knew how bad it would turn out to be.

Moments later, Jones joined the communications heads in the hallway, followed by Kalanick. Whetstone grabbed a laptop from the conference room and set it down on a chair in front of them. She opened a webpage to *Bloomberg News*'s website; they had just posted a story about Kalanick online. At the top of the article was a video clip.

The four executives huddled around the laptop, with Kalanick kneeling on the floor in front of the chair. They watched as a grainy dashcam video began playing. Shot from inside an Uber, the video shows a driver with three passengers: two women and a man, Travis Kalanick, sandwiched in between them in the back seat.

It begins innocuously, the tinny audio capturing snippets of the group's conversation and shared laughter—the giddiness suggested a tipsy ride home from a night out. As a Maroon 5 song plays on the radio, Kalanick starts shimmying his shoulders, swaying to the beat. As they watched their boss on camera, some in the room could only think of one word: "douchebag."

As Kalanick and his friends pull up to their destination, the driver strikes up a conversation, acknowledging that he knows who Kalanick is. Then the video takes a turn. Fawzi Kamel, the driver, presses Kalanick on Uber's dropping prices for customers, which in turn has hit the drivers hard. "I lost $97,000 because of you," Kamel tells him, "I bankrupt because of you. You keep changing every day."

"Hold on a second!" Kalanick interrupts. The conversation starts getting heated. "What have I changed about [Uber] Black?"

"You dropped everything!" Kamal pushes back.

"Bullshit. You know what?" Kalanick says, beginning to get out of the car. "Some people don't like to take responsibility for their own shit!" he shouts, now shouting over Kamel's protests and into his face.

Kalanick raises a finger and jabs it into the air as he finishes his thought. "They blame EVERYTHING in their life on somebody ELSE. Good luck," he jabs back. Kalanick exits the car to a shouting Kamel, disappearing from the frame of the video seconds before it ends. Someone closed the laptop.

Kalanick—the flesh-and-blood one in the hotel that Tuesday morning—already brought to his knees, began muttering to his lieutenants. "This is bad, this is really bad." He fell further forward, writhing around on the floor. "What is wrong with me?" he yelped.

None of the executives knew what to do. Seeing Kalanick squirm like this made them deeply uncomfortable.

Kalanick dialed the only person he felt he could turn to; he called Arianna Huffington. "Arianna, we need help," he cried into his phone. "How are we going to get out of this? This is so bad. I fucked up." Huffington cooed platitudes into the phone, attempting to calm down the distraught Kalanick.

Jones tried to offer some solace, suggesting talking to crisis PR firms[*] to help strategize and figure out what to do next to pull Uber out of its tailspin.

"There are experts who can help us here, Travis," Jones said.

Whetstone disagreed. "I don't think you're going to find better people than me and Jill," she offered. Whetstone believed the PR leaders could still pull him out of this disaster.

Kalanick lashed out, directing his anger toward Whetstone and Hazelbaker. "You two aren't strategic or creative enough to help us get out of this situation," he said. The room was silent as Kalanick's insult hung in the air. Whetstone and Hazelbaker had had enough. The two of them stood up, gathered their belongings, and walked out of the room.

Kalanick soon realized his mistake: he had pissed off the very people trying to protect him from a press corps that was about to tear him apart. As he chased his communications executives down the hotel hallway to try and convince them to stay, Hazelbaker confronted him.

"How dare you!" she yelled, inches from Kalanick's face, as the rest of the group watched in shock. "I've walked through fire for you and this company! You did this TO YOURSELF!"[†]

As the group split and the day wound down, Kalanick eventually managed to convince Whetstone and Hazelbaker not to quit their posts. Half of the group made its way back to Hazelbaker's townhouse, a twenty-minute Uber ride away in San Francisco's Cow Hollow district. Hazelbaker ordered takeout for the group.

---

[*] The group made a brief call to Steven Rubenstein, a crisis PR expert who regularly worked for the Murdoch family. Rubenstein ultimately decided not to take on Kalanick as a client, though he would cross paths again with Kalanick less than a month later. But as a parting gift, Rubenstein offered two pieces of advice: First, Kalanick had to "find his Sheryl," a reference to Mark Zuckerberg's relationship with Sheryl Sandberg, then widely considered a competent counterbalance to Zuck's leadership. Second, he said Kalanick needed to take a leave of absence. "You either shoot yourself in the foot, or the press will end up shooting you in the head."

[†] One witness to the confrontation between Hazelbaker and Kalanick recalled the communications executive using far more colorful vocabulary during the encounter.

Sitting on the sofas in Hazelbaker's living room, Uber's top executives shared pizza and beer and mulled their options. Meanwhile, Kalanick continued his theatrics, writhing around on Hazelbaker's carpet. Kalanick kept repeating the same thing over and over: "I'm a terrible person. I'm a terrible person. I'm a terrible person."

Whetstone tried to console him, halfheartedly. "You aren't a terrible person. But you *do* do terrible things," she said.

By the end of the day, Whetstone, Hazelbaker, and Kalanick had settled upon a statement to hand out to reporters. By then, the press and the public were frothing at the video, which had gone viral. Here was conclusive proof that Kalanick didn't care about drivers. That he partied like a douchebag. That Travis Kalanick was, in fact, an asshole.

Later that evening, Kalanick circulated an apology memo to his employees. They posted the memo to the company's public blog the next morning.

By now I'm sure you've seen the video where I treated an Uber driver disrespectfully. To say that I am ashamed is an extreme understatement. My job as your leader is to lead . . . and that starts with behaving in a way that makes us all proud. That is not what I did, and it cannot be explained away.

It's clear this video is a reflection of me—and the criticism we've received is a stark reminder that I must fundamentally change as a leader and grow up. This is the first time I've been willing to admit that I need leadership help and I intend to get it.

I want to profoundly apologize to Fawzi, as well as the driver and rider community, and to the Uber team.

—Travis

Chapter 25

GREYBALL

A WEEK AFTER SUSAN FOWLER'S BLOG POST EXPLODED ACROSS THE Valley and the front pages of newspapers worldwide, I got a telephone call from a number I didn't recognize.

"Hello, is this Mike? Mike Isaac?" a voice on the other end of the line asked. "Hi Mike, my name is Bob. I work for Uber. Can we talk off the record?"

A few days prior, my story "Inside Uber's Aggressive, Unrestrained Workplace Culture" had run on the front page of the *New York Times*. I had spoken to more than thirty current and former Uber employees, detailing life inside of the company. Since joining the *Times* in 2014, I had written dozens of stories about Uber, but Fowler's post was something different.

For every woman in tech who had deflected sexual advances from a superior, or endured an inappropriate comment on Slack; for every female founder who saw men land funding for their subpar ideas over better, woman-led startups—the Fowler post perfectly articulated the harassment, the bias, and the abuse built into the "meritocratic" systems touted so arrogantly by tech utopianists.

Fowler didn't know it, but her post signaled an early shot that later in 2017 would turn into a movement. That fall, the *Times* and the *New Yorker* would publish groundbreaking investigations into the systematic, widespread sexual harassment of Harvey Weinstein, the Hollywood mega-producer, which would eventually lead to his arraignment and spark the #MeToo movement. In the wake of the Fowler post, I joined the scrum of journalists reporting on the chaos and lawlessness inside of Uber.

Bob told me he appreciated my story, which was the first to dive into the details of Uber's Las Vegas bacchanal during the company's "X to the x" retreat, the multiple lawsuits waged by Uber's own employees against the company, the rampant drug use and sexual harassment beyond Fowler's initial claims. "It was the most accurate piece I've seen that attempted to capture what it was like inside," he said.

"But you only scratched the surface. Does the term 'Greyball' mean anything to you?" Bob asked. It didn't.

He suggested we meet to talk about it.

THE PARKING LOT OF THE ramshackle pizza joint in Palo Alto was mostly empty at eight o'clock on a Tuesday evening. The place was a dump, serving greasy slices and flat soda pop. That was exactly the point; Bob didn't want to be seen in public with me. We *definitely* wouldn't run into other Uber engineers in a dive like this.

As I sat in my car, I went over Bob's checklist. Before I left my house, I was to delete my Uber app and check the setting buried in the app submenu that deleted my contact information from Uber's servers. One of Uber's features requested users to upload their phone books to the cloud. If two friends or colleagues took a ride together, this feature allowed them to quickly split the fare. For most users, this was a nifty, convenient feature. For Bob and me, it was a liability; if Uber's information security team wanted, they could spy on the rides I'd taken, the names and numbers of my contacts and sources—any information I'd willingly given over to Uber. Better I delete Uber from my phone entirely. I was to leave my phone in the car, turned off, and bring nothing but a pen and notebook. He'd find me when I got there.

The place was dingy. It felt old, with shabby plastic booths and half-lit Budweiser ceiling lamps above the pool tables. I ordered a pizza for both of us, took a seat in a grungy booth and waited for Bob to show up. The only people there were two young guys playing pool, the cashier behind the counter, and another guy in the back, making pizzas.

Bob was nervous when he walked in, wearing a baseball cap and carrying a file folder bulging with paperwork. He wasn't used to meeting with reporters, and the risk he took was a big one; if Kalanick discovered what he was up to, Uber's lawyers could turn his life upside down. I appreciated the lengths he was going to just to meet me. I waved him over, looking as harmless as I could. My reporter trick is to play dumb and friendly; dumb and friendly is always more approachable than eager and prodding.

Over sweating cups of cold Pepsi and pepperoni slices, Bob and I went through a file of documents, bits of evidence from different Uber projects. One of Uber's local general managers had sent an email offering drivers a list of tactics to evade police capture:

> —*Keep your Uber phone off your windshield—put it down in your cupholder*
> —*Ask the rider if they would sit up front*
> —*Use the lanes farthest from the terminal curbside for pickup and dropoff*
>
> *Remember, if you receive a ticket while picking up or dropping off Uber riders at the airport, Uber will reimburse your costs for the ticket and provide any necessary legal support. Take a picture of your ticket and send it to XXXXXXXXXX@uber.com.*
>
> *Thank you and have a wonderful day!*

The email was written in a friendly tone, but it demonstrated that Uber was systematically schooling drivers to avoid detection.

After we finished our slices, Bob pulled a laptop out of his backpack. He opened up a web browser and punched in a URL. The page loaded to a three-year-old YouTube video posted by the *Oregonian*, the local newspaper in Portland. In the video, a transportation official, Erich England, was trying to hail a ride as a part of a sting operation to catch Uber illegally operating in the city. As we watched the video together, we saw England fail to catch an Uber. Two drivers agreed to pick him up and then quickly canceled, he explains, unsure what went wrong. "Must be high demand," England says, shrugging. After that, the app showed no Ubers available at all. Eventually, he gives up.

"That was no accident," Bob said. "That was Greyball."

The genesis of Greyball, a software tool Uber used to systematically deceive and evade authorities, occurred in Philadelphia, one of the hardest fought markets Uber ever tried to enter. In the fall of 2014, as the company tried to launch UberX in Philadelphia, the Philadelphia Parking Authority sent a stern message to drivers. "If we find a civilian car operating as an UberX, we will take the vehicle off the streets. We *will* impound the vehicle," a PPA official said at the time. The PPA started creating fake Uber accounts to conduct sting operations; when Uber drivers showed up, PPA officials impounded their cars and issued thousands of dollars in fines. It was effective; people became too scared to drive for Uber.

Uber city managers in Philadelphia were panicking. How could Uber convince people to drive for them if the police kept impounding their vehicles? Bob showed me an SMS text Uber's Philadelphia managers sent to all of its drivers, reassuring them of Uber's support:

UBERX: REMINDER: If you are ticketed by the PPA, CALL US at XXX-XXX-XXXX. You have 100% of our support anytime you are on the road using Uber—we are here for you, and we will get you home safe. All costs associated will be covered by us. Thank you [for] providing safe, reliable rides to the citizens of XXX-XXX-XXXX. Uber-ON!

Philadelphia's operations team was pushing engineers back at headquarters to come up with a solution. As other teams across the country encountered similar challenges, the pressure in San Francisco mounted. The operations and engineering departments at HQ wanted to know the precise laws around operating UberX vehicles in Philadelphia. The legal department, led by Uber's general counsel Salle Yoo, said it was a gray area; there were no specific laws about ride-hailing services, so technically, Uber would argue, it wasn't illegal to drive for one.

The term "gray area" was music to Travis Kalanick's ears. One talented engineer on the fraud team, Quentin, had an idea. Quentin's team had dealt with the widespread fraud in China, and dealt with the fallout when Kalanick had to explain to Eddy Cue why Uber had broken Apple's App Store rules. Quentin explained that when a rider opened the app there was a tool that controlled what cars the rider could see on the Uber map. They used the feature for all manner of things. If Uber was running a promotion, such as the popular "on-demand ice cream truck," the feature would only show the customer the drivers who were delivering ice cream nearby, and hide all other Uber cars on the road. The tool was nicknamed "Greyball," the idea being engineers were tricking customers—or "greying" over their eyeballs—to obscure or highlight specific vehicles.

What if, the engineers thought, they were able to "Greyball" the police or other parking enforcement officers who opened the app, hiding from them all UberX cars on the road? Authorities wouldn't be able to figure out which cars were Ubers, while drivers would be safe from impound and customers would still be able to catch their rides. Everybody wins—that is, everybody except the Philadelphia Parking Authority.

The big problem, Bob explained to me, was figuring out how to spot who the authorities were so they would know which customers to start "Greyballing." If Uber picked the wrong people, they could end up tricking a customer who wouldn't be able to catch a ride.

So Uber engineers, fraud team members and field operatives came up with about a dozen ways to spot authorities. One method involved "geofencing"—or drawing a digital perimeter around police stations in a city that Uber attempted to enter. City managers would closely watch which customers within that perimeter were rapidly opening and closing the Uber app—a behavior engineers called "eyeballing," or monitoring nearby drivers. City managers would also scan other details on new user accounts—personal information like credit cards, phone numbers, and home addresses—to check whether the data were tied directly to a police credit union or some other obvious giveaway. After Uber managers felt confident they had spotted police or parking enforcement, all it took was the addition of a short piece of code—the word "Greyball" and a string of numbers—to blind that account to Uber's activities. It worked extremely well; the Philadelphia Parking Authority never noticed the deception and car impound rates plummeted.

Quentin's fraud team created a new playbook for how city managers should use Greyball. The playbook was called "Violation of Terms of Service"—VTOS for short—and asserted that authorities using the app to fraudulently hail rides were violating Uber's terms of service agreement. That violation gave Uber the right to deploy Greyball. Any employee could find the playbook in Uber's internal, wiki-like information directory—alongside the dozens of other playbooks the company had created for other varied tasks.

With Uber facing opposition in almost every market it entered, the VTOS playbook and Greyball seemed like a godsend. In South Korea, for instance, local police were paying civilians to report drivers. A similar bounty program was conducted in Utah. The use of Greyball spread so quickly that members of the fraud team had to call a summit—attended by Uber general managers from more than a dozen countries around the world—to explain best practices.

As Bob explained the program to me at the pizza parlor, he started to relax. He felt relieved, he said, to finally explain it to someone after keeping it a secret for so long. Greyball might be illegal. Uber was potentially obstructing justice to make its numbers.

"I don't know what you're going to do with all of this," he told me, pushing aside our empty paper plates to stow his documents. "I don't know. Just meeting you, talking right now, whatever this is, I feel a little bit better.

"Maybe this will change something," he said.

After we said our goodbyes, I left the pizza shop and walked back to my car, my mind spinning. Months later, the phone number Bob used to call me was disconnected. It was the first and last time I'd ever see him.

ON THE MORNING OF MARCH 3, the *New York Times* sent out a push alert to the mobile phones of subscribers: "Uber has for years used its app to secretly identify and sidestep law enforcement officials where it was restricted or banned," the alert read.

The blowback was swift. Attorneys general across the United States began asking Uber whether or not it used Greyball in their cities. Days after the report, Joe Sullivan, Uber's security chief, prohibited employees from using the Greyball tool to target authorities in the future, and said Uber was reviewing the use of Greyball over Uber's entire history. The US Department of Justice opened a probe into Uber's use of Greyball and whether or not it was lawful; the inquiry widened to Philadelphia, Portland, and other cities where it had been used. Uber already had the reputation for being uncooperative and aggressive. Now, people were calling them potential criminals.

Attrition rates started climbing. Employees stopped wearing their Uber-branded T-shirts in public. During the Trump council revolts two months prior, protesters had handcuffed themselves to the front doors of Uber's headquarters; now there were demonstrators out front almost weekly. The company's reputation got so bad, employees stopped showing up to work. At one point, two of the remaining policy team employees in the office started rolling a ball down the entire length of Uber's cement corridors—thousands of feet from one end to the other—just to see if anyone would notice.

No one did; no one was there.

AFTER THE OFFSITE INCIDENT, the video of Travis berating a driver, and now the use of Greyball inviting federal investigation, Jeff Jones was done—he needed to get the hell out of Uber.

The entire reason he was hired was to fix the broken relationships between Uber and its hundreds of thousands of drivers. The driver video alone would have been enough to scuttle his efforts. When Uber cut rates in 2015, rather than worry about the effects lower income would have on drivers, Kalanick was giddy. To Travis, lowering prices

meant raising demand. Growth would explode again, and growth—
not the concerns of his drivers—was Travis's top priority.

It didn't matter to Kalanick that drivers were logging more trips
and picking up more people—basically doing twice the work—to make
the same amount of money. It didn't matter that drivers were com-
muting absurd distances to busy cities like San Francisco—often from
places two hours away, but occasionally as many as six hours away—
sleeping in their cars overnight on side streets and empty parking lots
for the chance at more rides per hour. It didn't matter that San Francisco lacked sufficient public bathrooms for drivers, forcing them to
find coffee shop bathrooms, or, more often, make do elsewhere. And it
certainly didn't matter that drivers pulling daylong shifts were over-
worked and under-slept.

Kalanick had no sympathy for drivers and their bills—vehicle wear
and tear, medical insurance, among many others—and classified them
all as 1099 freelance workers. The entire business model was based on
Uber minimizing the company's responsibility over its drivers.

Drivers did find ways to push back. They formed unofficial unions,
and used forums like UberPeople.net to congregate, share information, and organize walkouts and other protests. Harry Campbell, an
aerospace engineer who drove for Uber and Lyft on the side, started
a personal blog to document tips and insights. He called it *The Rideshare Guy*. Drivers were starving for more help and support from Uber;
instead, they found it amongst themselves.

Reporting to Kalanick, Jeff Jones was powerless to help them. As
he surveyed the wreckage of the past six months, Jones decided to pull
the ripcord. On March 19, 2017, *Recode* ran a story saying Jeff Jones,
Uber's president of ridesharing, had resigned from Uber, with sources
claiming his departure was directly due to the string of controversies
that plagued the company.

Kalanick tried to fight back in the press. He had his communications staff leak a memo to *Recode*. In it, Kalanick said Jones left
after being passed over as a potential chief operating officer. But Jones
wasn't going to let his former boss trash him without a fight. After

Uber's statement, he sent an on-the-record comment to *Recode*, in which he directly blamed the company's leadership culture for his departure:

> I joined Uber because of its Mission, and the challenge to build global capabilities that would help the company mature and thrive long-term.
>
> It is now clear, however, that the beliefs and approach to leadership that have guided my career are inconsistent with what I saw and experienced at Uber, and I can no longer continue as president of the ride sharing business.
>
> There are thousands of amazing people at the company, and I truly wish everyone well.

In the world of carefully worded corporate communiques, this was Jones giving it to Kalanick with both barrels.

Jones's maneuver worked. After time off, he would eventually go on to be hired as president and chief executive officer of H&R Block, the tax preparation giant. He moved to Kansas City, Missouri, home of H&R Block's headquarters, and still lives there with his wife.

The month had not gone well for Uber. But the worst days were yet to come.

ALSO IN MARCH 2017, Gabi Holzwarth was still trying to get over her ex-boyfriend, Travis Kalanick. She was working her new job at an automotive startup—her first full-time gig since she split with Kalanick—when she received a call from Emil Michael.

After dating Kalanick for the better part of three years, the two had split towards the end of 2016. The breakup had been difficult for both of them. Friends and co-workers believed they both genuinely cared for one another. As Kalanick skyrocketed to fame, the two became each other's support systems. When he wasn't working, Kalanick spent most of his time with Holzwarth.

But there was a more difficult side to their relationship. Later, as Holzwarth reflected on her time with Kalanick, and would later tell reporters, she realized he could sometimes be emotionally insensitive. He never yelled, but Kalanick knew how to be cruel and cutting, both in his bullying and put-downs to employees at work and to Holzwarth at home. At his request, Holzwarth would help organize events like Kalanick's birthday parties, flying models in to join them as scenery. Holzwarth went along with it at the time, but looking back she felt bad about herself, about how Kalanick treated her and the other women in his life.

"You go to an event and there's just a bunch of models they've flown in," she later said. "That's what they like to play with. That's pretty much it."

One event, however, stood out in her mind years later. In mid-2014, Holzwarth was on a business trip with Kalanick in Seoul. Uber had faced numerous problems with Seoul officials while trying to launch UberX in the region. One evening, a group that included Kalanick, Holzwarth, Emil Michael, and another female Uber employee joined a handful of South Korean Uber managers for a night of drinking. Things got wild, and the group ended up at what they thought was just a standard karaoke bar.

As the group from Uber stumbled in, another group of women who worked for the establishment sat in a circle in front of the bar's patrons, each with a different numbered "tag" attached to their miniskirts. The men were allowed to look over the women, sizing them up. Customers could then pick a number, and the woman would follow them to a separate room to sing karaoke together, or the woman would serve the patrons drinks. Sometimes, after a few rounds of karaoke, the women would go home with the customer.

At least some in the group quickly surmised that the numbered bar girls were escorts for hire. Holzwarth and the female Uber employee were uncomfortable, but went along with things so as not to make the situation awkward for the others. Four of the South Korean male Uber managers picked out women to join the group to sing karaoke. After a few minutes, the female Uber employee left, appearing visibly shaken. After a round or two of karaoke—Michael sang "Sweet Child O' Mine"—Holzwarth, Kalanick, and several others departed, leaving the local Uber managers with their escorts at the bar.

By all accounts, Kalanick, Michael, and others had done little more than sing karaoke in a public room at the bar and order drinks from the servers. Nevertheless, the outing could have landed everyone in deep trouble. Kalanick and Michael had appeared indifferent as their subordinates cavorted with sex workers. Months afterward, the female employee complained to human resources, and later told Kalanick she was uncomfortable about the situation. Little was done; human resources briefly raised the issue with the executives, but it seemed like everyone had chosen to forget the entire thing.

Holzwarth hadn't planned on saying anything about it to anyone, either. That is, until she got a telephone call from Emil Michael.

On March 1, Michael texted Holzwarth out of the blue. She hadn't spoken to him since before she and Kalanick had broken up late in 2016. He asked to call her, and she agreed.

They made small talk initially; Holzwarth was never close with Michael, though she often spent time with him and his girlfriend, since he was Kalanick's wingman. "Things have been really rough out here," Michael said. They both acknowledged the difficulties Uber had been going through. Then, Michael got down to business.

"Remember that night in Korea?" Michael asked. "Well, there are reporters digging around, trying to break the story. I just want to go over things with you," he continued. "We just went to a karaoke bar and that's all that happened, right?"

Holzwarth was getting upset. To her, Michael sounded like a thug, like some mafia consigliere, trying to tie up loose ends. He would tell others that he was simply trying to warn her that the incident might appear in the press.

"Can you just leave me out of it?" Holzwarth said. She was trying to move on with her life, and was having a difficult enough time doing it with her ex-boyfriend's face plastered across every website and news-paper. Now she had Michael elbowing his way back into her life, bul-lying her.

He wouldn't drop the issue, pressing her on the events. The kara-oke bar, he insisted, was the only thing they did that evening in South Korea. "Right?" he said. "That was it, right?"

Holzwarth started to cry. "I'm dealing with my own shit!" she sput-tered, in between sobs into her iPhone. "Just please, please leave me out of this!"

Eventually, Holzwarth agreed to keep quiet and stay away from reporters if they tried to call her. Michael feigned support at the end of the conversation, attempting his best impression of a concerned friend as she sniffled through tears on the other end of the line. "I hope every-thing's OK," Michael said.

"Thank you, take care of yourself," Holzwarth responded. They said their goodbyes and hung up.

After she got off the phone, Holzwarth burst into violent sobs. Everything about her failed relationship—the way Kalanick treated her, the way she felt like she debased herself by being with him—she would later tell others that it all came flooding back.

Soon after, she called Rachel Whetstone, Kalanick's head of com-munications, completely distraught. Holzwarth told Whetstone about the Korea incident, about Michael contacting her about it—she told her everything. Whetstone, aghast, apologized over and over to Holz-warth. Whetstone asked Holzwarth a few questions, tried to console her further, then the two hung up.

Whetstone congregated with other members of the executive lead-ership team at Uber—including Salle Yoo, the general counsel, Liane Hornsey, the chief of human resources, and Arianna Huffington, who was increasingly involved in managing damage control—to figure out how to handle the situation, and pray that it wouldn't leak. The entire group was furious with Emil Michael; what he did was stupid, reckless, like something out of *The Godfather*.

Michael must have realized his error. The day after the executives met, Holzwarth started receiving texts from Michael, clearly trying to cover his ass. He texted:

I am so sorry for being cold the other day on the phone. I was super panicked. I should have asked about you and how you were doing. I care about you and consider you a friend. We shared some amazing times together. I hope you believe me. I would love to see you at some point too.

Michael tried dispatching other women to help him. He had his girlfriend text Holzwarth. Later, another female Uber employee texted Holzwarth to ask how she was doing. And yet another woman—a friend of Michael's—contacted Holzwarth later in the day to invite her to a birthday party.

Holzwarth was sad, confused, scared. Most of all she was angry— angry at Emil Michael for putting her in this position. She didn't want to be silenced, she would later tell a reporter. Michael was a bully, plain and simple, and thought he could bully her the same way Uber bullied those who got in the company's way.

Holzwarth thought back to one of her earliest gigs: She had played violin at the launch party of *The Information*, a tech journalism startup that debuted a few years back. The site had covered Uber aggressively in the past, and she felt close to some of the people there.

She still had the reporter's cell phone number.

Chapter 26

FATAL ERRORS

STARTING WITH #DELETEUBER IN JANUARY 2017, AND CONTINUING through the Fowler blog post, the *Bloomberg* video, the Trump council fiasco, and the Greyball revelations, Kalanick's reputation had plummeted.

The Korea story reinforced what the public already suspected: that Travis often turned a blind eye towards Uber's toxic culture, which went to the very top of the company. Executives Ed Baker and Amit Singhal were toast. Jeff Jones had exited in a blaze of bad press. Things looked dire for Emil Michael, his right hand and confidante. Looming above all of this was the Holder report, which had yet to be presented to the board of directors. Around Uber HQ "the Holder Report" took on a mythic air, like a corporate sword of Damocles, ready to fall at any moment. There were eight years of executive behavior to investigate. Who knew what else Holder would dig up?

Worse for Kalanick, his board of directors was putting new pressure on him to fire Anthony Levandowski. By late March, the dirt that had come out on Levandowski turned him into a major liability.

In December 2016, Levandowski had launched a self-driving-car test program in San Francisco without a permit and in direct defiance of the California transit authorities, who called the maneuver illegal. Almost immediately, the test program went awry. One of Uber's test cars blew through a red light in broad daylight, an event captured on the dashboard camera of a nearby motorist. As the clip went viral online, Uber issued a statement: "This incident was due to human error. This vehicle was not part of the pilot and was not carrying customers. The driver involved has been suspended while we continue to investigate. This is why we believe so much in making the roads safer by building self-driving Ubers."

But three months later, the *New York Times* published a story, citing internal documents, that claimed Uber's narrative was false; it was the self-driving software that missed the red light, not the driver. Uber had lied to reporters, on the record, about an illegal program it was running in its hometown.

Levandowski made it look like Uber had something to hide in other ways. He refused to cooperate during the civil case filed by Waymo; towards the end of March, he pleaded his Fifth Amendment right not to incriminate himself should the government pursue a separate criminal prosecution.

Kalanick knew he needed to fire Levandowski, but he couldn't bring himself to do it. Levandowski, like Kalanick, was a born charmer, full of charisma and showmanship. Their long walks along the Embarcadero together, the ongoing jam sessions, the "brother from another mother" mentality—all of it had enraptured Kalanick. The men shared a dream of a fully autonomous future, one where one software program piloting a fleet of automated vehicles could do the work of millions of drivers. But nevertheless, Kalanick had to act. To improve the optics, Kalanick and Levandowski dreamed up a series of internal demotions, claiming Levandowski would work on self-driving cars but remove himself from discussions around lidar, the key technology at the heart of Waymo's lawsuit. The ridiculous gymnastics fooled no one; Levandowski was still running the show.

Kalanick had always struggled to fire people face-to-face, even when it was painfully clear—like in Levandowski's case—that only Kalanick could swing the axe. Ultimately, it took pressure from Bill Gurley and David Bonderman on the board to force Kalanick's hand. Towards the end of the spring of 2017, Anthony Levandowski was unceremoniously terminated. Kalanick was sad to lose such a close ally and friend. The rest of the company, from the top down, was not.

Weeks later, Judge William Alsup, who presided over Waymo's civil case against Uber and Levandowski, would refer the case to the US

Attorney's Office in San Francisco for "investigation of possible theft of trade secrets." If they decided to pursue the matter, it could mean bringing criminal charges against Levandowski, raising the possibility he might even spend real, hard time in prison.

"Uber regrets ever bringing Anthony Levandowski on board," one of Uber's lawyers later told the jury as the Waymo lawsuit went to trial. "All Uber has to show for Anthony Levandowski is this lawsuit."

AS EACH DISASTER HIT in the first quarter of 2017, the company's communications team scrambled to fix the damage. Some described it as walking through a field of landmines; every step forward brought them closer to the next explosion.

One tactic was to demonstrate transparency. Days after the Korea incident surfaced, Uber unveiled its first-ever diversity report, detailing the gender and ethnic breakdown of the company's workforce. To discuss the findings, Liane Hornsey, the new head of HR who had started just weeks before Uber's February scandals broke, expressed contrition in a press interview. She tried to soften Kalanick's hard edges, and to acknowledge that while Uber had more work to do it was indeed up to the task of creating lasting internal change.

Hornsey's interview seemed to dampen public outcry for the moment. Arianna Huffington also spoke out, saying the company would no longer hire "brilliant jerks." Huffington soon began to assume a larger leadership role in the brand rehab campaign. She sensed a power vacuum, a crisis of leadership in the corporation and a personal crisis—of Kalanick's ability to trust the people around him. She owned shares in the company, yes, but there was a special status and power being the person to help right the ship in a time of crisis—especially if that ship was a $69-billion Titanic like Uber.

THE DIVERSITY REPORT and Hornsey's interview landed in late March, quelling public ire for just a little over two weeks. Then another bombshell hit.

By this time, the competition between Uber and Lyft had become a famous rivalry. Kalanick didn't just want to beat Lyft; he wanted to bankrupt them. On April 13, 2017, it became clear how ruthless Uber had become. A report in the tech press unearthed the existence of Uber's program "Hell," the one that illicitly repurposed iPhone technology to target Lyft drivers and lure them to Uber. But that was just the beginning.

Hell had been created by a group called the "competitive intelligence" team—COIN for short—established to keep tabs on competitors. Uber engineers set up special computer servers which were unconnected to the company's primary infrastructure and kept "unattributable" to Uber. On those servers, Uber stored, processed and analyzed information Uber engineers had "scraped" or harvested from Lyft's apps, websites, and code repositories.

The team kept tabs on overseas competitors like Ola in India and DiDi in China. Another entity, the Strategic Services Group, the SSG for short, employed the most clandestine tactics of the bunch. It was made up of ex-CIA, Secret Service, and FBI operatives, and hired subcontractors on special anonymous contracts with Uber so that their names couldn't be traced back to the company. This outfit of black-hat spies engaged in a wide range of activities, some of which eventually spun out of Uber's control.

Led by Nick Gicinto, SSG operatives would carry out espionage and counterintelligence missions using virtual private networks, cheap laptops, and wireless hotspots paid for in cash. Undercover operations could include impersonating Uber drivers to gain access to closed WhatsApp group chats, hoping to gather intelligence on whether drivers were organizing or planning to strike against Uber.

They conducted physical surveillance, photographing and tracking competitors at DiDi and Lyft, and monitoring high-profile political figures, lawmakers, and police in contentious cities. They followed people on foot and in cars, tracking their digital activities and movements, and even took photographs of officials in public places. They impersonated Lyft drivers or riders to gain intelligence on the competing company. SSG operatives recorded private conversations between opponents at

DiDi and at Grab, their Southeast Asian competitor. One Lyft executive grew so paranoid about being followed by Uber that he walked out onto his porch, lifted both middle fingers in the air and waved them around, sending a message to the spies he was absolutely sure were watching.

Internal communications within SSG were carried out over an enterprise version of an app called Wickr. Because of its architecture, Wickr end-to-end encrypted every message, meaning only the sender and recipient would be able to read them. All messages were automatically deleted after a certain period of time, undermining any future legal discovery. Craig Clark and Sullivan, both licensed lawyers, would often designate documents as attorney-client privileged, another safeguard against potential legal threats.

The budgets for these black ops departments were obfuscated; Kalanick had purview over them. With nearly unlimited resources, Kalanick and other members of his A-Team could dispatch SSG operatives to go on covert missions—"some real spycraft shit," as one member of the team described it—and gather intelligence against those Kalanick perceived as threats. The mission to photograph Jean Liu, the DiDi president, at the Code conference was an undertaking by the SSG. It was unclear how much of this intelligence was actionable or even valuable. Nevertheless, Kalanick okayed budgets that spun into the tens of millions for surveillance activity, global operations, and information collection.

Kalanick wanted to know his competitors' every move. He was fighting a war of inches, across multiple countries, and used the SSG to gain intelligence about the other side. But Kalanick's motives went beyond utility. Photographing Jean Liu at the exact moment she learned of Uber's $3.5-billion Saudi investment—that was revenge for the pain DiDi had caused him in China. People close to Kalanick said DiDi's infiltration of his own ranks had changed something in the Uber CEO. After China, he harbored a constant, creeping suspicion that others were trying to blind him or trick him. Kalanick believed his spies could gather the data he needed to make sense of the ongoing fights.

Joe Sullivan, Kalanick's security chief, didn't see anything wrong with the practice. As he and his deputies, Mat Henley and Craig Clark, would later tell investigators, Uber's activities hardly differed from information-gathering that *all* companies engage in. It was called market research. Buying intelligence from third-party firms to gain an edge was normal. Anyone who criticized Uber for running a slick spy unit should have seen things before Sullivan had arrived, when Uber's systems were in utter disarray, every employee had access to "Heaven," thieves defrauded Uber's incentive system at will, and drivers were literally being murdered in their vehicles. When Sullivan assigned SSG operatives in the field in South America, India, and other locations, he did it to save lives, he said. And his efforts produced results quickly; Sullivan's new law enforcement outreach division helped police investigate threats against Uber drivers, and fraud had fallen by more than thirty-two basis points, an enormous drop.

Still, the SSG and COIN made many people uncomfortable. Employees were unnerved by mass deletion of internal emails, group chats, and company data, carried out under an internal initiative to "eliminate data waste" throughout all levels of the company. Internally, many believed executives wanted to cover Uber's tracks, anticipating a subpoena for some unknown future court case.

There were also the bribery problems in some Asian markets. Local employees considered bribery a necessary evil, a cost of doing business for an American company operating on foreign soil.

The never-before-reported details of a case in Indonesia, for example, would grow into an enormous problem. As Uber set up shop to compete with Grab in Indonesia, Uber would open "green light hubs," which were makeshift checkpoints for drivers in the area to receive vehicle inspections, register complaints with district managers, and other activities. The problem was that the hubs were set up in suburban districts zoned for residential use only. Almost overnight, the green light hubs began attracting hundreds of drivers, which clogged the suburban streets and angered the locals. When the police found out, they threatened to shut Uber's hubs down.

Instead of moving the company's hubs, local Uber managers decided to pay off the cops. Every time a police officer would show up, an

Uber manager would fork over a cash bribe—usually around 500,000 rupiah, around the equivalent of $35 USD, and the officer would leave. Unsurprisingly, the police became regular visitors.

Uber employees were known to take money from the petty cash bin to pay off bribes, or forge receipts in the amount of the bribe and enter them into the expense account management system for reimbursement, behavior that, as of this writing, the Department of Justice is still investigating as a potential violation of the Foreign Corrupt Practices Act.

Problems like this began to emerge throughout the company that spring as Eric Holder's law firm, Covington & Burling, conducted hundreds of interviews with employees in preparing their report. Rachel Whetstone had warned Kalanick that bringing in an outsider was the fastest way to lose control. Now her warning seemed prophetic.

Kalanick was vulnerable, but his colleagues weren't sure if he truly appreciated the gravity of the situation. Once investors found out the depths to which employees had carried out murky, even potentially criminal behavior, it would surely impact the company's valuation. In just three months, Uber had gone from the world's greatest investment to a $70 billion time bomb.

JUST WEEKS AFTER firing his close partner, Anthony Levandowski, Travis Kalanick was forced to fire another old friend. Eric Alexander was supposed to be Travis Kalanick's fixer. Instead, he became a wrecking ball.

If Emil Michael was Kalanick's number two, then Alexander was number three. That didn't mean he was third in line to the CEO seat; Alexander's official title was president of business in APAC, or the Asia Pacific region, responsible for maintaining relationships across all of Asia. Alexander was the guy who knew how to connect you to other guys. He was a valuable asset in the knock-down, drag-out Asian-market transportation wars, a man who could fix things that were broken.

Over time, he became more than that—he became a friend. When Kalanick and Michael would head out for a night on the town in Korea or Southeast Asia, Alexander would inevitably join them. Like his boss,

Alexander devoted his life to Uber, spending hours on planes every week, traveling from one country to the next.

When Uber's rape controversy exploded in India in December 2014, Eric Alexander was one of Kalanick's first calls. Alexander parachuted into the region and immediately did effective damage control with Indian politicians and the press. Eventually, Uber was able to settle a lawsuit the woman had initiated against the company. And though Uber's operations were temporarily suspended in the Delhi region, Uber was back up and running in India by early 2015. The company seemed to have come through the worst of it.

In the summer of 2017, however, the tech press discovered that Alexander, as a part of the investigation into the India rape case, had been given the victim's personal, private medical files through a law firm that had obtained them; the records detailed her examination by doctors hours after her sexual assault. Alexander brought them back to the United States with him. Kalanick and other executives were briefed on the investigation by the legal team as it was unfolding.

In the immediate aftermath of the attack, a theory was floated that the rape possibly may not have taken place at all and was in fact part of a plot against Uber perpetrated by executives at Ola, Uber's major Indian ride-hailing competitor. According to a review conducted on the driver's and victim's accounts, the driver held multiple Uber accounts and the victim's account was used by several different people. Investigators were having trouble reconciling the different identities attached to each account. They raised the possibility that they had been created to stage the attack.

There was another sticking point. According to the medical file, the young woman's hymen was still intact—a fact that stuck in Kalanick's head as investigators tried to verify the claims. Kalanick would occasionally raise the fact among colleagues.

In April 2017, the *New York Times* had approached Uber about Kalanick's comments about the India rape incident, but the executive denied that he ever had doubts about the truth of her claims to members of their communications team. Uber's communications team members then swatted the story down as false, and the story never ran.

By the summer, *Recode*, the technology news site, reported discovering that Alexander had carried around the files with him; *Recode* said it was planning to publish the story the next day. It didn't help that Alexander was also present during the infamous South Korean karaoke bar incident with Kalanick and Michael. Once the public found out, it would be devastating.

Members of Kalanick's executive leadership team were disgusted, interpreting his raising the details of the medical file as a way to cast doubt on the woman's claims. Aghast, at least two people flung back in Kalanick's face the possibility that the victim could have been anally raped during her sexual assault.

If there were any lingering doubts held by the executives, they didn't get in the way of the company assisting law enforcement in India. Soon after the attack, Alexander shared GPS records with Indian officials, which showed that the account went offline in proximity to the attack's location and around the time it had been reported to take place. Alexander also later testified in the driver's criminal trial.

Nevertheless, carrying around the private medical files of the victim was questionable, and Kalanick knew he had to fire his friend before the *Recode* story ran. So he called Alexander, told him the situation, and apologized for what he had to do. By June 7, it was over for Eric Alexander. The executive was gone, a last-ditch effort to save face. The effort failed. After the story dropped, employees were beyond outraged.

Members of the executive leadership team had reached a crossroads. The events up until this point were bad enough, but apparent rape denialism was going too far. By then, the head of communications, Rachel Whetstone, was gone. After she had threatened to quit several times, and Kalanick had convinced her to stay, the CEO had finally had enough. Kalanick accepted her resignation in April. About a half dozen of the remaining executives drafted a letter addressed to Uber's board of directors. Uber, they wrote, desperately needed an independent chairman to counter Kalanick's all-encompassing power. They pleaded with the board for the termination of Emil Michael, who leaders saw as an accelerant to Kalanick's worst impulses.

Most of all, they wished Kalanick would take a leave of absence. For them to even begin repairing the company's reputation they would need resources and commitment from the board, and Travis's continued presence made their work impossible.

TRAVIS KALANICK WAS IN New York at the end of May when he got the call. His parents were involved in a freak boating accident. Kalanick needed to fly to Fresno immediately.

As he contracted a private plane from Manhattan to Fresno, his mind was on his parents, the only two people left in his life that he could truly count on. Now, five months into the worst year of his life, his father, Donald, was in critical condition, badly hurt during the accident. His mother fared far worse.

In the weeks before the crash, Travis was considering joining his parents on the outing to Pine Flat Lake for Memorial Day weekend. The family spent summers there when Travis was young, playing on a dust-covered campground and putting in hours repairing his father's broken-down motorboat. "The last note I got from her was a gorgeous picture of the lake as you approach it from the campground, still cajoling me to cancel my meetings on the East Coast and join them," Kalanick wrote in a post to his Facebook page, just days after the incident. "But I didn't."

During those summer trips as a kid, Travis and his family would motor north twenty miles to the source of the lake at Kings River, a trip his parents repeated together that Friday at the end of May. As they approached the riverhead, Bonnie said she wanted to take the wheel, "a switch I've seen them do dozens of times before," Kalanick wrote. The family dog got in the way at the last moment, and the wheel turned sharply, steering the boat directly toward a cluster of rocks. Before Donald could swing the wheel back in time, the boat struck the rocks, pitching him over the edge of the boat and into the cold lake water. Bonnie was still in the boat when it hit.

With five fractured ribs, a cracked vertebra, a broken leg, and one collapsed lung, Donald Kalanick swam back to the sinking boat to rescue his wife before she was carried under the water with the sinking craft. He wrapped his wife in life jackets, swimming for nearly two hours to pull both of them to shore. Once they reached the beach, Donald tried performing mouth-to-mouth resuscitation on his wife, but to no avail. Bonnie Kalanick was killed in the crash, having died immediately upon impact. Eventually a fisherman found the two of them, carting them to safety.

Travis was devastated. Bonnie had been the person in the world to whom Travis felt closest. His parents had supported him through everything, even the lean years when he was a loser living in his parents' house after college, too poor to hack it on his own in startup world. Bonnie doted on him whenever they were together. Now she was gone.

For a moment, the tech world stopped pummeling Travis as he grieved his mother. The CEO sat by his ailing father's bedside with his brother, Cory, as the two waited to see if Donald would pull through. As the news reached the public, emails began pouring in, offering words of consolation. Even Tim Cook—someone who Travis had battled with in years past—sent him a note expressing his condolences.

Kalanick, not knowing where to turn, called Arianna Huffington. She was on the next flight out to Fresno. As the news spread, others reached out to Kalanick to see if they could help. Angie You, an ex-girlfriend who remained a close friend, asked if he wanted her to come out and join him as he sat by his father's bedside; she knew Kalanick's parents well from their years of dating. Huffington would later tell friends she expressed genuine care and concern for Kalanick during the darkest moment of his life. Onlookers said she took on a maternal role for Kalanick, caring for him in the absence of his actual mother. Yet some Uber executives close to the incident couldn't help but feel that Huffington was controlling the situation to get closer to Kalanick.

WHEN KALANICK WASN'T AT his father's bedside, he was back at the Holiday Inn across the street from the hospital, trying to salvage what was left of his career. He had rented out a conference room in the hotel—circumstances much reduced from his usual Four Seasons suites—and used it as a makeshift war room away from Uber headquarters. The idea, to take his mind off his parents, was to write a letter to his staff expressing true contrition, something that proved he was listening to their grievances and that he wanted to change. He paced as he dictated his thoughts aloud to himself, alternating between the conference room and a huddled position in the dim, carpeted hallway of the hotel, working on the letter. Following Arianna's advice, he was trying to strike the right tone, somewhere between humble, apologetic, and inspiring—something befitting of a leader willing and able to make a comeback and steer the company through a difficult time.

After many drafts, they settled on a version that, to them, sounded like Kalanick was taking full responsibility for his actions. The letter said things that he believed his staff would appreciate hearing from him. It was an apology, the first he had ever uttered in writing about his shortcomings. He realized he should have owned up to his mistakes long ago, and thought the letter might even save his job.

> *Team,*
>
> *Over the last seven years, our company has grown a lot—but it hasn't grown up.*
>
> *I've been an entrepreneur my whole life. Most of the time, I've been on the brink of imminent failure and bankruptcy. I was never focused on building thriving organizations. I was mostly just struggling to survive.*
>
> *When Uber took off, for the first time in my life I was leading an organization that wasn't on the brink of failure each day. In just the last three and a half years, our service and our company has grown at an unprecedented rate . . .*
>
> *Growth is something to celebrate, but without the appropri-*

ate checks and balances can lead to serious mistakes. At scale, our mistakes have a much greater impact—on our teams, customers and the communities we serve. That's why small company approaches must change when you scale. I succeeded by acting small, but failed in being bigger. . . .

. . . Over the last few days, as I'm sure you can imagine, family has been on my mind a lot.

My mother always encouraged me to stay as connected as possible with the wonderful, talented, inspiring people that make Uber everything that it is. She always put people first, and it's time I live her legacy. My dad taught me that actions speak louder than words, and to lead by example. So I felt it was important to be very candid here about the challenges we face at Uber—but also how we're taking action without delay to make things right.

I hope you will join with me in building an even better Uber.

Hunched over his laptop in the hallway of the Holiday Inn, Kalanick looked at the letter, which included the line, "sometimes it's more important to show you care than to prove you're right." He was tired, having gone days without any real sleep, but felt like this was solid work. It was something he could deliver in the next few weeks, something to re-instill faith in his leadership after the Holder report was finished.

At that moment, Kalanick would have no way of knowing he would never deliver the letter to his employees.

PART V

Chapter 27

THE HOLDER
REPORT

OVER THE FOUR AND A HALF MONTHS SINCE TRAVIS HAD HIRED ERIC
Holder and his partners at Covington & Burling to investigate Uber, the forthcoming report had been elevated to mythical status among employees and outsiders alike. Some saw it as a Necronomicon, an almost occult document filled with the company's dark secrets. Others saw it as a chance to clean house, to acknowledge and admit wrongdoing, and begin reframing the debate. Either way, on Tuesday June 13, at Uber's internal all-hands employee meeting, the company planned on presenting the recommendations of the Holder report.

Everyone knew the report would contain new bad news. The question was, how much? Executives at Uber decided to do some damage control in the run-up to the big day. At an internal meeting on June 6, Uber announced it had already fired twenty people as a result of the findings. That group included Josh Mohrer, the general manager from New York who had spied on a reporter and toyed with staff. Mohrer was given a soft landing; he said he was leaving to become a managing partner at Tusk Ventures,[*] a firm founded by a political operative and early Uber supporter and advisor. Others had similarly comfortable exits. Aside from the firings, the company announced that thirty-

---

[*] Bradley Tusk, founder of Tusk Ventures, was an early advisor and aide to Kalanick as he attempted to conquer Manhattan. Tusk, whose fee for political consulting at the time was in the tens of thousands, opted instead to take shares in Uber as a form of payment. Those shares are said to be worth more than $100 million today.

one employees were in counseling or additional training, while seven employees had received written warnings for their behavior.

On Sunday, June 11, Uber's board of directors met at Covington & Burling's offices in downtown Los Angeles to discuss the findings of the report, and the firm's recommendations. Each of the seven board members walking into the firm's offices that afternoon had a different agenda. Bill Gurley, the venture capitalist, needed the drama to end. David Bonderman, the private equity magnate, wanted Uber to pull out of this awful press cycle. Both men wanted Uber to make its way to an initial public offering so that their firms could reap billions in returns on their initial investments.

Garrett Camp, the man who had invented the company in the first place, had been checked out for years, an absentee founder who was happy to allow Kalanick to take the reins. After all, Kalanick had made Camp very, *very* rich by now, and was only going to make him richer. Ryan Graves, Uber's first, brief-tenured CEO and operations chief emeritus, felt loyal to Kalanick. Graves, too, believed that the press was unfairly targeting Uber and in turn Kalanick. Graves didn't think Kalanick should be ousted, but Graves *did* believe that some temporary time away from Uber would serve Kalanick—and the company—well.

Yasir al-Rumayyan, the representative of the Saudi Public Investment Fund, had sided with Kalanick from the very beginning. The Saudis were looking to diversify the royal family's holdings and eventually move away from being an oil dynasty. And Kalanick was the one who brought the Saudis into the Uber fold. Al-Rumayyan liked Kalanick; there was no reason, he believed, for him to leave the company. He would follow Kalanick's lead.

Arianna Huffington, the independent board member, was far from impartial. Even as the Holder investigation was under way, Huffington was expressing public shows of support for Kalanick. "He definitely has my confidence, he has the board's confidence," Huffington said of Kalanick at a conference in March. That made other board members and executives nervous; it was clear to those at the top of the company that Huffington was on "Team Travis," and would vote to keep Kalanick at the helm of the company. The two had grown close over the years since she joined Uber's board. Importantly, she also knew that Kalanick still held majority voting power as long as his allies Camp, Graves, and al-Rumayyan were with him. While Huffington professed independence in public, everyone inside Uber knew her allegiance.

And finally there was Kalanick. He hoped the delivery of the Holder report would bring Uber a much-needed reprieve from public scrutiny. No matter what the report recommended, he had no intention of ever leaving his position as CEO.

To forestall unintended leaks, the group settled on a secure method. Each board member was required to read a printed copy in the Covington & Burling office, leaving all electronic devices outside the room. No digital copies would exist outside of the Covington & Burling office hard drives.

Those who read the report in its entirety were shocked. It was hundreds of pages long, a winding, repetitive list of infractions that had occurred across Uber's hundreds of global offices, including sexual assault and physical violence. The company had numerous outstanding lawsuits against it, and would likely face many more. After Ryan Graves read the report, he felt he needed to vomit.

In a marathon meeting that Sunday, June 11, Uber's seven-person board met to discuss the pages they had just read. No one outside of the room was ever going to see the investigation findings, but still there was great fear of leaks. Up to that point, reporters had found willing sources at every level of the company. It seemed inevitable that the Holder report would also hit the press. Graves asked, from the beginning, for everyone to keep the meeting's contents between themselves. Then he started begging. "Please. Please don't talk to the press."

With the report came a series of recommendations from Holder and partner Tammy Albarrán. The final list would span a dozen pages and include a number of serious structural alterations, and different versions of the recommendations would later be distributed. But Holder and Albarrán had put their most important action items at the top:

Travis Kalanick needed to take a leave of absence from his own company, relinquish his control of Uber's business, and hire a proper chief operating officer to help him do so. The second order of business, the report recommended, was to fire Emil Michael. And finally, the company badly needed to appoint an independent board chairperson, someone completely unconnected to Uber, to give perspective and balance to executive deliberations.

The room was split. Gurley and Bonderman worried Holder's recommendations might not have gone far enough, since they stopped short of pushing Kalanick out for good. Some members of the executive leadership team were convinced Huffington had leaned on Holder and Tammy Albarrán, Holder's partner, to convince them not to recommend Kalanick's termination in the final report. But ultimately, Gurley and Bonderman were still satisfied with the proposed reforms; it was time to clean up the company, and those changes started from the top.

Though they didn't want him axed, Huffington, Camp, Graves, and al-Rumayyan all believed Kalanick needed some time away. Public scrutiny of Uber was growing too intense; the press wanted blood, Kalanick leaving the spotlight—if only momentarily—would alleviate the pressure.

Travis already knew he was going to be asked to leave, the question was whether he could ever return, but the decision to fire Michael was painful for Kalanick. The CEO had watched the world turn against his friend in just six months, but Michael had stood by him to the end, the only person connected to Uber he felt he could trust. Even his ex-girlfriend had betrayed him at that point. But Kalanick knew the board had to act in unison to make Uber's turnaround appear legitimate and earnest. By the end of the day, all seven board members voted unanimously to accept all of Holder's recommendations. Though no one on the board—perhaps not even Travis himself—knew what Kalanick was going to do on Tuesday, when they were expected to present the report to the entire company.

Michael got the call that evening. Many employees who worked for Michael still supported him. Even Michael's biggest detractors admitted he was a talented executive, whose work ethic and ability to forge relationships and close deals was equaled by few.

"Uber has a long way to go to achieve all that it can," Michael wrote in a letter to his team, full of self-congratulatory praise. "I am looking forward to seeing what you accomplish in the years ahead." Michael also dialed in, unannounced and uninvited, to one final conference call with his former business team employees. Michael was grief-stricken. He had devoted the past four years of his life to Uber and had tried to be a mitigating influence on Kalanick. Instead, he had let Kalanick bring out the worst in him. As Michael commandeered the conference call, he again told employees that he was proud to have helped build a world-changing company.

And that was the end of Emil Michael's career at Uber.

EVERY TUESDAY MORNING, each Uber employee earmarks the 10:00 a.m. Pacific time slot on their calendar for the company's all-hands meeting. Workers from across the world dial in to Uber's video conference line to watch company leaders give an update on the state of affairs, from vice presidents to board members to Travis Kalanick himself. As employees filed into Uber's spacious conference area, they saw some of the executive leadership team and a handful of board members, ready to give a presentation.

Uber employees were anxious. The negative articles and upheaval over the past six months had impacted their jobs and affected their personal lives. Throughout the spring, Arianna Huffington had appeared on television—CNN, CNBC, and other outlets—to discuss the report. On those shows, Huffington would claim the report was coming in a week or two weeks, in hopes of holding off the press. Bonnie Kalanick's unexpected death in a boating accident had further delayed the presentation. But now Arianna stood just offstage, waiting for everyone to take their seats.

On the morning in question, Kalanick was nowhere to be seen. As it would turn out, he wasn't in the building at all. Over the weekend, a

news story appeared saying that Kalanick may take a leave of absence
from the company. But even the topmost executives had no idea what
Kalanick was going to do that day. As employees dialed in to the all-
hands, Kalanick was offsite typing furiously at his keyboard, deciding
what to say to them. At 9:59 a.m., all Uber inboxes received a note from
Kalanick, just as Arianna Huffington was taking the stage, flanked by
fellow board members Bill Gurley and David Bonderman.

"Good morning everyone," Huffington said into the microphone.
A few people in the audience chirped a half-hearted "good morning"
back at Huffington, the morning's apparent emcee. "Before we begin, I
want to address the elephant in the room. Where is Travis?" Huffington
asked, rhetorically.

The answer was in Kalanick's cryptic email, which some employees
had opened just as Huffington began her spiel. It read:

Team,
For the last eight years my life has always been about Uber. Recent
events have brought home for me that people are more important
than work, and that I need to take some time off of the day-to-day
to grieve my mother, whom I buried on Friday, to reflect, to work on
myself, and to focus on building out a world-class leadership team.

The ultimate responsibility, for where we've gotten and how
we've gotten here rests on my shoulders. There is of course much
to be proud of but there is much to improve. For Uber 2.0 to suc-
ceed there is nothing more important than dedicating my time to
building out the leadership team. But if we are going to work on
Uber 2.0, I also need to work on Travis 2.0 to become the leader
that this company needs and that you deserve.

During this interim period, the leadership team, my directs,
will be running the company. I will be available as needed for the
most strategic decisions, but I will be empowering them to be bold
and decisive in order to move the company forward swiftly.

It's hard to put a timeline on this—it may be shorter or longer
than we might expect. Tragically losing a loved one has been diffi-

*cult for me and I need to properly say my goodbyes. The incredi-
ble outpouring of heartfelt notes and condolences from all of you
have kept me strong but almost universally they have ended with
"How can I help?." My answer is simple. Do your life's work in
service to our mission. That gives me time with family. Put people
first, that is my mom's legacy. And make Uber 2.0 real so that the
world can see the inspired work all of you do, and the inspiring
people that make Uber great.
See you soon,
Travis*

So there it was. Kalanick was going to step away from the company.[*]
On one hand, it was difficult to imagine an Uber without Travis Kalanick at the helm. The man lived and breathed the business. But on the
other, employees now recognized how toxic a symbol he had become.

Kalanick's last-minute additions to his letter terrified some of the
executive leadership team. The phrases "see you soon" and the suggestion he would be away "shorter or longer" than people anticipated did
not inspire comfort. Still, they were relieved that he was willing to step
aside for *some* period of time, whatever that ended up being.

Huffington continued. The recommendations stemming from the
report were the result of an exhaustive, months-long process, she said.
Holder and Albarrán interviewed more than two hundred people personally, while also fielding tips and holding anonymous conversations
with hundreds of other current and former employees through an
anonymous hotline. The firm reviewed more than three million documents, turning the company inside out. Huffington did not note that
the undertaking had cost Uber tens of millions of dollars; management
considered the money well spent if it allowed them to purge Uber of
its problems.

---

[*] Nine months later, Alex Trebek asked *Jeopardy!* contestants the name of the Uber CEO
who "took a leave of absence to work on Travis 2.0." Kalanick tweeted a photo of the question on TV, adding the hashtag: "#bucketlist."

"The recommendations are going to be posted on the Uber news site momentarily," Huffington said, as employees began scanning Kalanick's email. Uber had formed a special committee to oversee the report, composed of Huffington, Gurley, and Bonderman. All three of them adopted the recommendations before passing them on to the rest of the seven board members. "On Sunday, at the board meeting—which certainly was the longest board meeting I've ever been at—the full board adopted them unanimously," Huffington said.

As Huffington spoke, the recommendations from the report were posted online. At the meeting, employees breathed a collective sigh of relief. At the very top of the report was the news they'd just heard: Kalanick's role would be diminished, and he would be subject to far more oversight. No one outside the board would see the report's raw text. It read like a repository for every grievance and complaint employees had filed against Uber. After months of waiting, some in the audience felt Uber owed it to employees to publish the report itself, if only to come clean entirely. Huffington noted it would be improper, citing privacy and legal issues.

Speaking to the point about diversity, Huffington continued: "I just want to say that for me, a personal and stated goal since I joined was to increase the diversity of the board, much as I do love my white male colleagues," she said. "Today, I'm delighted to announce the addition of Wan-Ling Martello to Uber's board," Huffington said, accompanied by faint applause. Martello, a career executive in the food industry who spent time at Kraft, Borden, and most recently as executive vice president of Nestlé in Asia and Africa, was supposed to be an independent director, a voice of reason who could vote to serve the best interests of the company and its shareholders.

"She's someone I know you'll love to get to know," Huffington continued. Many in the audience didn't know who Martello was or what to think, but the addition of another female board member was probably good. Martello would tell others that she intended to be "Switzerland" on the board between constantly warring factions, but she entered the company at a time when acrimony was at its peak. Huffington went on to highlight the diversity she would bring to a board that was still very white and very male. "There's a lot of data that shows when there's one woman on the board, it is much more likely there will be another on the board," Huffington said.

From her side, David Bonderman piped up. Until that moment, he and Gurley had been quiet, letting Huffington present her section of the report. But a thought popped into his head.

"I'll tell you what it shows," Bonderman said. "It's that it's much likelier to be more talking on the board."

The room froze. Had one of Uber's board members just made a sexist comment about women talking too much?

The audience was stunned; Bonderman, a seventy-five-year-old white billionaire hedge funder from Fort Worth, Texas, was dunking on women in the middle of the board's company-wide presentation about changing Uber's misogynist culture. Bill Gurley, who was standing behind Bonderman, shook his head.

Huffington tried to recover, playing off the moment and moving on. "Oh, come on David," she said, chuckling. "Don't worry everybody, David will have a lot of talking to do as well." The room was dead silent.

"So, the final category," Huffington announced, trying to move away from the awkwardness of the moment. "The final category is culture."

Someone in the audience laughed aloud.

FOR MONTHS, Bonderman had been driving Kalanick crazy.

Bonderman, a career financier and no-nonsense businessman, had sat on the boards of plenty of companies during his time as co-owner of Texas Pacific Group, the private equity firm he had helmed for a quarter century. Born an Angeleno but now a Texan, Bonderman had moved to Fort Worth and made his fortune working for the enormously wealthy Bass family, who controlled substantial oil and gas concerns in the Dallas–Fort Worth metroplex. It was there Bonderman met his partner, Jim Coulter, working for Robert Bass. Coulter and Bonderman struck out on their own to found TPG in 1992. Coulter was the conservative, sensible partner; Bonderman loved to take risks. When TPG invested in Uber, its growth trajectory made it a sure bet. But with Kalanick at the helm, Bonderman would have less control over the company than he enjoyed on other boards.

To most, Bonderman was a shuffling giant, tall, white and unkempt in ill-fitting suits. He did *not* look like the 239th richest man in the world (which he was), and certainly didn't appear in the ostentatious cowboy garb of other Texan energy tycoons. Balding, with a gruff, high voice and utter contempt for small talk, Bonderman had no qualms about speaking up against Kalanick during boardroom meetings. He agitated for change around Kalanick's Ahab-esque pursuit of the Chinese market, chafed at firing of Brent Callinicos, Uber's first and only chief financial officer. And Bonderman was furious that it had taken so long for Kalanick to fire Anthony Levandowski, someone who was a clear liability to the entire company.

Bonderman didn't care about Kalanick's feelings. He didn't care about the feelings of the legion of bros at the company. What he cared about was his money, and that Uber became as successful as everyone hoped it would be. TPG had billions riding on it.

So when Travis Kalanick saw his opening, he took it. Kalanick was tired of the older man's prodding and complaining. After Bonderman's slip-up onstage, Kalanick started working the phones. Even before the presentation had ended that Tuesday morning, Kalanick had text messages out to board members and others on the executive leadership team.

Kalanick's message was clear: Bonderman needed to go.

AS ARIANNA HUFFINGTON'S AUDIENCE shifted in their seats, unsure how to handle Bonderman's comment, the presentation continued. Huffington announced a few symbolic changes. For example, workers wouldn't have to wait to eat dinner at the office until eight o'clock at night anymore, a practice long espoused by Kalanick in order to keep employees in the office for longer workdays. And the famed "War Room" in the middle of the office was given a new name, courtesy of Huffington herself: "The Peace Room." Though this last change seemed cheesy, the room seemed to accept it.

It was Gurley's turn to take the stage.

"I wanted to make a few comments just to put all of this in perspective," Gurley began, his towering frame always at odds onstage with the awkwardness of his personality. "This company is undoubtedly the most successful startup in the history of Silicon Valley. It grew faster, bigger, it touched more people customers, countries cities faster than ever before.

"But I want to bring up a phrase you hear pretty often but I think is applicable," Gurley continued, his tone turning grave. "With great success comes great responsibility. We are no longer considered a startup by the outside world. We are considered one of the largest, most important companies in the world. And our behavior, our corporate behavior, has to begin to equal and parallel that expectation or we're gonna continue to have problems."

Audience members nodded along.

"We're in a reputational deficit," Gurley continued. "You can read something and say that's not fair, but that's not going to matter. Because, it's gonna take us a while to get out of this, and people are not going to give us the benefit of the doubt."

"No one thinks since we announced the Holder recommendations here that everything will be fine," he said. "Don't pay attention to that right now. Let's just do our best work and help get to Uber 2.0," Gurley said, handing over the microphone.

The audience cheered. Perhaps it was possible, they believed, to finally turn the company around.

DESPITE BONDERMAN'S GAFFE, it appeared that the Tuesday all-hands had been a success. Shortly after Arianna Huffington took the stage, a *New York Times* reporter had somehow infiltrated the meeting and began live-tweeting the event.* Kalanick was apoplectic; members of the security team scrambled to try and find the reporter. But fortunately for Uber, the *Times* seemed to have missed Bonderman's comment.† Perhaps the company could deal with the situation privately.

They had no such luck. Hours after the presentation ended, another website published the entire contents of the presentation, highlighting Bonderman's sexist comment. This latest blow was absolutely crippling. After months of waiting for the report which would launch serious change, a member of the company's *board of directors* had suggested to more than six thousand workers that women talk too much. Employees were outraged, while journalists felt validated; Uber's culture was poisoned from the very top.

For Kalanick, it was different. He finally had the ammunition he needed to take out Bonderman. After a day of texting and emergency board deliberations, Bonderman knew he had to fall on his sword. His note to employees was sent out by the end of the day.

> *Today at Uber's all-hands meeting, I directed a comment to my colleague and friend Arianna Huffington that was careless, inappropriate, and inexcusable. The comment came across in a way that was the opposite of what I intended, but I understand the destructive effect it had, and I take full responsibility for that. . . .*
>
> *I do not want my comments to create distraction as Uber works to build a culture of which we can be proud. I need to hold myself to the same standards that we're asking Uber to adopt.*
>
> *Therefore, I have decided to resign from Uber's board of directors, effective tomorrow morning. It has been an honor and a privilege to serve on Uber's board, and I look forward to seeing the company's progress and future success.*

---

* Uber was not happy with me.

† I was not happy with myself.

And with that, Bonderman was gone, and Kalanick had one less enemy on the board. With the day behind them and Kalanick claiming he was heading out on his leave of absence, it was time for Uber to heal—to become "Uber 2.0."

At least, that was what was supposed to happen.

Chapter 28

THE SYNDICATE

WHEN DAVID BONDERMAN OPENED HIS MOUTH AT THE ALL-HANDS meeting, Bill Gurley had one thought pop into his head: "You've got to be kidding me."

Gurley was optimistic heading into the all-hands that Tuesday morning. He was disgusted with the report's contents; it read like a lewd magazine, a racist, sexist Silicon Valley bachelor party. But with the board unanimously accepting the report's recommendations, Gurley had hope.

Everyone looked to Gurley as the one to somehow fix the mess. Gurley had known Kalanick for years. He was on the board. Along with his physical height, Gurley had an air of authority. Now, the adult in the room was expected to do something, and fast.

But the pressure was getting to him. By the middle of June, the lanky and trim Gurley had started to gain weight. Earlier in the year, Gurley flew to San Diego for an extensive restorative surgery on one of his knees. In the weeks leading up to the June 13 meeting, Gurley had been fielding calls with his injured leg propped in an anti-swelling machine. Benchmark's offices were in Woodside, California. There, leaning back in his desk chair, he would complain to partners in his South Texas baritone about Kalanick's inflexibility. His knee hurt—a lot—but it was nowhere near as bad as the pain this Uber situation was causing him and his firm.

Gurley did have support. Benchmark had historically operated as a true partnership. Every Monday morning at Benchmark's partner meeting, the close-knit group of VCs would spend hours reviewing each of the companies in their portfolio. That meant input from partners like Matt Cohler, an early Facebook employee and customer growth savant. Peter Fenton, the "high EQ partner," helped Twitter juggle the ousting of two founders and the installation of a third CEO at the social network. Eric Vishria and Sarah Tavel, the most recent Benchmark additions, could give the perspective of what it was like to be a founder or an executive inside a closely scrutinized startup.

Yet Gurley shouldered the brunt. His cell phone was constantly buzzing with calls from Benchmark's limited partners—the group of enormously wealthy investors, from college endowments to pension funds, who put up the hundreds of millions of dollars that Benchmark used to invest in other companies. They were terrified Uber would tear itself apart, evaporating the billions of dollars in returns they expected. Through every frustrated email, every anxious phone call, Bill Gurley was there to soothe them, assuring his LPs that he had everything under control.

To friends, it didn't look like it. One evening in 2017, David Krane, a partner at Google Ventures who helped lead the quarter-billion-dollar investment in Uber four years earlier, threw a party at his house to benefit a scientific research foundation. Peter Fenton showed up full of his usual bubbly friendliness. In tow was Bill Gurley, who spent most of the evening nursing a drink and moping in the corner of Krane's living room or propped against the outdoor bar on Krane's back porch. Gurley was so tired and stressed that he could barely stand upright. He had tried to take care of himself, he told friends. The six-foot-nine Texan started doing yoga and meditating. Yet he still couldn't sleep. Gurley was exhausted.

The Holder report event was *supposed* to contain the fallout. Travis Kalanick was stepping away, the company would take steps to rebuild its brand—there was a possibility everything would shake out just fine.

But the meeting had been a disaster, derailed by David Bonderman's sudden sexist remark. And Kalanick had no intention of laying low. The very next day, Kalanick was on the phones, calling up department heads and members of the executive leadership team, running the business as if he hadn't just pledged to the entire world that he intended to stand down. Within the week, Travis began working with Uber engineers during his so-called "leave," who would carry out his orders without informing the board.

The simple solution would be to step in and insist Travis's leave be made permanent. But Kalanick wasn't going anywhere; he had completely ignored his promise to go on leave. Travis would fight any further attempt to sideline him, and Gurley knew Travis well enough not to underestimate him in a fight.

Gurley's qualms were as much philosophical as practical. Benchmark's image was based on its reputation as a "founder friendly" venture capital firm. When hedge funds or private equity firms invested in a founder's company, the founder often had to accept a more heavy-handed approach to governance. Bonderman, for instance, had no qualms about criticizing Uber's burn rate. Bonderman was a private equity man, though, and venture firms wanted to be seen as "founder friendly." Benchmark was there to support its portfolio company, to help in recruiting top executives, to contemplate strategy, and give welcome advice. If they ousted Kalanick permanently, would the next Uber, the next Facebook, the next big thing, ever let Benchmark invest again?

Beyond Benchmark's reputation, there were other practical matters: money. Uber's valuation by then had swollen to an enormous $68.5 billion. Uber was worth more than even Facebook at its private valuation peak, and Benchmark had invested at the ground floor. The firm's initial $11-million stake was now worth *billions* of dollars, easily one of the greatest venture capital investments in Silicon Valley history. Now, Benchmark's Uber shares were in serious jeopardy. Every new negative press story chipped away at Uber's valuation, which tarnished Bill Gurley's incredible play, and meant less money in the end for shareholders.

Some investors were turning on the company publicly. Mitch Kapor and his wife, Freada Kapor Klein, both early investors in Uber, had long been active in so-called "impact investing," a socially conscious approach to capitalism. "We feel we have hit a dead end in trying to influence the company quietly from the inside." the two wrote in a public blog post. "We are speaking out publicly, because we believe Uber's investors and board will rightly be judged by their action or inaction. We hope our actions will help hold Uber leadership accountable, since it seems all other mechanisms have failed."

It was a message from a founder that made Gurley realize just how bad things were. One afternoon that summer, as Gurley was checking his inbox, a new email popped up on his screen. It was from Katrina Lake, the chief executive of Stitch Fix, a much-loved, successful e-commerce company that sold personally styled outfits to customers over the internet.

Gurley knew Lake well. Benchmark had led a $12-million round of venture funding in 2013, back when the young company was showing promise—Lake had created Stitch Fix in her bedroom, during business school—but was still far from a sure thing. By 2017, Stitch Fix had gone public in a successful IPO, netting Benchmark hundreds of millions in returns. As a board member of Stitch Fix, Gurley and Lake had grown close over time. He believed in her company, and she trusted his advice.

Lake's email was brusque. "It's demoralizing and sad that something like Uber can even exist and even thrive," she wrote. "And I'm disappointed that someone I respect so much has had a part in it."

For Lake, the Uber story was deeply personal. Lake stood out as one of the most prominent female chief executives in Silicon Valley. As she traveled the path from running a small startup to a multimillion-dollar enterprise, Lake had dealt with her share of sexist scumbags. At one point during Stitch Fix's ascent, Lake was sexually harassed by one of her own venture capital investors, Justin Caldbeck. Caldbeck sat on her board of directors as an observer until Lake insisted he be removed after the incident. She knew how awful the bro-friendly culture of tech companies and venture capital could be for women.

But after she read about Susan Fowler, about what happened in India, about the torrent of other scandals flooding out about Kalanick, Lake felt ashamed for her company to be spoken about in the same breath as Uber. To think that her mentor was idly standing by—even abetting it—bothered her.

To Lake, being an entrepreneur in Silicon Valley wasn't just about

doing novel things with the latest in tech. It was about building companies that lived by the values that founders wanted to see in the world.
"I'm hopeful that Stitch Fix can be a living, breathing counter-narrative,
a company that is successful because of its values and not in spite of it."

Gurley responded quickly, thanking her and expressing his gratitude. "It's been a nightmare," Gurley wrote. Gurley's investment in
Uber had made his reputation, but Lake's email was a kick in the ass.

When Benchmark held its next partner meeting after the Holder
presentation, the group agreed: Benchmark needed to do the "right
thing." Kalanick needed to go.

But Benchmark couldn't do it alone. Gurley needed help.

THERE WAS A REASON Kalanick had kept such tight control on his
investors: if the day ever came where the venture capitalists turned on
him—just like Michael Ovitz did with Scour—he wanted to be able to
defend himself.

He had done well. Over time, Kalanick slowly eroded shareholder
power and influence. Kalanick withheld as much information as possible, hindering investors' basic ability to understand the company's
finances. Investors grumbled. They had invested a great deal of money
in Kalanick's company, and they felt they had a right to know how the
company fared and what decisions Kalanick was making with that capital. One investor said Kalanick treated them like mushrooms: he fed
them shit and kept them in the dark.[*] Kalanick thought they should be
grateful even for that.

Investors seemed to intuit as much. As Uber's valuation rose, few of
them attempted to meddle. Legally, Kalanick's position was indefen-
sible; investors who owned a high percentage of a company had rights
to information about that company. His response, according to at least

---

[*] The investor cribbed the line from Mark Wahlberg's character in *The Departed*, the Academy Award–winning Martin Scorsese crime drama from 2006. Wahlberg, a cop, was referring to his combative relationship with the FBI.

one investor, was to call their bluff: "So sue me," he told this person. "What's your rep going to be in this industry if you sue your own company?" He was right.

What's more, over time Kalanick had amassed so-called "supervoting shares," a more powerful class of stock that gave him more votes per share than the one-vote-per-share common stock most other investors held. Kalanick held an enormous amount of supervoting shares, a scenario he engineered in Uber's earliest days, as did two of his allies, Garrett Camp and Ryan Graves. And his pile of common stock shares was only getting bigger by the day. If Uber employees wished to cash out some of their stock through an internal repurchase program, Kalanick required Uber employees to sell those shares back to him. With each passing day, as employees left or sold shares through normal turnover and attrition, Kalanick's voting power grew stronger.

Supervoting shares wouldn't save him in every situation. Some decisions—like voting an executive out of the company, for instance—were only decided by full board vote.

There, Kalanick had another advantage: he effectively controlled the board of directors. Of the eight-person board, most were aligned with him: Arianna Huffington, Wan-Ling Martello, Yasir al-Rumayyan, Ryan Graves, and Garrett Camp all followed Kalanick's lead. And in 2016, during the $3.5-billion Saudi investment round, he negotiated for himself another ace in the hole. The terms of that round, all of which were unanimously agreed upon by the board, gave Kalanick the ability to appoint three additional board members whenever he desired.

In the summer of 2017, as outside scrutiny intensified, Graves and Camp began to worry. But both the men felt indebted to Kalanick. Camp had never wanted to run Uber in the first place, and had been content to be a back-seat passenger. Graves had spent much of the past few years at Uber partying and traveling, yet the CEO never hung him out to dry. Graves evidently felt like the CEO truly cared about him— like they were "bros."

Whenever Camp, Graves, or anyone else in Kalanick's orbit began to chafe at his actions, he usually responded with some version of the same placative sentiment: "Do you know how much money I'm going to make you?"

The line almost always worked.

BY MID-2017, everyone who had money tied up in Uber felt helpless. Kalanick wouldn't have asked for their help anyway; over the years, he had managed to alienate key investors through subtle betrayal and financial skullduggery. His goal was to undermine his investors before they ever had a chance to do the same to him. To accomplish that goal, he executed a series of preemptive strikes over a period of eight years.

Shawn Carolan, a partner at Menlo Ventures, had negotiated a board observer seat during his firm's early investment. Kalanick made sure it came with no voting power. Rob Hayes, a VC from First Round Capital, was fortunate enough to invest in Uber during its "seed round," one of the earliest stages of fundraising. Along with a sizeable stake in Uber, Hayes secured himself a board seat.[*] But during the Series B round of funding, Kalanick altered the contracts in a way that stripped Hayes of his voting seat and limited his access to information. Chris Sacca, an ex-Google lawyer turned investor and founder of Lowercase Capital, once considered himself a friend to Kalanick. Sacca's $300,000 early investment in Uber also gave him a sizeable portion of the company. But when Sacca began attempting to buy up shares of Uber from other early investors—a practice known as "secondary share purchasing"—Kalanick turned on him. The CEO stopped allowing Sacca to attend board meetings as an observer; the two rarely spoke afterwards.

Gurley knew all this. He had secretly been talking to the spurned VCs for months, all of them back-channeling, worried about whether their investment was going to implode. As he rallied the group of investors, Gurley began reaching out to other people for advice. He

---

[*] Hayes's seed investment of $500,000 bought First Round Capital a 4 percent ownership stake in Uber. Eight years later, that investment was worth more than two billion dollars. Similar to Gurley's bet, it would prove to be one of the most successful tech company venture capital investments of all time.

contacted law professors at Stanford, whose backgrounds included expertise in corporate governance and white-collar crime. He spun up lawyers from Cooley and Paul, Weiss, two top-tier Valley firms that regularly consulted tech companies and VCs. He hired a crisis public relations firm. And he proposed a plan that would rely on all of them to work together. Gurley knew Kalanick would never step down of his own volition. They had to force his hand.

The plan Gurley devised was simple. He would lead a syndicate of Uber's largest shareholders—Benchmark, First Round, Lowercase, Menlo—all of whom collectively held a more than a quarter of Uber's stock. They would approach Kalanick with a letter that put forward a simple request: Step down from your position as chief executive for the sake of the company. If Kalanick refused to do so, the group would go public. They would call up the *New York Times*, tell the reporter the entire plan, and their letter to Kalanick would land on the front page of the paper the next morning. That was strategic, too; going public would help rally more of Uber's dozens of investors to the cause.

Gurley assumed that Kalanick would dismiss their letter and decline to step down even after they confronted him. For that moment, Benchmark hired Steven Rubenstein, the crisis communications expert, who would handle outreach to the press once the reporter for the *Times* went live with the story.* Gurley knew it was important for the syndicate, not Kalanick, to control the public narrative. With Arianna Huffington's help, Kalanick could try to gain sympathy from outsiders and paint the venture capitalists in a horrible light.

If everything went sideways, the syndicate had a secret weapon. The lawyers discovered a flaw in Uber's company charter. Currently, the group all held a significant amount of Class B stock, classified as "supervoting shares," which carried ten votes per every one share they held. But if the syndicate deployed its "nuclear option," it could force everyone to convert all supervoting "Class B" shares to Class A shares, which carried only one vote per share. While that would severely cur-

---

* Rubenstein, ironically, had nearly been hired by Travis Kalanick months ago, after the video showing him screaming at a driver went viral.

tail Benchmark's supervoting power, it would also rein in Kalanick's supervoting power as well. The result would be a scramble to build coalitions of shareholders that could seize power. But the group didn't want to go there quite yet; giving up its supervoting power was a last-resort technique.

The most important factor was time. Gurley's syndicate needed to give Kalanick a strict deadline to answer the syndicate's demands. Gurley knew Kalanick, like a rock climber looking for a toehold, would search for any weakness in the syndicate's attack. With enough time and effort, Kalanick would find one, exploit it, and sink them all. He was a survivor; they needed to box him in.

On the day they decided to confront Kalanick, Gurley orchestrated a conference call with the syndicate and its advisors. Gurley was down at Benchmark's Woodside office, holding court in the firm's main conference room, an airy space with a dozen black leather and metal Steelcase chairs encircling a long, polished hardwood table. Benchmark had been pitched by a who's who of valley founders at that table. He had signed high-profile term sheets there, and held countless discussions about portfolio companies like Uber, Snap, and Twitter. But on June 21, 2017, Gurley would use the room as mission control for the syndicate's attempt to oust Travis Kalanick from his position as chief executive. The call that morning covered logistics of the day's events. At one point, Gurley took a moment to explain why they had to move so quickly and decisively, and what they were risking if they moved ahead. A cadre of other investors, lawyers, and associates listened in.

"Did you ever see the movie *Life*?" Gurley asked everyone on the conference call. "The one with Ryan Reynolds in space, with that black goo alien they captured? Once they find the alien, they place it in an indestructible box inside a lab in their spaceship to keep themselves safe while they perform tests on it. Then, eventually, the alien escapes. It gets out of the box somehow, and ends up killing everyone on the spaceship. It heads to earth to kill everyone there, too. All because it got out," he said.

The syndicate members listened quietly on the line, wondering where Gurley was going with this. Some of them chuckled to themselves; Gurley loved analogies.

"Well, Travis is exactly like that alien," he said. "If we let him out of the box—at any point during the day—he'll destroy the entire world."

Chapter 29

REVENGE OF THE VENTURE CAPITALISTS

THE DAY BEFORE GURLEY'S CONFERENCE CALL WITH THE SYNDICATE, Travis Kalanick was supposed to be in San Francisco. But on June 20, he wasn't home in his apartment at the top of The Castro. Nor was he pacing in his bunker at Uber's headquarters on 1455 Market Street. Instead, Kalanick was two thousand miles away, working on his laptop.

It was eighty degrees in Chicago that Wednesday, warm and humid but not yet the sweltering blanket of deep summer in the Midwest. Kalanick was there to interview Walter Robb, the former co-CEO of Whole Foods. The CEO thought Robb a potential candidate to become Kalanick's new chief operating officer. For the interview, Kalanick had rented a private conference room on one of the uppermost floors of the Ritz-Carlton Chicago, downtown, off Michigan Avenue. Kalanick liked flashy hotels and nothing was more baller than working from the top of the Ritz.

Travis's trip to Chicago threw a wrench in the syndicate's plan. They all knew he was still working around the clock—people kept calling Gurley to tell him Kalanick hadn't taken leave—but they didn't know he was interviewing people to be his second-in-command[*] in a different state. For their plan to work, they would have to travel to Illinois.

By the summer of 2017, Gurley and Kalanick weren't speaking. Gur-

---

[*] Not that hiring a second-in-command would have mattered much. Kalanick had once said to a different candidate in an interview that the job would be to carry out his marching orders. Gurley was furious.

ley knew he couldn't be the person to fly to Chicago that day and per-
suade the CEO to tender his resignation. Kalanick had come to resent
Gurley's nagging, his worrying, his insistence that Kalanick accept
change. The minute Gurley walked into the hotel room to negotiate a
surrender, Travis would tell the six-foot-nine Texan to go fuck himself.
They needed a neutral emissary.

The syndicate picked Matt Cohler and Peter Fenton. Cohler, a brilliant early Facebook employee who joined Benchmark in 2008, was
practical, realistic, and frank; he could deliver the news to Kalanick
in a sober way that the CEO could understand. Thin and fair, with
curly brown hair, wide eyes, and rosy cheeks, Cohler had just turned
forty, but he looked at least a decade younger. After Gurley, Cohler
knew Uber the best. He was there in the beginning when Gurley first
approached Kalanick to source the deal. At the very least, Cohler was
a familiar face that Kalanick wouldn't immediately want to punch.

What Cohler lacked, though, was a high level of emotional intelligence. That's what Peter Fenton brought to the table. Fenton was one of
the most charismatic Benchmark partners, able to meet a young startup
founder and put them at ease with a soft touch and bright smile. Like
Cohler, Fenton looked young despite pushing forty-five; dirty green
eyes, a high forehead, and sandy blond hair gave him more of a "boy-
next-door" vibe than the driven, experienced venture capitalist he was.
He was a tough negotiator, but reasonable, able to yield when necessary, making the other party feel like they were being heard. It was a
great quality for someone who needed to close deals or, in this case,
break tough news to someone.

The group had been discussing the situation for weeks. Gurley had
spent hours on the phone with the other VCs—Josh Kopelman and
Rob Hayes over at First Round, Doug Carlisle and Shawn Carolan at
Menlo, Chris Sacca from Lowercase. He spent even more time agonizing over the situation with his own partners at Benchmark. The
group was so paranoid someone would overhear their plans that they
used codenames for Kalanick when discussing the matter in public;
"Travis" was always "Bob" or "Jeff" or a dozen other names picked at
random if a partner happened to be talking about Kalanick in the back

of an Uber. After the driver video of Kalanick surfaced, it was assumed someone might be recording or listening at all times.

Most of Gurley's meetings with Benchmark staff took place around the long, wooden conference room table in Benchmark's Woodside offices. The group—Gurley, Cohler, Fenton, and partners Eric Vishria, Sarah Tavel, and Mitch Lasky—worked out the details of their plan together, over and over, until they had it memorized. They worked through every possible permutation of what could happen when Cohler and Fenton approached Kalanick. (Would Travis throw a fit? Would he accede immediately? Would he lunge over the table and murder them?) They typed up a dozen different versions of the letter they would deliver to Kalanick, one for each possible scenario of the coming showdown. Lawyers at Paul, Weiss—the venerable white-shoe law firm—vetted each draft.

The group met at Benchmark's office the Tuesday before the big day. They worked out their plans once more, gearing up for what promised to be an ugly, public fight. There was no way Kalanick was going to yield. An Uber executive recalled that Kalanick said he was "ready to take Uber's valuation to zero" before he would ever leave the helm. (Later, through a spokesperson, Kalanick denied he ever said this.) The syndicate needed to steel themselves for what was to come.

For all their agonizing, the firm made peace with what the partners were about to do. This was about saving a company—about saving the firm's legacy—and they couldn't sit idly by while Travis Kalanick single-handedly drove a $68.5-billion behemoth into the ground. Over the six weeks they had been discussing the plans together, their mantra to outsiders, to each other, was the same: "We have done all we can." It had come to this only because they had exhausted all other options.

As daylight waned in the conference room, Gurley looked around the room at his partners and nodded. He was anxious. But he was resigned.

"I really think we're on the right side of history here," Gurley said.

MATT COHLER AND PETER FENTON chartered a private plane that morning, a direct flight from San Francisco International to O'Hare. The two arrived and checked into another swank hotel—down the street from the Ritz-Carlton—to prepare for the day ahead. After walking off the tarmac and taking an Uber to their hotel, the two met up with Steven Rubenstein, the crisis communications expert who had flown in from the east coast to handle press outreach after Kalanick inevitably turned down their proposal. Rubenstein was a fixture in the world of crisis PR firms, having done damage control for Rupert Murdoch during the infamous phone hacking scandal that engulfed News Corp in the early 2000s. A wiry, sarcastic New Yorker with thick black glasses, Rubenstein would begin spinning the events to journalists once the confrontation had occurred.

Though they all knew Kalanick was at the Ritz, Cohler, Fenton, and Rubenstein couldn't escape the sneaking suspicion they would accidentally run into the CEO somewhere in the city. Benchmark's paranoia had been running on high for months now; after the firm's investigation into some of Uber security's more clandestine activities, Gurley couldn't shake the feeling that people were following him, or that Kalanick had cameras placed outside of his home.

Back in Woodside, Gurley sat at the head of the table, swiveling back and forth in a black leather armchair. Benchmark partners trickled in and out of the room. To keep things running smoothly, the syndicate started an internal group chat using WhatsApp, the text messaging service. At this point, more than a dozen people were involved in planning Kalanick's ouster. They needed a way to keep everyone on the same page. Besides the WhatsApp group, numerous other texting threads sprang up between members of the syndicate. But Gurley remained the point man.

Cohler and Fenton left their hotel and began their journey to the Ritz-Carlton to see Kalanick. Rubenstein stayed behind, waiting to hear from the partners in case they were forced to go public.

Though not everyone in the syndicate knew it, one person aware of the plan had called a *New York Times* reporter over the weekend. The investors were rumbling, this person said, and the *Times* needed to be ready to write a story in case something dramatic happened. The source's words were cryptic and intriguing.

That reporter was me.

AT NINE O'CLOCK in the morning on June 20, I was sitting in the Virgin America terminal of San Francisco International Airport, when my phone started buzzing in my pocket. I was flying down to Los Angeles to interview an executive onstage at a tech conference, and planned to use the rest of my week there to meet up with my other industry contacts. I thumbed the mute button and checked my iPhone—it was one of my key Uber sources calling.

The source had contacted me over the weekend practically out of the blue, warning me that something big was about to go down. I've received plenty of bogus tips over the years that have amounted to nothing. But that morning, as I was boarding my plane, the source told me Kalanick's clock was ticking. There was a chance that he might be forced to leave Uber that day. "He's gonna go. It's gonna go down today," the source said.

I was caught off guard. "What? What the fuck?" I stammered. "I'm literally about to get on a plane. Is this happening right now? Do I have to cancel my flight?"

The flight attendant started calling out boarding groups. Even if I had Wi-Fi access in the air, I couldn't take a phone call at 30,000 feet. All across Silicon Valley, after months of scandal and public outcry, every tech worker was watching to see whether or not Kalanick would be able to keep his job. More so now, after the botched delivery of the Holder report and the sudden ousting of David Bonderman. If today was the day Travis Kalanick was going to be pushed out, I needed to be ready.

"Be by your computer, and keep your phone on all day," the source said. "I'll call you." And then the source hung up.

WHEN COHLER AND FENTON walked out of the gold elevator doors and onto the black and white marble of the Ritz-Carlton's twelfth-floor lobby, they didn't expect to be swarmed by suits. The hotel was hosting a real estate conference that week; white guys in boxy suits from national realtors like RE/MAX and Coldwell Banker flooded the main foyer. The two VCs politely pushed their way through the scrum.

The Ritz had been open in the city for more than forty years without a refresh, and locals and regulars had noticed the stale feeling over the years. But as Cohler and Fenton walked in that morning, they saw a hotel transformed; in a few weeks the Ritz would unveil a grand remodel, $100 million and eighteen months in the making. A Roy Lichtenstein painting hung above the suited throngs. Across the lobby, a wall of windows faced north, overlooking Lake Michigan. Were the two men not brimming with anxiety, they might have appreciated the view.

Cohler and Fenton crossed the lobby to the second set of elevators, those leading to the guest rooms, private offices, and business suites on the topmost floors in the building. Fenton had contacted Kalanick that morning, telling him that the two VCs were in Chicago, and needed to speak with him urgently. Kalanick, caught unawares, knew something was up. But he told them to come find him at the Ritz; he was working, waiting, upstairs by himself.

As the two VCs entered the private conference room, the men gave each other a solemn greeting. Gurley's absence made it slightly less difficult—Kalanick hadn't stormed out of the room, at least—and Cohler and Fenton gently eased into the matter at hand. They had a request. More than a request, really. They wanted Kalanick to step down, "immediately and permanently," for the good of the company.

As Kalanick sat there, stunned, Fenton slid a letter[*] across the table. Kalanick looked down and read the note in front of him.

Dear Travis:
On behalf of Benchmark, First Round, Lowercase Capital, Menlo Ventures, and others—which collectively owns more than 26% of Uber's economic stock, and over 39% of Uber's voting shares—we

---

[*] The syndicate drafted many versions of the letter. This is not the final version Kalanick ultimately received that day, but it is very similar to the final, which included a last-minute addition to the investor syndicate: Fidelity Investments, the mutual fund giant. This version of the letter has never been made public until now. Some identifying information has been removed to protect the sources connected to the document.

are writing to express our profound concerns about Uber's direction and to propose a way forward.

Please know that we are deeply grateful for your vision and tireless efforts over the last eight years, which have created a company and an industry of which no one could have dreamed. Unfortunately, however, [the] series of recent revelations have deeply affected us. . . . [A]ll of these issues are causing tremendous damage to Uber's brand and threaten to destroy Uber's value for its shareholders and stakeholders. We believe the issues stem from deep-seated cultural and governance problems at Uber and from the tone at the top. . . .

We must take concrete steps to address these issues and strengthen Uber's brand and governance. If we do not adequately address these issues now, Uber's brand and market share will continue to erode, to the detriment of the company and all of its shareholders, including you.

. . . With these changes we firmly believe Uber can regain its place as one of the most important companies Silicon Valley has ever produced. We hope you will agree to move forward with us on this path.

"Moving Uber Forward": Investor Demands

First, you must immediately and permanently resign as CEO. We strongly believe a change in leadership—coupled with effective Board oversight, governance improvements, and other immediate actions—is necessary for Uber to move forward. We need a trusted, experienced, and energetic new CEO who can help Uber navigate through its many current issues, and achieve its full potential.

Second, Uber's current governance structures, including the composition and structure of the Board of Directors, are no longer appropriate for a $68 billion company with over 14,000 employees. The new CEO must report to an independent Board that will exercise appropriate oversight. . . . Further, as you know,

*the Holder Report calls for the appointment of additional inde-
pendent Board members. To that end, you should fill two of the
three Board seats you control (retaining one for yourself) with
truly independent directors who comply with the Holder Report's
recommendations for qualification for service. . . .*

*Third, . . . [y]ou should support a board led CEO search com-
mittee, with an independent chairperson, and the inclusion of a
representative of senior management and a representative of the
driver community. . . .*

*Fourth, the company should immediately hire an adequately
experienced interim or full-time Chief Financial Officer. The
company has intentionally operated without a properly qualified
executive in the top finance [role] for over two years. The investor
group broadly believes that this specific executive hire needs to be
addressed urgently.*

*We hope you will agree to move forward with us on this path,
and look forward to your response.*

Kalanick was furious. He stood from his chair, unable to finish the
letter at first. He was livid that the two men had sprung this on him
just weeks after the sudden, tragic death of his mother. How could
Benchmark do this to him?

Kalanick immediately recalled a board meeting that had occurred
just a week before. Graves, Gurley, and other board members had been
discussing Kalanick's leave of absence, and were openly wishing him
well. Gurley, who said he was on the fence about Travis returning after
his leave, added that he'd support the outcome either way. Kalanick
was heartened by that comment; in the wake of all the personal turmoil
he had been through, Kalanick felt supported and comforted by his colleagues on the board. Now, as Cohler and Fenton slid his death letter
across the table, Kalanick wondered how long they had been plotting
his demise. It was the ultimate betrayal.

He started pacing, the way he always did. He shouted at the two
investors—people he once considered his allies, his supporters—while

Cohler and Fenton sat there, stone-faced, and took it. Kalanick was lashing out like a cornered animal. He wouldn't take it lying down. He wouldn't accede to their demands. He was going to fight.

"If this is the path you want to go down, things are gonna get ugly for you," Kalanick said. "I mean it."

The two investors knew Kalanick meant it. But so did they. When Cohler and Fenton presented him with the letter, the syndicate started a ticking clock. It was currently close to noon. They gave Kalanick an ultimatum: They needed an answer by the end of the day, around 6:00 p.m. If he declined to answer, if he took too long, if he tried to stall them further—if he did *anything* that seemed like an underhanded maneuver—the venture capitalists would walk out of the room, text members of the syndicate and immediately spin up the communications machine to take the fight public. By the next morning, the fight would be on the front page of the *New York Times*. Word would spread fast, and other investors—emboldened by the hard-line stance taken by Uber's largest shareholders—would ultimately join them. At least one heavy hitter already had; at the last minute, Fidelity Investments, an enormous stakeholder and powerful ally to Benchmark, also signed on to the letter demanding Kalanick's removal. Others—Glade Brook Capital Partners, Wellington Capital Group, angel investors like David Sacks—were already privately agitating to remove Kalanick. The rest would join them, the two VCs threatened, once Benchmark spread the word.

Kalanick knew he was pinned. After a prolonged back-and-forth between the two sides, he asked the two of them to give him some time to think. Cohler and Fenton agreed, and the two sides parted for the first time that day.

After Cohler and Fenton updated Gurley at Benchmark's headquarters back in Woodside, Gurley typed in a text message updating the syndicate on what was happening.

"He's stalling," Gurley texted.

THE CHIEF EXECUTIVE OF UBER was doing more than just stalling. After Cohler and Fenton left the room, Kalanick began frantically calling people, beginning with Arianna Huffington, one of his few remaining allies. Huffington said she was as shocked as Kalanick that the cabal of investors would pull such a move. The two discussed Kalanick's options.

Kalanick trusted Huffington. But what he didn't know was that Huffington was already helping prepare a draft of his resignation statement. As Kalanick's world was crumbling down around him, Huffington jumped into a recording booth to tape a podcast with Ashton Kutcher.

Cohler and Fenton kept updating the group back in Woodside. Gurley wasn't panicking. He had anticipated this would happen. Kalanick had lost many friends inside Uber, and would likely reach out to allies, to make a play to save himself. All Benchmark had to do was keep Kalanick under pressure—to keep him boxed in like the alien in *Life*.

Another investor texted Gurley, wondering what the latest was. Gurley texted back: "dancing around."

KALANICK DIDN'T JUST CALL Arianna Huffington. He rang up top business development executives still at Uber like David Richter and Cam Poetzscher, both of whom had sway within the organization and could perhaps find a way to help Kalanick out of this mess. He started calling board members and old allies like Garrett Camp and Ryan Graves. He called other investors who might have a plan to help him take on the syndicate.

Suddenly, Kalanick saw a way out. If he rallied enough Uber shareholders to his side, Kalanick could potentially amass enough voting stock to fight back against the syndicate if it came to a public shareholder battle. To that end, he started calling up individual members of the syndicate, figuring he could charm and eventually flip them back over to his side.

"Shawn!" Kalanick yelped into his iPhone, using his best puppy-dog, pity me voice. "I can't believe it's come to this! I can change! Please let me change!" he said.

For Shawn Carolan, the Menlo Ventures partner and early backer of

Uber, it was hard not to believe him. Carolan told friends that he had always found Kalanick persuasive; the founder's confidence, his scrappiness, his smarts, and his charm were all the original reasons that Carolan had invested in him in the first place. Now he heard the pain in Kalanick's voice as he fought for his job. It sounded like the CEO was crying on the other end of the line. Though he knew he shouldn't, Carolan felt guilty. He was trying to annihilate one of his own founders, a cardinal sin of the venture business. But after a beat, and some hemming and hawing over the phone, the venture partner shook it off, steeling himself for a proper response.

"I'm sorry, Travis, I really am," Carolan told him. "As much as I want to trust you, I just can't. I cannot, in my right mind, support you any longer as the chief executive of this company." And then, the venture capitalist hung up on Travis Kalanick.

AFTER SOME TIME HAD PASSED, Cohler and Fenton returned. They leveled with Travis. If he agreed to step down, peacefully and without a fight, they would give him the dignity of a graceful exit. Often, the VCs knew, when executives were replaced or demoted from the companies they helmed, there was usually a bit of Kabuki theater and posturing on the announcement. A half-truth like "I'm stepping back to be an advisor," or "I've decided to step down to spend more time with my loved ones"—these were the platitudes of corporate coups d'etat. Benchmark was happy to allow Kalanick that luxury.

It was getting later; the sun began to dip outside the Ritz conference room. Fenton had put himself in a bit of a bind; he couldn't wait Kalanick out with perfect calm as Fenton had to catch a flight across the Atlantic later that evening to see his children, who lived with their mother in France. At around 4:00 p.m., Kalanick's clock was still ticking. But Kalanick still resisted, asking for another break. The group parted again.

Eventually, Kalanick started dispatching proxies to talk to the venture capitalists as he tried to figure out what to do. Arianna Huffington, who by this time had been in close contact with Kalanick for hours, began talking and texting with Fenton.

Up until this point, Huffington had been in lockstep with Kalanick, supporting him and playing defense for him against the press and angry employees. Huffington had been on CNN earlier that year, claiming that Kalanick had "evolved" in his behavior and defending his capabilities as chief executive. Travis Kalanick, she claimed on live television, was the "heart and soul" of Uber. Huffington had aligned with Kalanick for years until this very moment—she suggested to him that maybe he should consider an exit.

Were it any other time in Kalanick's life, he would have immediately cast the idea aside. Kalanick never, ever stopped fighting for what he wanted, and more than anything in the world, he wanted to recover from a terrible year and continue Uber's quest for global domination.

But things were different. Kalanick was shattered by the sudden death of his mother. The syndicate's coup had caught him still shaken and absorbing the fact that he would never see her again. It had only been a couple of weeks since he had flown to Fresno after the boating accident and sat by his father's hospital bed, hoping he would recover. Even less time had elapsed since he buried his mother in Los Angeles. For the first time in his life, Travis Kalanick realized he was tired of fighting. Maybe, for once, stepping down and walking away to mourn was the right thing for him to do.

Gurley texted an update to a member of the syndicate: "he's leaning towards backing down," he wrote. The VCs and advisors chattering over the WhatsApp group together couldn't believe it actually might happen.

BY THE TIME THE SUN went down in Chicago that night, Kalanick had stalled for hours. The investors had had enough. After Huffington's first hint to the syndicate that Kalanick might step down, the group had made little progress. Fenton and Cohler had been periodically talking with various people representing Kalanick and his interests throughout the day. Travis had scrambled and found a handful of allies to help advocate on his behalf.

But as hours passed with no definite word from Kalanick, they were fed up.

At 9:19 p.m., Eastern Standard Time, Peter Fenton texted Arianna Huffington to let them know they were going to call up the *New York Times*.

"Forgive my anxious tone, I'm leaving for Europe in fifteen minutes," Fenton wrote, noting the plane he soon had to catch. "But there's no way I can stop the group from going public. I know you're moving heaven and earth, but we are out of time."

Huffington responded quickly. "Calling him right now," she said. Fenton shot back a "praise hands" emoji text as a sign of gratitude.

The syndicate's final bit of prodding—coupled with Huffington's last-minute counsel that Kalanick should step down—had finally worked. The CEO was exhausted, out of options. He had failed to convince any of his former allies to join him in battling the venture capitalists. He agreed to meet again and sign the paperwork.

The final meeting at the Ritz-Carlton was a flurry of ink and negotiation. Kalanick pulled out a pen and started tearing through the letter, crossing out things he wouldn't agree to, amending stipulations he thought overreaching.

Even if he was no longer going to lead the company he created, he was damn sure still going to have a say in the future of the business at the board level. The investors agreed; allowing him to stay on the board was the least they felt they could do.

It became untenable for Bill Gurley to remain on the board. Gurley may have won the fight, but Kalanick never wanted to see or deal with the venture capitalist ever again, much less work with him on Uber's board for years to come. After some haggling, the sides agreed on a compromise; Gurley would step off Uber's board and be replaced by his partner, Matt Cohler, instead.

The group promised Kalanick his soft landing. The exit would be simple and graceful; outsiders would understand.

Over text messages, Fenton showered Huffington with effusive praise:

My deepest and most heartfelt thank you. Today, you made the impossible happen. i'm in awe. I would love to work with you anytime, anywhere. Just think of what we can do when the circumstances aren't so unduly stressed. I badly want all the energy going forward to be towards the positive, fresh Uber. This company has a brilliant future.

Gurley texted the syndicate a final update. "We have a signed resignation letter."

AT 9:30 P.M. PACIFIC TIME, I got a final tip from a source as I walked back to my hotel in Downtown Los Angeles. I had been relayed a copy of the letter, and learned a general sense of Kalanick's day facing off against investors in Chicago. I was told to call Kalanick and Huffington, and ask for Kalanick's statement.

What I didn't know was that the two sides had negotiated a peaceful exit for Kalanick. I had no idea that they were going to tell the press that Kalanick decided, of his own accord, to step down from his position.

All I was told was that Kalanick was being pushed out in an investor coup, and that I needed to hurry up and file a story before someone beat me to it. What I wouldn't know until much later: at least one source in the syndicate full of people plotting Kalanick's downfall wanted to make sure Kalanick would *never* return to his spot at the top. While most of the syndicate expected the soft landing story to be carried out as planned, there were a select few who wanted it to look as messy as it all really was. And they used me, an unwitting participant, to make that happen.

As I received word from my contacts, I scrambled upstairs to my hotel room, furiously typed out a thousand words on Kalanick's ouster, and called the CEO and Huffington for comment.

"I love Uber more than anything in the world," Kalanick wrote to me in a final emailed statement. "At this difficult moment in my personal life, I have accepted the investors' request to step aside, so that Uber can go back to building rather than be distracted with another fight."

The story hit the web at just after 1:30 a.m. Eastern Time, as a push notification from the *New York Times* smartphone app was sent out to the home screens of hundreds of thousands of subscribers simultaneously. "Travis Kalanick resigned as chief executive of Uber after investors began revolting over legal and workplace scandals at the company," it said.

Kalanick was blindsided. He was supposed to get a soft landing, to tell the story of his departure on his own terms. Instead, he was utterly humiliated. Someone had betrayed him.

Back in Woodside, the members of the syndicate were all in shock. Someone had leaked the entire story to the *Times*. In the end, all they wanted was Kalanick's resignation, not his embarrassment. Somehow, in the scrum of the past forty-eight hours, things had gone sideways. There was a sense of guilt among the syndicate. But outweighing the guilt was something greater: a sense of relief. Travis Kalanick was no longer the chief executive of Uber.

Now, the company could start to rebuild.

LESS THAN TWENTY-FOUR HOURS LATER, a front-page story dropped in the *Times*, laying out in excruciating detail what happened to Travis Kalanick in Chicago. Seeing the tick-tock of events on page A1, followed by an enormous graphic of Kalanick's face shattered—like pieces of glass—across the front of the business section, was too much for him to bear. He was livid; the venture capitalists screwed him, like he always suspected they would.

They made a fool out of him in front of the entire world.

Chapter 30

DOWN BUT
NOT OUT

THE MORNING AFTER THE SYNDICATE DEPOSED HIM AS CEO AND A source sold him out to the *New York Times*, Travis Kalanick flew home to California, unclear what to do next. The two things he loved most in the world—his mother and his company—were gone. The press wouldn't stop hounding him. The majority of his workforce was cheering his departure.

What does a founder do when he has been fired from his own company? Kalanick, a man of immense energy and urgency, suddenly had nowhere to direct it. The fight was over, and he had lost. What now?

Kalanick decided he would travel to paradise. It was Diane von Furstenberg, the luxury fashion designer, who suggested he recuperate on a faraway island. Her husband, Barry Diller—the Manhattan media mogul and chairman of InterActiveCorp—had space on his yacht in the South Pacific. Diller and von Furstenberg were known for throwing killer parties in Tahiti; this was the kind of invitation Travis had always relished, but he was in no mood to party. Still, he felt punch-drunk, and followed the judgment of those around him. Despite her last-minute shift in allegiances in Chicago, Travis still trusted Arianna Huffington, who agreed with the celebrity couple.

At the end of June, Kalanick boarded a plane to Pape'ete, the capital of French Polynesia, where he spent a week off the coast recuperating on Diller's yacht, the *Eos*. The yacht—the second-largest of its kind in the world—was named for the Greek goddess who opens the gates of heaven for the sun each morning. Celebrities and friends cycled on and off Diller's boat, which sleeps sixteen (served by twenty crew), and others moored nearby. The visitors came and went, but the ex-CEO stayed for weeks. Kalanick's only consolation was an empathetic von Furstenberg, who tried her best to cheer him up.

Maybe, if details of his ouster hadn't leaked into the public immediately, Travis Kalanick would have spent more time in Tahiti. He needed more than a quick island getaway after eight years of working eighteen-hour days, seven days a week. In that time, he might even have made peace with what had happened, and learned from the cataclysmic conclusion to his career at the company. At that moment, Travis Kalanick had a chance to grow.

But after he saw the minute-by-minute tick-tock of his ouster splashed across the major newspapers in the world, he abandoned thoughts of a peaceful surrender. In Tahiti, at the end of June, Kalanick began preparing for war.

BILL GURLEY THOUGHT his headache was over after the showdown in Chicago.

For a few weeks after that day—one of the most stressful of Gurley's life—things seemed to quiet down. After the initial flurry of coverage, the media spotlight began to turn elsewhere. The board was ready to begin interviewing candidates to be Uber's next chief executive.

In the absence of a full-time CEO, the fourteen-person executive leadership team took responsibility for running the company while the board searched for a permanent replacement. It was an inelegant, over-sized solution; a committee of fourteen people was hardly a replacement for a fast-moving, decisive executive.

Worse, they quickly found themselves barraged with requests from Kalanick, who tried to isolate each of them individually, and coax them over to his side. Each day, a different member of the team would get a flurry of texts and phone calls from their former boss, all attempts to involve himself in daily decision-making—as if the Ritz-Carlton showdown had never occurred. Kalanick kept pinging one employee to discuss fallout from the infamous driver video incident, something that continued to plague him long after he had left.[*] He peppered them with questions on the present health of the business, while trying to guide decisions over its future. Kalanick was supposed to have stepped back—he hadn't.

Some members of the team were conflicted. Executives like Andrew Macdonald, Pierre-Dimitry Gore-Coty, and Rachel Holt—the triumvirate of leaders who collectively oversaw Uber's operations in hundreds of cities globally—had worked only for Kalanick most of their professional lives. Daniel Graf, a math and logistics wizard, felt close to Kalanick after he elevated Graf into a more senior role working on the core Uber product. Thuan Pham, Uber's chief technology officer, had been recruited personally by Kalanick and worked in lockstep with him for years. Now, they were all forced to shut him out.

But members of the executive leadership team weren't blind devotees. Many had floated in and out of kinship with their boss over the past eighteen months. Their positions against him hardened when Kalanick took his cajoling one step too far. In July, Kalanick began calling people—including Ryan Graves, a key shareholder and board member—asking for their allegiance and voting power, if he were to require it. At his most vulnerable and unpredictable, Kalanick was seeking sworn allies. The behavior scared members of the ELT; they didn't know what Kalanick was going to do next.

In the end, all fourteen members drafted and signed a letter to Uber's board of directors, urging it to take further action against Kalanick to stop his meddling. On Thursday, July 27, they sent the following message:

Dear Board of Directors:
In fulfilling our obligation to surface issues we feel are significant, we call your attention to three examples:
*   1. Travis recently reached out directly to an employee, first asking if he would talk to a reporter about an upcoming negative*

---

[*] Kalanick ended up paying the driver, Fawzi Kamel, approximately $200,000 of his own money, to keep Kamel quiet and prevent further damage. Some wondered if it was worth it, considering the video was already public.

*story related to the Fawzi Kamel incident (Kamel was the driver
in the March video). Previously, Travis' personal lawyer had also
reached out to this employee on the same topic.*

*Travis also requested that the employee produce private, inter-
nal emails for him, and said that if he refused to send them he
would exercise his right as a board member to get them directly
from the Security team. The employee did not produce the emails
and Travis subsequently asked the Security team to produce the
emails. The Security team also declined to produce the emails and
reported the incident to Salle, who subsequently advised the ELT
that we should stand firm against any requests that may violate
an employee's right of privacy, and that a Director on their own
cannot conduct independent investigations.*

*Travis also asked whether the employee had spoken with the
Covington investigators about the issue in question. The employee
was very troubled by this given the confidential nature of the Covington process and reported his concerns to the Legal team.*

*2. Travis recently called an ELT member to ask if he could
count on their votes (the particular purpose/vote was not identified). Current and former employees have reached out to the ELT
with similar reports. This has put the ELT in a difficult position,
wondering what Travis might be up to and whether or not it is a
cause for concern.*

*3. Travis continues to reach out to employees beyond the ELT
for business purposes. Regardless of the intention of the outreach,
it is disruptive to the daily work at Uber. There is also cause for
concern in that the outreach often comes with a request to conceal
the conversation from management.*

With deep respect,

The ELT

The letter came with an even firmer demand: If he continued to try
to regain power, all fourteen members of the ELT would resign from
their positions.

The move shook Gurley and the rest of the board. A mass exodus at the top could send the company into a death spiral. They had to do something to neutralize Kalanick.

Joe Sullivan, Uber's chief security officer, had an idea. With others lacking the courage to halt Kalanick, Sullivan knew what he needed to do. He would strip Kalanick of his electronic access to Uber.

One by one, Sullivan revoked all of his old boss's permissions to Uber's most sensitive information. Kalanick's Google drive access was shut down. His ability to enter chat rooms, internal wiki pages, and employee discussion forums were gone. In just a few keystrokes, Sullivan had defanged Kalanick.

It helped—for the moment.

THE BOARD HAD TAPPED an executive search firm to quickly find Uber's next CEO. Board members were convinced naming the next leader would prevent Kalanick from trying to claw his way back inside. But that person had to be strong enough to bring the hammer down on Uber's prodigal son.

Benchmark thought it had found that person: Meg Whitman. The firm had deep ties to Whitman, a career executive who had climbed the ladder at one Fortune 500 company after another. A Princeton grad and Harvard Business School alum, Whitman was tough, hard-charging, decisive. She held positions as a consultant at Bain & Company, a strategy executive at The Walt Disney Company, a general manager at Hasbro, and other positions. Her expectations on staff were exacting; low-performers were demoted, cut loose, or pushed out of the way.*

But her biggest break was in March 1998, when Bob Kagle, a founding partner of Benchmark, brought Whitman in to become

---

* Literally, in some instances. In 2007, Whitman was accused of shoving a subordinate in front of multiple employees. Whitman and the employee, Young Mi Kim, eventually dealt with the matter privately, with Whitman reportedly forking over a settlement of around $200,000.

the new CEO of eBay. The online auction site was a crown jewel in Benchmark's investment portfolio, though its founder and leader at the time, Pierre Omidyar, was no professional CEO. Kagle, who sat on eBay's board of directors, saw the growth potential in the young company and placed a young Whitman at the helm. By the time Whitman left eBay a decade later, the company was an industry titan, with more than 15,000 employees and a market capitalization of more than $40 billion.

After a failed campaign to become California's governor, Whitman eventually landed the chief executive position at Hewlett-Packard in 2011. Though she had supporters who cited her success at eBay, some pegged her from the start as the wrong CEO to reverse HP's declining hardware business. So, Whitman was open to offers when she was approached as a potential chief executive candidate for Uber—a skyrocketing, high-profile business emblematic of this generation's "unicorn" era.

Whitman had been acquainted with Uber since the beginning, and had even made an angel investment in 2010. At Benchmark's request, she had provided mentorship to the young executives at Uber while they were still getting their bearings, chatting over the occasional dinner at Whitman's house or office over the years. She liked Graves—he was a hard guy not to like—but kept a watchful eye from a healthy distance on the brash and unruly Kalanick. She advised them on possible board member additions, and was considered a possible candidate for Uber's board herself. Her first real counsel and business advice to Kalanick was about his obsession with China, a market she knew he would never conquer. "You will never have more than 30 percent of that market," Whitman once told Kalanick. "Because of the Chinese government, you could be Mother Teresa herself and still never gain more share."

As soon as Kalanick was ousted, Heidrick & Struggles, the executive search firm, reached out to Whitman as a possible candidate. At first, Whitman equivocated. She was still CEO of HP, after all. And even from the sidelines she could see Uber was an enormous mess. "I suggest you talk to everyone before me," Whitman told the recruiters. "If you run through that list and still find that you want me, call me back."

Benchmark had made its mind up already: they were Team Meg. She had the professional acumen they wanted, and experience scaling a software-based business globally. Most of all, she had a hard-and-fast rule: if Whitman was going to run Uber, Kalanick had to be completely gone. No meddling, no interfering, no nothing. She said that an Uber under Meg Whitman meant the end of Travis Kalanick—music to Benchmark's ears.

With Kalanick scratching at the windows, Benchmark had to act fast. Whitman was rushed to interviews with all the sitting board members—Matt Cohler of Benchmark, David Trujillo (who had replaced David Bonderman) of TPG, Ryan Graves, Garrett Camp, nearly everyone.

On the afternoon of Tuesday, July 25, Whitman was driving her car in downtown Palo Alto when she got a frantic phone call from Henry Gomez, her top communications and marketing strategist at HP. A story was about to be published that claimed she was among the candidates for the top job at Uber.

Whitman was beyond livid. One of Kalanick's allies, who knew of Whitman's hard-line stance against him, planted the leak with the press to smoke Whitman out. As the CEO of a public company, Whitman would be forced to bow out, lest she risk revolt among her employees and shareholders at HP.

In the weeks that led up to the courtship, Whitman had stressed that there could not be any leaks about her participation or consideration. Any hint of her departure could devastate HP, which was already in dire financial straits. She made clear she would deny everything if the press caught wind of the situation. For two days, Whitman's spokesman gave the same statement over and over: Whitman was "fully committed" to Hewlett-Packard, and planned to stay with the company until her work was done.

The press speculation continued, however, and then escalated on the afternoon of Thursday, July 27, after another leak: Jeff Immelt, the outgoing CEO of General Electric, was also a top candidate for Uber CEO.

Much of the board speculated about the purpose of the leak—some believed while Benchmark was trying to rush Whitman through, Kalanick and his allies were pushing Immelt, a candidate who was much more amenable to Kalanick's presence at the company than Whitman. Whitman didn't want Kalanick allowed in the building, but with Immelt, Kalanick saw a pathway to a comeback.

This was Gurley's worst nightmare. It wasn't clear Immelt had a vision for his current company—GE's stock price and business had cratered during Immelt's tenure—much less a coherent one for Uber. But Gurley worried more about something else: if Immelt left even a crack for Kalanick to push his way back in, who knew what would happen next?

The crescendo of media coverage intensified the pressure on Whitman. HP's board felt her first statement was not unequivocal enough; employees and shareholders agreed. So Whitman did what she believed had to be done: she pulled the ripcord.

At around seven o'clock in the evening that Thursday, just as Uber's board of directors was beginning a quarterly meeting to discuss the progress of the CEO search, Whitman sent out three brief tweets to the world.

"Normally I do not comment on rumors, but the speculation about my future and Uber has become a distraction. So let me make this as clear as I can. I am fully committed to HPE and plan to remain the company's CEO. We have a lot of work still to do at HPE and I am not going anywhere," she said. The eight board members' cell phones began lighting up and buzzing, one after the other, as Whitman's tweets began circulating throughout the Twittersphere.

Her final sentence was unequivocal: "Uber's CEO will not be Meg Whitman."

BILL GURLEY was crestfallen.

His firm had spent weeks grooming Whitman to take the CEO job, only to have her nuke her own candidacy in public at the last moment.

Kalanick, refreshed and geared for war, had begun to play dirty—just as Gurley had feared.

Now it was Gurley's turn to fight back. On August 10, Benchmark partner Matt Cohler—who was on safari in the middle of the African outback, surrounded by elephants, lions, and hippopotamuses—began calling other board members to notify them of Benchmark's next move: the firm filed a lawsuit against Travis Kalanick accusing him of defrauding Uber's shareholders and breaching his fiduciary duty, a stunning act of open warfare between board members at a high-profile company.

Gurley's move was strategic, but also desperate. Kalanick was laying siege; he had reneged on his deal to name two independent board members to the seats he controlled. If Kalanick added two puppet directors to the board, their support could clear a pathway for him to return.

Gurley's idea in suing the former CEO was to invalidate Kalanick's rights to those board seats entirely. In the lawsuit, Benchmark claimed Kalanick lied to Gurley and the rest of the board, all of whom would never have given such power to Kalanick were they to know how poorly Kalanick had been running his company.

This was an ironic case for Benchmark to make. Gurley hadn't blinked when Travis barreled past regulators and operated illegally in cities, or launched self-driving cars in San Francisco against the wishes of transportation authorities. He invested in a transportation disruptor, and disruptors by definition don't play by the rules. Some inside Uber thought Gurley and Benchmark expressing shock—*shock!*—at Travis's behavior was disingenuous.

Still, a VC firm suing one of its own CEOs was a big deal, showing just how far Benchmark was willing to go to rid Uber of Travis Kalanick. If the coup hadn't already done so, the lawsuit damaged the "founder friendly" image Benchmark had worked so long to cultivate.

Shervin Pishevar, an early Uber backer, came to Kalanick's defense. In the war of investors versus Travis Kalanick, Pishevar sided with Kalanick. On August 11, Pishevar sent a letter to Benchmark, asking the firm to step down from Ubers' board of directors.

"We do not feel it was either prudent or necessary from the standpoint of shareholder value, to hold the company hostage to a public relations disaster by demanding Mr. Kalanick's resignation," the letter said, claiming to represent a group of shareholders. The move came with an offer: Pishevar and his coalition said it would buy out 75 percent of the shares held by Benchmark—an action that would require Benchmark to step down from the board.

Gurley and his allies didn't believe it. To them, Pishevar was nothing but talk. He made such enormous boasts, so frequently, that he had become a Silicon Valley in-joke among VCs, so over-the-top that not even his friends took him seriously. Now he claimed, with no evidence, that he represented a group with billions of dollars in capital to purchase Benchmark's stake. They believed Pishevar was clearly a front for Kalanick, who was trying to force Benchmark off the board.

On the contrary, if Kalanick were able to pick who would wage battle on his behalf in a war against Benchmark, Pishevar would not have been his first choice. Nonetheless, Pishevar's parry made a certain amount of strategic sense. Based on the rules of the company, the board and Benchmark were forced to seriously consider his proposal. If Benchmark were off the board, it would give Kalanick the room he needed to return as CEO. More than a clever stalling tactic, it might have actually worked.

AS KALANICK, GURLEY, and their allies traded blows in public, a new figure was circling like a vulture over the company, drawn by the smell of money. That figure was Masayoshi Son.

Known more colloquially by the business world as "Masa," Son was the founder and CEO of SoftBank, a Japanese mega-conglomerate with stakes in some of the world's most successful finance, telecommunications and technology companies.

He also happened to take a "madman" approach to the world of business; rivals could never predict Masa's strategy, could never guess his next move.

A short, vivacious Korean who grew up in Japan, Son was always an outsider; his childhood Japanese classmates threw rocks at him for his heritage. After his idol, the founder of McDonald's in Japan, told him to study in the United States, Son made his way to California and enrolled at the University of California, Berkeley, where he majored in economics. He bankrolled his college years by importing Pac-Man video arcade machines and then leasing them to bars and restaurants around the Bay Area.

He returned to Japan to make his fortune, starting SoftBank to disrupt the telecommunications industry in 1981. Over twenty years, Son grew his fledgling startup into a corporate behemoth with a $180-billion market capitalization, based on Masa's maverick style of making big bets on world-changing companies and industries. He invested widely across Silicon Valley in the height of the dot-com craze, spreading SoftBank's capital across dozens of risky bets. In 2000, the crash erased billions of market value overnight, and SoftBank's value plummeted. Masa himself lost some $70 billion in personal wealth. Webvan, one of Son's biggest losing investments, also happened to be a portfolio company of Benchmark's.

Masa wasn't down for long. Over the next decade he continued making big, bold bets and built SoftBank back to the force it once was. By the early 2010s, SoftBank owned stakes in more than a thousand internet companies; his acquisition of Sprint made SoftBank the world's third-largest telecommunications company. Friends and colleagues considered him fearless. Son said he hoped to be remembered as a "crazy guy who bet on the future."

By 2017, SoftBank had been making serious turbulence in Silicon Valley by slinging money from the "Vision Fund," an enormous $100-billion pool of capital formed by the Public Investment Fund of Saudi Arabia, the Abu Dhabi Investment Authority, Apple, Qualcomm, and SoftBank itself, along with a few others. Masa's mandate was simple: by focusing the fund on technology investments—something he had done practically his entire career—SoftBank would finance the global tech infrastructure that would undergird the future. He designed the investment vehicle for speed; Vision Fund was required to invest all of its capital within five years of its closing date. That meant parking truckloads of cash in startups, fast.[*]

As Uber's management and morale eroded, Masa sensed opportunity. The feud between Kalanick and investors had surely knocked enormous sums off Uber's potential market value. If SoftBank could buy shares at a lower valuation than Uber's last $68.5-billion round, Masa had a chance at making *billions* by the time Uber had re-stabilized and went public—if that day should ever come.

An IPO was still a big "if." At the moment there was open warfare between members on the board of directors, employees continued to leave, and users were fleeing the product for Uber's main competitors. There was a real chance the company would continue to stumble, and perhaps even implode.

For Masa Son, that made an investment in Uber that much more attractive. He needed to find a way in.

---

[*] This strategy disturbed the dynamics of venture investing in Silicon Valley. None of the funds in the Valley had the money to compete with SoftBank. A $100-million investment from SoftBank could make a startup overnight, while getting shut out from SoftBank could break one.

Chapter 31

THE GRAND
BARGAIN

BEGINNING ON FRIDAY, AUGUST 25, AND THROUGH THE REST OF THAT
weekend, Uber's board of directors planned to make a final decision on
who would become the company's new chief executive officer.

By the end of the summer, just a few weeks after Meg Whitman had
pulled herself out of consideration, the recruiting firm had produced
a list of five potentials, then whittled the group to three. Over the last
weekend in August, each of the three candidates was asked to give
individual presentations to the board of directors. It was a test run, a
demonstration of their skills and an opportunity to present a roadmap
of how they, if chosen, would run Uber.

Jeff Immelt, General Electric's outgoing chief, was still Kalanick's
top choice. The sixty-year-old executive was winding down a terrible
run at GE. The storied corporation had lost billions in market cap
during Immelt's tenure, and his board asked Immelt to "retire" early in
2017. Bringing Uber out of its darkest period and, eventually, across the
finish line of an initial public offering, would certainly rehab Immelt's
image and cement his business legacy. Most important to Kalanick,
though, was that Immelt was malleable, open to Kalanick's continued
influence at the company. Immelt was the best possible pick for the
ousted founder who wasn't ready to relinquish control.

Then there was the dark horse candidate: Dara Khosrowshahi. A
career executive and the current chief of the travel and logistics site
Expedia.com, Khosrowshahi made plenty of sense on paper. He was a
Brown undergraduate in bioelectric engineering who later turned investment banker at Allen & Company. With thinning hair balancing a thick

brow line and a full, steeply-bridged nose, Khosrowshahi was hand-some, charming, even cool. Like someone's dad, who also happened to look good while wearing black skinny jeans. Westerners often found his Persian surname tricky; everyone ended up calling him "Dara."

Khosrowshahi's family fled Tehran in the late '70s in the midst of the revolution that brought the Ayatollah Khomeini to power, escaping to the south of France before eventually settling in Tarrytown, New York. His parents, trying to usher their sons into American culture as pain-lessly as possible, enrolled young Dara and his two brothers at Hackley, a K–12 private prep school in the area, where they quickly assimilated. Khosrowshahi worked hard in high school to gain entry into the Ivy League. "There's this chip you have on your shoulder as an immigrant that drives you," he later said of his childhood.

After Allen & Company, Khosrowshahi joined Barry Diller's Inter-ActiveCorp where he worked for years until moving over to Expedia and eventually rising to the top spot. Expedia's travel business was all about logistics, getting people around the world via an online market-place. It was, as it turned out, a business not terribly different than the one he was asked to consider running at Uber.

But where Kalanick had the kinetic energy of a pinball machine, Khosrowshahi was calm and collected, a perpetual Zen that, to the uninitiated, sometimes came off as boring, even passive. Uber's direc-tors were used to Kalanick, the vibrant, world-changing visionary—a true showman. The understated Khosrowshahi made perfect sense as an executive, though he lacked some of the punch and panache the board was used to. It was clear that everyone on the board liked Dara. But it was also clear that none of them quite *loved* Dara. As a result, he became an emergency back-up candidate, a safe pick for the group. His identity was kept well hidden from the press through the process.

The last thing the Benchmark partners wanted in the next CEO was a passive weakling. Give Kalanick even an inch, and he'd claw his way back inside. Jeff Immelt wasn't going to control Kalanick, and they were unsure whether Khosrowshahi would have the mettle, either. They needed someone truly unflappable: Meg Whitman.

Gurley believed that they might be able to convince Whitman to get back in the game. It would be difficult; Whitman's statement on Twitter had been ironclad. She would look enormously hypocritical were she to take the position. So it would be on them to convince Whitman it wouldn't matter. Become CEO of Uber first, repair the fallout later.

It was Ryan Graves who ended up coming through for Benchmark. Whitman had coached Graves on executive leadership. Graves, the affable, bear-sized mascot of Uber, had grown closer to Whitman than Kalanick had during their time together. In the days leading up to the final weekend of demonstrations Graves called up Whitman, begging her to reconsider. "We're down to the short strokes here," he said. And this time, Graves swore there would be no more fuckups. "I promise you, Meg. This *Will. Not. Leak.*"

Whitman was still stewing from the last go-round. A public company CEO entertaining an offer from another company wasn't just a bad look, it was a material problem for shareholders. Whitman didn't want to court a risky situation just to get publicly screwed again. She needed reassurances.

"This is what you need to do," Whitman said. "Talk to the other two, and make *sure* you want *me*, not them, after that."

Graves responded that the only person on the board who liked Immelt was Kalanick, and though everyone certainly liked Khosrowshahi, no one was 100 percent sold. Graves was upfront: "We want you, Meg." He all but guaranteed Whitman would have the job if she came back and presented to the board over the last weekend in August.

Meg Whitman had made up her mind: She wanted to be the next CEO of Uber. "Okay then," she told Graves. "Let's talk."

ON FRIDAY, AUGUST 25, Jeff Immelt and Dara Khosrowshahi both made their way to 345 California Street, through the concrete and gold-colored entryway, and up to the offices of Texas Pacific Group, the private equity fund founded by David Bonderman. Most of Uber's board of directors[*]—including Travis Kalanick—had gathered in an airy, well-lit conference room in TPG's office on the thirty-third floor. For most of Uber's lifespan it had been led by a man with unparalleled hustle, a super pumped visionary who pushed himself and his company to the very edge until, ultimately, he had hurtled over it. Now Uber needed something different in a leader: Uber needed a grown-up. The two male candidates would lay out their visions that Friday, while Meg Whitman would give her presentation the following day.

Immelt went first, and he was a disaster. He seemed completely out of touch and unprepared. He seemed not to understand what went into running a sophisticated, heavily regulated three-sided marketplace. One board member called the entire spiel "a bad joke."

Immelt's presentation looked even worse after Dara Khosrowshahi arrived. As Khosrowshahi fired up his laptop and began walking the board through slides, it was immediately clear to the room that he understood the fundamentals of Uber's business. Khosrowshahi came from logistics and the world of online marketplaces; over his twelve-year run as the CEO of Expedia, he grew annual revenue from $2 billion to $10 billion. He understood the intricacies of the ride-hailing market, the complicated economics of balancing riders' desire for cheap fares with drivers' need to earn enough to keep them on the road. While Khosrowshahi knew Uber was a company rooted in its on-the-ground operations prowess, he appreciated the technical chops and importance of a strong engineering culture. Most of all, Khosrowshahi understood the importance of brand—and at the moment there were few brands in the business world in worse shape than Uber's.

At one point during the presentation, Khosrowshahi pulled up a slide on his PowerPoint deck that made everyone in the room tense up: "There Cannot Be Two CEOs," the slide read. As Khosrowshahi looked across the room, directly at Travis Kalanick, he made clear that were he to become Uber's new leader, Kalanick would truly have to

---

[*] Two board members were out of the country, and had to dial in to a conference line that morning.

take a hike. The former chief's only involvement would be his board duties, but no more, Khosrowshahi said.

Having finished for the day, the board went out for dinner together that evening to discuss the candidates' performance. Over bottles of wine and farm-to-table entrées, board members talked about how impressed they were with Khosrowshahi. For someone who hadn't attracted much attention over the weeks-long search, Dara had nailed his presentation, which was a pleasant surprise. Even if things went completely askew with Whitman and Immelt, they knew they had a backup candidate they could all feel comfortable with.

One thing everyone could agree on was that Immelt was not cut out for the job. No one, in good conscience, could vote for him to be the next chief of Uber—even Kalanick and his supporters.

On Saturday morning, August 26, Meg Whitman walked out of the elevator at the Four Seasons Hotel on Market Street, through the spacious fifth-floor lobby and into a private executive suite to meet the rest of Uber's board members for her presentation. Whitman wore a cap pulled down tightly to shield her eyes and face as much as possible, in case there were any press camped out at the hotel restaurant or by the elevators. It was common for Silicon Valley executives to dine at MKT, the hotel eatery, and Whitman's face was well known in the Bay Area. If anyone saw her at Uber's offices or even at TPG, things could get ugly for her in public again.

Whitman's presentation was an upfront, no-bullshit display. If she were chosen for the job, she meant business. "If you think I'm the right person for the job, we need to settle a few things immediately," Whitman said, commanding the attention of the room. "This lawsuit?" Whitman said, referencing the battle between Benchmark and Kalanick. "We need it settled and done."

Worse for Whitman were the never-ending cascade of leaks from the boardroom. She said it reminded her of early days at Hewlett-Packard, when a dysfunctional board fed stories to the press that rattled directors and their trust in one another. "We need to seal these board leaks," she said. "No one—*no one*—on this board should just be taking unilateral actions on their own. The board is *splintered*," she said, the word hanging in the air for effect. "We need to be cohesive, we need the board to be as one. What we *cannot have* are random acts of violence against this company."

Whitman was tough, especially on Kalanick. She made it clear that Kalanick would not have an operational role. Kalanick was a founder and a director, but *not* the CEO. And for Whitman, it was going to stay that way. Moreover, if the board chose her, they would also have to enact a wholesale restructuring of company governance. Kalanick's overwhelming power over board seats would not be tenable under her watch.

On Sunday morning, as board members prepared for a day of deliberation, a tweet began circulating virally: It was from Jeff Immelt, who decided to take himself out of the running in public. "I have decided not to pursue a leadership position at Uber," Immelt wrote. "I have immense respect for the company & founders—Travis, Garrett and Ryan." People close to Immelt immediately tried to spin the move as a decision Immelt had made on his own, to avoid a dysfunctional situation. The board knew better; on Saturday evening, one board member reached out to Immelt as a courtesy. They let him know he did not have the support needed to win the position. Hours later, in order to save face, Immelt bowed out in a tweet.

With the choice now down to two candidates, it was time for the board to deliberate and vote. It quickly became clear where the lines were drawn. As soon as Immelt dropped out, his four supporters quickly switched to backing Khosrowshahi. That left the remaining board members all stumping for Whitman. Throughout the day, the group voted anonymously using a creative method: each of them texted their choices to Jeff Sanders, a partner at Heidrick & Struggles, who had been helping the board during its search. The votes kept coming in deadlocked; neither side would budge.

As the meetings dragged into the afternoon with little progress, Matt Cohler of Benchmark stood up in the conference room to make a speech. Benchmark had already prepared for Whitman's ascendancy and practically assured her the job. Uber's communications team had already prepped a memo ghostwritten for Whitman to announce her acceptance of the job to employees internally. The rollout plan was ready, all they needed to do was vote Whitman in.

Then, some believe, Cohler made a miscalculation. The Benchmark partner gave the table an ultimatum: If the board voted for Whitman, Benchmark would drop its lawsuit against Kalanick. It read to the room as an ultimatum. This was the price of peace.

For once, Kalanick wasn't the only one who directors felt was acting childish. Cohler's brinksmanship dismayed almost everyone in the room. Instead of following a fair process to determine the best candidate, Benchmark was effectively holding the board hostage to approve the candidate of their choice.

Cohler's speech may have cost Whitman the job. After the next secret ballot, the votes came in again, but this time it was not deadlocked. It was five to three in favor of Khosrowshahi. The group had chosen Uber's next CEO.

To give the process some semblance of cohesion, the board had agreed that whoever won, they would cast one final ballot and all unanimously vote for the same candidate. That way, when they finally announced the decision to the public, the board could pretend it had been united the entire time.

In the end, none of it went according to plan. As the group contacted a spokeswoman to prepare a final statement announcing it had made its decision, word of the decision began dribbling out to the press. Shortly after five o'clock in the afternoon, reporters published stories that confirmed Uber's new chief executive even before the board was able to call Khosrowshahi and give him the good news. Huffington was the one tasked with calling Khosrowshahi to officially offer him the position.

"Hello, Dara?" Huffington said, in her unmistakable Greek accent. "Dara, I have good news and I have bad news for you." Khosrowshahi listened, chuckling into the phone.

"Dara, the good news is that you are Uber's next CEO. The bad news is that it has already leaked."

THE THING THAT FINALLY SEVERED Kalanick from his leadership of Uber wasn't the vote to name Dara Khosrowshahi as CEO, nor was it Gurley's lawsuit against him for defrauding investors. It was a deal that would come months later, in the form of what Gurley called "the Grand Bargain," courtesy of SoftBank and Masayoshi Son.

In December, Son reached a deal with Khosrowshahi and Uber's board, in which SoftBank would purchase some 17.5 percent of Uber's overall shares in what was called a tender offer, a way for outsiders to buy stock from existing shareholders in the company. The Soft-Bank shares would come from a group of people, including employees who were long waiting to sell but couldn't due to Kalanick's restrictions. They would come from investors like Benchmark, First Round, Lowercase, Google Ventures, and other early Uber investors. Most importantly for Son, SoftBank would purchase those shares at a steep discount from Uber's valuation earlier in the year. Son and Khosrowshahi settled on a purchase price of $33 per share, pegging Uber's valuation at about $48 billion—a steal for SoftBank. That meant that the scandals of the previous twelve months had knocked about $20 billion off Uber's private market value.

To keep the price propped up on paper, investors did some sleight-of-hand maneuvering. SoftBank would purchase $1.25 billion in additional, newly issued shares at Uber's previous existing valuation of $68.5 billion. The premise was absurd; the secondary market clearly valued Uber's shares at far lower than they were before Uber's 2017 from hell. Yet in the eyes of the market, the maneuver worked; Uber's valuation would remain at $68.5 billion.

The grand bargain would also add an additional six new seats to the company's board, two of which would go to SoftBank, while the remaining four would go towards independent directors and a new board chairperson. Six is an enormous number of seats to add to a company's board of directors, but most observers felt it was necessary. Adding that many new independent seats provided a counterbalance to Kalanick in case he began another fight for control.

As the terms of the "Grand Bargain" were being negotiated, Kalanick tried exactly that. In September, he used an old portion of Uber's charter that allowed him to name two new directors—Ursula Burns of Xerox and John Thain of Merrill Lynch—to the company's board. It was a pure preemptive strike; he had given the rest of the board just five minutes' notice before announcing the move to the public.

Gurley could only laugh. He knew if the board could negotiate its way to completing the deal with SoftBank, Kalanick's stunt would be in vain.

"Today's move is a futile, last-ditch effort as the door closes on Kalanick's dark reign," Gurley texted, a few drinks in, to a close friend in the hours that followed Kalanick's board appointments.

Gurley had added a final, important provision in the SoftBank deal. For years, Kalanick held an enormous amount of stock that carried with it ten-to-one voting rights. By enforcing a "one share, one vote" rule, the agreement severely curtailed Kalanick's influence over the company and his ability to use his stock to sway the company's direction. The revocation of supervoting power, combined with additional board seats for neutral directors, finally gave the board the leverage it needed to unlock the death grip Kalanick had held on the company for nearly a decade.

By December 28, 2017, Gurley's grand bargain was signed. Kalanick had lost. Gurley had finally won.

AFTER THE CEO DECISION that last Sunday in August, the next forty-eight hours were a blur for Khosrowshahi and the board. The leak was devastating. The board had worked so carefully to keep its weekend deliberations secret. This was supposed to be a moment to turn over a new leaf, yet instead it fractured trust at the top.

It took two days to finalize the negotiations. While the whole world knew Uber wanted Khosrowshahi as its next chief, he still hadn't agreed to take the job; he could extract steep concessions from the company as conditions for his acceptance. He wound things down at Expedia, enhanced his new contract, and prepared to meet his new company. Khosrowshahi negotiated himself quite a package as a result; if he was able to take Uber public by the end of 2019—roughly two years from his hiring—at a valuation of $120 billion, Khosrowshahi would net a personal payday of more than $100 million.

After the requisite contracts were signed and he said his goodbyes to employees at Expedia's offices in Seattle, Khosrowshahi boarded a flight to San Francisco on Tuesday to visit his new employer.

Huffington immediately began to orchestrate the succession. She proposed that she would introduce Khosrowshahi at an employee all-hands meeting on Wednesday, and interview him on stage, live-streamed to all of Uber's 15,000 employees so that they could begin to get to know their new leader. And to make it seem like the executives were beginning to settle their disagreements, she asked Kalanick to appear on stage with him the next day. Huffington was good at this kind of thing; she loved the pageantry of it, passing the torch of Uber from one leader to another. Kalanick agreed to the meeting.

The board knew Wednesday would be a feeding frenzy for the press. Before the big event, leadership thought it would be a good idea to have one last supper together, a chance for everyone at the top of Uber to get to know one another.

On Tuesday evening, Uber's board of directors and executive leadership team met to eat at Quince, the Michelin-rated restaurant in the Jackson Square neighborhood of San Francisco. In a spacious private room at the back of the restaurant, the group of twenty peppered their new CEO with dozens of questions. Months of anguish melted as executives allowed themselves to laugh and loosen up around one another. Wait staff poured glasses of Bordeaux and Riesling from enormous glass decanters.

By the end of the night, many in the group were hammered. People were giving impromptu speeches, opening up about grievances they'd had and intended to settle that evening. It was the last night they might have together before an entire new cycle of scrutiny from the press and the public, as Kalanick faded into the background while Khosrowshahi rose to prominence.

Kalanick, to his credit, spent much of the evening being magnanimous, choosing not to dominate the conversation, as was his tendency. He was reserved but not quiet, friendly though not overjoyed. Whatever anguish he harbored he was able to hide it well, spending the night in good spirits with his former subordinates.

Apropos of nothing, Joe Sullivan, Uber's chief security officer, stood up at the table to make a toast. With a glass of red in one hand, the tall, awkward Sullivan turned toward his new boss and told him what he believed they were all feeling at that moment.

"Dara, I just wanted to say we're glad you're here, on behalf of all of us," he said. "We came to this company because we thought it was changing the world. We wanted to be a part of that. And we still do—we want Uber to be iconic," Sullivan continued. He was drunk, he later admitted, but with that came honesty that Sullivan hadn't shared with executives in some time, given the back-biting, cutthroat atmosphere over one incredible, horrible year.

"We hope you're not a two-year CEO," he said, looking directly at Khosrowshahi. "We hope you're the one."

Sullivan raised his glass. "Cheers," he said.

The room answered: "Hear, Hear!"

EPILOGUE

OVER THE NEXT EIGHTEEN MONTHS, UBER'S NEW CHIEF EXECUTIVE Dara Khosrowshahi would systematically undo nearly everything his predecessor had stood for.

Gurley's "Grand Bargain" had worked. Kalanick's power over the company had been diminished. In exchange for permanently removing Kalanick, Khosrowshahi had effectively accepted a $20-billion hit to Uber's valuation.

Khosrowshahi's first job was to repair relationships with its hundreds of thousands of regular drivers after years of abuse and neglect. By the time Khosrowshahi was voted in, Uber was halfway through its "180 days of change" campaign to improve relations. Led by two executives from Kalanick's reign—Rachel Holt and Aaron Schildkrout—the campaign involved an extended listening tour and apology, as well as new features and improvements drivers had been requesting for years. One of the most meaningful changes was something Kalanick had personally prevented: the ability to tip drivers. After Kalanick was ousted, the feature was implemented, earning the company an enormous amount of goodwill.

Khosrowshahi also began recruiting his own lieutenants. Where Kalanick had Emil Michael, Khosrowshahi hired Barney Harford, an Expedia executive and longtime trusted colleague, as Uber's chief operating officer.[*] Where Kalanick managed all of the company's

---

[*] Harford's tenure has not been all sweetness and light. Months after he began his role at the company, I reported what appeared to be Harford's persistent problem with making sexist and racially insensitive comments to subordinates. Harford was reprimanded and forced to undergo sensitivity training and executive coaching, but was not let go.

finances himself, Khosrowshahi hired Nelson Chai, a former Merrill Lynch executive, as the new CFO, who investors hoped would help get Uber back to a place of fiscal responsibility. Ronald Sugar, a former chief executive of the defense contracting firm Northrop Grumman, joined Uber's board of directors as its independent chairman, another position that was vacant until Khosrowshahi started at the company. And by hiring Tony West, a former associate attorney general at the Department of Justice, Khosrowshahi made clear that Uber was going to take its legal and compliance obligations seriously. For the first time in its nine-year history, Uber had installed proper corporate governance—mechanisms and officers that Bill Gurley had long desired.

Khosrowshahi then revamped Uber's overarching philosophy. Uber's fourteen values, once sacrosanct, were replaced with a list of eight simple maxims. Items like "super pumped" and "always be hustlin' "—values sprung from the mind of an arrogant young man—were discarded. Instead they were replaced with a set of fairly bland platitudes that touched on "customer obsession," à la Jeff Bezos, and a celebration of employee differences. The most important norm was the one Khosrowshahi repeated at nearly every press appearance and television interview during his extended 2018 apology tour: "We do the right thing. Period."

The new values were a full-throated repudiation of Khosrowshahi's predecessor; where the *old boss* was a ne'er-do-well, the *new boss* had integrity. With a warm smile, balding hairline and fuzzy beard, Dara Khosrowshahi was dubbed the new "Dad" of Silicon Valley.

Nearly overnight, "Dad" was everywhere. Over the airwaves, in magazines and newspapers, and across YouTube, Uber blanketed media with ads featuring Khosrowshahi's face. The company earmarked half a billion dollars for 2018 solely to repair and rebrand its battered image. They bought ads widely over the NBA playoffs and finals games, prime-time popular TV shows, and major publications like the *Wall Street Journal*.

But beyond the positive media blitz, executives at the company were happy to make as little news as possible. After spending nearly a full year with negative Uber headlines plastered across print, TV, and the internet, the company intentionally tried to stay as dull as possible.

"Khosrowshahi has been perfecting his brand of boring for 365 days," *WIRED* said of the CEO's one-year anniversary at the company in the fall of 2018.

Image and public relations were no small tasks, but Khosrowshahi had a much larger and thornier challenge ahead: reining in Uber's profligate spending and creating a path to profitability. For years, Kalanick had no checks on his decisions. He had burned through billions of dollars, for example, in money-losing wars with other ride-hailing companies across multiple continents. Khosrowshahi, a CFO for years under Barry Diller at InterActiveCorp, was a number-cruncher, an executive who met budgets. As he looked at Uber's balance sheet, awash in red ink, he began to cut losses. That meant selling off Uber's business in Southeast Asia to Grab, the local ride-hailing competitor, for a 27.5 percent ownership stake in the Singaporean company. Where Uber was notorious for poaching employees from competitors in earlier years, Khosrowshahi stopped flinging around enormous pay packages to compete with Facebook and Google for engineering talent.

Uber's self-driving division, one of the biggest drains on Uber's finances and once considered an "existential" area of development for the company, is in limbo at the time of this writing.

After his days at Uber, Anthony Levandowski, now disgraced in Silicon Valley, would not go quietly into the night. Levandowski would go on to form *another* self-driving trucking startup, Pronto.ai, that provided an off-the-shelf kit to bring autonomous driving to long-haul truckers for only $5,000. "I know what some of you might be thinking: 'He's back?'" he wrote in a blog post announcing his new company. "Yes, I'm back."

When not working on his startup, Levandowski was busy founding his own religion: a church devoted to Artificial Intelligence as a god-head. It was called the "Way of the Future."

Employees have found solace with the state of things under Khosrowshahi, largely happy that their employer has emerged from being one of the most hated companies in America, as it was in 2017. Cocktail parties are safe again. But for some, there remains a lingering, persistent concern: Is Uber under Dara Khosrowshahi still going to swing for the fences? Or has Uber lost its appetite for moonshots and world domination—the alluring, Travis-like quests that attracted them to the company in the first place?

As one former employee put it: "Is Uber going to be an Amazon, a company that dominates in every sector it branches into? Or are we going to be another eBay?"

LIFE FOR BILL GURLEY has grown quite a bit easier.

At the end of 2017, Gurley stood on the floor of the Nasdaq stock exchange with Katrina Lake, the entrepreneur and CEO of Stitch Fix who counseled him earlier in the year. That day in November, Lake addressed her newfound public investors for the first time with her fourteen-month-old son in tow. Gurley towered behind her, tall and awkward in a black suit and sky-blue tie, his fast-graying dark hair neatly parted to the left, as he smiled and applauded his entrepreneur, the youngest female founder ever to take a company public. It was protégés like Lake that made him the proudest. Lake was thirty-four. She had fought for every inch of ground that her startup had gained since she founded it at age twenty-eight. Lake valued Gurley's advice and direction but was confident following her own instincts, building Stitch Fix into a public company with integrity.

He wouldn't be able to do the same with Kalanick. The SoftBank deal had been a sad coda to the relationship between a founder and an investor who once considered themselves the closest of allies and friends. And the damage to Benchmark's "founder friendly" image was going to last indefinitely.

Still, things had changed. Back during Uber's darkest times, Gurley wondered if the entire business was destined to fail. He had truly believed tens of billions of dollars in value might evaporate in an instant, all because Gurley had failed to save the company from a risk-taking madman named Travis Kalanick. It was a thought that kept him up late into the night.

Gurley didn't have to worry about that anymore. He was sleeping much better lately.

AT THE END OF NOVEMBER 2017, Joe Sullivan was in the mountains near Lake Tahoe with his family. Sullivan and his daughters were preparing for Thanksgiving dinner the next day, a tradition they spent at the family's cabin just a few hours north of San Francisco. As Sullivan was cooking and half-listening to a football game playing on television in the other room, he got a message from Uber's human resources department, asking him to join a conference call later that day.

Sullivan wasn't stupid. Executives don't just get surprise emails from human resources asking to arrange an emergency phone call the evening before a holiday weekend. He replied, refusing the offer and demanding to know what was going on.

The HR rep responded: Uber was firing him. More than a year ago, when Uber suffered a security breach that resulted in the theft of millions of drivers' identities, Sullivan had not sought outside legal advice or counsel, nor did he inform the authorities it had occurred. Sullivan and his team had spent millions in an operation to find the hacker, pay a bounty for them to delete the data, and keep the incident quiet.

For Sullivan, the payment was considered part of a "Bug Bounty" program, a common Silicon Valley tactic by which corporations paid out so-called "White Hat" hackers—or "good guy" hackers—to find security flaws in a company's systems and point them out or exploit them, then notify the company. For their trouble, companies would pay the hackers a bounty; the bigger the "bug," the heftier the bounty. Sullivan paid this hacker, known as "Preacher," $100,000 for his bounty. The operation had been successful, Sullivan believed, as he managed to thwart a potentially catastrophic incident.

New management at the company saw it differently. Tony West, now Uber's chief legal officer, was furious that Sullivan or Kalanick hadn't informed authorities immediately of the breach. West was baffled as to why Sullivan spent millions to track down "Preacher"—the hacker who turned out to be a twenty-one-year-old man named Brandon, living with his mother and brother in a trailer park in Florida. Sullivan should have turned Brandon in to the authorities, West said, because Uber was required to notify consumers in the event of a data breach. Not doing so would ultimately result in millions of dollars of payments. Instead, Sullivan paid "Preacher" off and sent the hacker on his way.

For the better part of an hour, Sullivan tried to explain to Uber's legal and human resources representatives that they had it wrong, that the way he and his team handled the breach was all aboveboard.

His efforts were in vain. The best they could offer was a lump-sum payment if Sullivan agreed to sign a non-disparagement agreement. Sullivan was too pissed off to even consider it; he declined.

He didn't ponder the decision he had made for long. Forty-five minutes later, Sullivan got a phone call from a reporter, requesting comment on his handling of the 2016 data breach and hacker payments. Uber executives had sanctioned a leak to the reporter that claimed Sullivan had led a cover-up operation to pay off hackers and hide the evidence of the incident from consumers. Fifteen minutes after that, the reporter's story went live to the world.

Before he even had time to react, Sullivan's electronics stopped working. His company laptop was remotely wiped of all its contents from Uber headquarters. Shortly thereafter, his Uber-issued iPhone was "bricked," rendered just as useless as his laptop and wiped of its data.

As he sat in the living room of the cabin, stunned and angry, Sullivan tried to figure out what he was going to do next. He had done his job trying to protect the company, and he believed he had done it well. Before he arrived in 2015, Uber's security systems were in complete shambles, practically nonexistent. They had him to thank for cleaning up the first data breach, not to mention the second one, and the nightmare of privacy violation issues Uber was dealing with at the time. Now, Uber's executive, legal, policy, and press leaders had effectively damaged his reputation and career in Silicon Valley. For at least a time in 2017, federal prosecutors would look into his actions for any possible criminal violations.

He didn't believe what he had done was wrong. But after spending the past three years of his life inside Kalanick's orbit and nebulous approach to ethics, Sullivan realized something: his life, for the foreseeable future, was going to get a whole lot worse.

AS JOE SULLIVAN'S LIFE WAS falling apart, new billionaire Travis Kalanick's was just beginning.

By 2018, Kalanick left his full-time home in San Francisco with its enormous 13.3 percent state income tax rate and spent time in Miami—a haven for the fabulously wealthy, with no state income tax. Kalanick was joined by two fellow Silicon Valley outcasts, his former lieutenant Emil Michael, and early Uber investor and friend Shervin Pishevar, both of whose reputations in the tech press had been tarnished. Kalanick's friends quickly established residency in the state, protecting their new fortunes from the government.

In the months after their ousters, the fallen Uber executives are convinced they fell victim to the ineptitude and trickery of Uber's communications team. Eric Alexander, who was pushed out after it was reported that he retained the private medical file of the rape victim in India, has sued Rachel Whetstone, Uber's former head of policy and communications; Alexander, Michael and others believe Whetstone conspired to knife them by soliciting information to reporters while she was still at the company. In a response to the suit, Whetstone strenuously denied the assertions. As of this writing, the lawsuit remains unresolved.

Kalanick, still a bachelor, quickly found his place in the Miami nightlife scene. Flocking from one club to another with friends, he had a habit of informing dates and female acquaintances of his new status as a member of the revered "three comma club"—a reference to the three commas present in the number 1,000,000,000. When he wasn't in Miami, he could be found aboard yacht parties in the French West Indies, or at one of his two houses in Los Angeles—one for both the East and West sides of LA, depending on where traffic kept him that day.

His entrepreneurial days far from over, Kalanick is in the midst of working on his next startup: a real estate play, purchasing underutilized buildings and creating so-called "micro-kitchens" inside them, which will serve food delivered by Uber Eats. His plan, should it succeed, rests largely on the continued success of Uber.

Some are worried what that will mean for Uber in the long run.

TRAVIS KALANICK SHOWED UP on the steps of the Phillip Burton Federal Building and US Courthouse at 450 Golden Gate Avenue on February 6, 2018, steeled and prepared for an afternoon of testimony.

Wearing a solid black suit, a lavender shirt, and a black and white tie, the former chief executive looked good for his day in court. Later that day he would take the stand in *Waymo v. Uber*, which after months of deliberation and discovery had finally gone to trial. A gaggle of paparazzi huddled around the building entrances, waiting to snap photos of the fallen billionaire. Reporters lined the hallway of the nineteenth floor. They had camped out since five o'clock in the morning in hopes of getting a seat in the courtroom.

Confident yet measured, Kalanick would go on to deliver compelling testimony later that day, assuring jurors that his actions surrounding Uber's Otto deal were on the level. He testified that when Uber began developing its own autonomous vehicle research, Google CEO Larry Page grew increasingly "unpumped"—the polar opposite of "super pumped." Sipping from a bottle of water, Kalanick was intense and charming, and his demeanor seemed be having a positive effect on some of the jurors, to the frustration of Waymo's trial lawyers.

"He answered every question, cool and calm," Miguel Posados, one of the jurors, said of Kalanick to a reporter after the trial had concluded. Steve Perazzo, another juror, said Kalanick "really seemed like a good guy," like "a guy who took this idea and was pretty aggressive with it and wanted to be the best in the world."

The jury wouldn't get a chance to render a verdict. Shortly after Kalanick's testimony, Waymo sensed its case going sideways. Uber ultimately ended up settling with Waymo for $245 million in company equity, abruptly ending the trial. However, there were some strings attached to the deal. As part of the terms of the settlement, Uber agreed that it would not use any of Waymo's trade secrets in developing its autonomous vehicle program. Additionally. Uber's self-driving division would be subject to independent, third-party reviews to be sure it no longer used any of Waymo's proprietary information. Nevertheless, Waymo too would have a stake in Uber's success.

I HAD SHOWED UP TO the courtroom on the morning of Kalanick's first day of testimony, assuming my editor at the *Times* might want me to write a piece for the paper. I was on leave, in the process of writing this book, but the ending had yet to come—the case was still playing out.

After a midday recess from the court, Kalanick was expected to testify next going into the afternoon. As a stable of lawyers on both plaintiff and defendant's sides filed into the courtroom, followed closely by the press, I headed down the hallway to quickly pop into the bathroom before proceedings began again.

By the time I got back to the courtroom, I was too late. They had closed the doors and the trial was back under way. I missed my chance to report on Kalanick's testimony from the witness stand. In the long wood-and-granite-lined hallway outside of the courtroom, I stood cursing myself quietly, hoping the armed US marshals would eventually let me back inside. The way things were going, I wasn't holding my breath.

Then I realized: Kalanick wasn't in the courtroom yet. He was walking briskly down the hallway behind me, towards the courtroom doors, waiting to be called to the stand. The guards held the doors shut, motioning for Kalanick to wait in the chambers outside, with me and a few others, before he was called in to testify.

Kalanick waited, quiet and unaccompanied, outside the courtroom. I hadn't seen or talked to him in months, and didn't expect he'd be particularly chatty right before one of the most important moments in his professional life. Our last real interaction had been in June 2017, when I had contacted him for comment before I had run the story of his ouster. I imagined he hated my guts. Some ten feet separated us in those quiet chambers outside of the courtroom. As he backed away from the doors, he walked to the other side of the room, putting the bodies of three idle lawyers between us.

A minute passed, and his head snapped up, as if he had decided something. Kalanick walked directly over to me, meeting my gaze, and stuck out his hand. "Hey man, how are you doing?" he asked me in a hushed voice in the quiet hallway, shaking my hand and putting an arm around my shoulder. I'm sure he considered me a bitter enemy, and yet here he was, leaning into his charm. I smiled, returning the handshake.

"Are you gonna be okay in there, sitting down?" I said, trying to break some of the tension. "They won't let you pace back and forth!"

"God, I don't know!" he said, chuckling but clearly nervous. By the end of his testimony later that day, which would last just under an hour, he would finish nearly four bottles of water.

Kalanick caught himself, as if someone had reminded him he was talking to a reporter. "Can we go off the record now?" he asked, wanting to chat but clearly not trusting me to keep things between us.

I agreed to go off the record, an agreement I will continue to honor here. We spoke in that hallway for about ten minutes, as if things were normal between us, as if the multibillion-dollar company he built wasn't at risk of a legal defeat that could bring about technological and financial ruin. Despite everything that had happened the past year—his ouster, the death of his mother, the loss of practically every friend he had—Kalanick was still capable of conjuring up his charming, bubbly self. Kalanick was still here, still standing.

I wondered if he had learned anything from the last nine years of his life. He was rich—filthy, stinking, three-comma-club rich. And he was famous, or infamous, now. He was trying to rehab his image—to truly become a "Travis 2.0" version of himself. I was told that two months before today, he spent his Christmas in St. Bart's. Days were spent with his father, who had recovered from his injuries since the accident in Fresno, and Kalanick's siblings; for Christmas, they all dressed up in holiday pajamas and posed for photos on Kalanick's Instagram account. Nights were spent on yachts, drinking and partying with friends and models.

I had heard that he was building his next startup, focused on food delivery and logistics. My sources told me he was working just as hard—if not harder—at his new company, cracking the whip on his employees as much as he did at Uber. And to build it, he was recruiting many of the employees he had fired from Uber—the employees who were forced out because of the Holder report.

Kalanick was a billionaire. Garrett Camp and Ryan Graves were rich beyond their wildest dreams. The venture capitalists would soon reap enormous rewards from their investments. And by the time Uber made its debut on the public markets in 2019, there would be plenty more newly minted millionaires in Silicon Valley who would join them, ready to christen the next wave of innovation, to fund the next era of new startups. I wondered if there would be a new generation of Travis Kalanick protégés soon. What would they think of the founder's rise, and the path he took to get there?

Kalanick and I shook hands again, ending the conversation. He walked away from me, peering into the courtroom through the glass panes in the closed doors.

"God," Kalanick said, still staring inside the chambers, speaking out loud in the hallway, to everyone and to no one. "It feels like we're in the tunnel in a stadium, right before the Super Bowl," he said, laughing quietly to himself.

He began slowly raising his arms above his head, his eyes still locked on the witness stand in the courtroom, ready to jog down the aisle to his seat. He smiled, waiting for the guards to open the doors and let him into the room.

"I'm ready," Kalanick said.

POSTSCRIPT

AFTER MONTHS OF SPECULATION, UBER ANNOUNCED THAT IT WOULD hold an initial public offering in May of 2019. Lyft had debuted on the public markets just a few weeks prior at $72 per share. On the opening day the price spiked at first, until settling around $78. Uber set its sights much higher.

As Uber prepared itself for its coming IPO, the company had hired Morgan Stanley and Goldman Sachs to market its public shares to investors, floating the sky-high valuation of an Uber worth $120 billion, nearly twice its last round of private funding.

In pitching themselves to Uber, the bankers had not ignored the particulars of the CEO's compensation package. Before Khosrowshahi left Expedia, he had been the highest paid CEO of a publicly traded American company. In taking the Uber job he left behind tens of millions of dollars in unvested Expedia stock. To balance that loss, Khosrowshahi negotiated a hefty perk into his Uber gig: if the CEO were able to take Uber public at a valuation of more than $120 billion and keep it above that market cap for more than 90 days, he would receive an enormous payout of more than $100 million. Bankers at Morgan Stanley and Goldman Sachs internalized that number and (if only tacitly) aimed to achieve that splashy $120 billion market cap.

But in the months leading up to Uber's big coming-out party, the bankers' high expectations met reality. SoftBank, a supposed Uber ally, began funding competitors in some of Uber's highest growth areas like Latin America and various food delivery industries. Uber's stats began to look less rosy as investors scrutinized the details. As the weeks wore on during Uber's "roadshow"—the process by which investment firms decide if they wish to buy Uber's stock—it became clear that the company was not going to fetch $120 billion.

Bankers and traders crowded the floor of the New York Stock Exchange as Khosrowshahi and his entourage arrived on the morning of May 10, 2019. Early Uber employees and even some of Uber's longest-tenured drivers joined them at the invitation of the company. Staff were passing out black, Uber-branded hats and T-shirts for securities traders to wear as they typed their first buy and sell orders into the NYSE computers that lined the floors. Caterers brought in rounds of Big Macs, french fries, and hash browns, a nod to Uber's big deal with McDonald's for UberEats food delivery. Everyone was ready to start trading shares of $UBER.

There was some tension leading up to the big day. Khosrowshahi had asked Travis Kalanick not to join him on the balcony for the ceremonial bell-ringing that morning, something that pissed Kalanick off. Word leaked to the press that the two men were at odds, and there was a question as to whether Kalanick would end up showing at all. But Kalanick showed, arriving in time for an early breakfast that morning where he and Uber's current CEO publicly buried the hatchet. Khosrowshahi called Kalanick a "once-in-a-generation entrepreneur"; everyone in the room agreed.

At the breakfast, Khosrowshahi called Kalanick up to the front of the room along with Garrett Camp and Ryan Graves as the men stood for a round of applause. In just a few hours, Graves's shares would be worth $1.6 billion, while Camp's would net a cool $4.1 billion in value. Kalanick would be worth the most; after the bell, his stake in the company would be worth $5.4 billion. The three men who built Uber into what it was over the past decade were all billionaires. They also no longer happened to be on close speaking terms. Kalanick left shortly thereafter without incident, hours before the shares would hit the market, leaving the spotlight to Khosrowshahi.

As Khosrowshahi gathered with his executive team to execute his first trade, a scrum of employees, securities traders, photographers, and the press surrounded him. The CEO looked up at the monitors, waiting to see what the price would be. The night before, Uber had set its IPO price at $45 per share, below the value it had initially sought but still carefully calibrated for a healthy first-day "pop," an initial surge in price bankers like to promote to clients as an incentive to buy-in early. At $45 per share, the bankers expected the stock to immediately open at at least a few dollars above that number.

That didn't happen. As the minutes ticked forward, the number started falling—$44, $43, and finally to $42, the price of its first official public trade. Khosrowshahi's face sank. The floor, once brimming with excitement, fell to hushed whispers. Uber would be opening at *below* its initial pricing target. That was unheard of, especially for tech stocks that normally see a healthy first-day pop. By the end of the day, Uber had lost more in dollar terms than any other American initial public offering on Wall Street since 1975. Uber's coming-out party was a disaster.

Almost immediately, questions began swirling about how private market valuations had spun out of control, and whether Uber—the king of the unicorns—was finally tamed by the realities of Wall Street. It wasn't tenable to be losing billions of dollars with no definite path to profitability when you were trying to convince public market investors to purchase your stock. Silicon Valley investors wondered if Uber's disappointing debut would be a harbinger of many difficult technology IPOs to come.

Khosrowshahi, for his part, tried to stay upbeat. Later that evening, at a party on the stock exchange floor, he would give a toast to employees holding Big Macs and glasses of champagne, attempting to inspire his team—many of whom owned a great deal of the declining stock— even as brutal headlines posted about the IPO.

"Now is our time to prove ourselves," Khosrowshahi said to the room. "Five years from now, tech companies that come IPO after us will stand on this very trading floor and see what we've accomplished." The mood was sober, but Khosrowshahi was doing his best to rally the troops.

"They'll say, 'Holy shit. I want to be Uber.' "

AFTERWORD

AFTER FINISHING THIS BOOK IN 2019, I HAD IMAGINED THAT UBER'S
rise and fall would be enough of a public-facing disaster for investors
and startup founders to avoid Travis Kalanick's hubristic blueprint.
Silicon Valley viewed the rapid decline of Uber as an example of management committing cardinal startup sins, a corporate cautionary tale,
something that young techies could point to as a clear example of how
not to run a company.

I couldn't have been more wrong. At the time of this book's release in
the fall of 2019, WeWork was imploding. The headlines looked familiar; it was another multi-billion-dollar startup in nosedive, backed with
big money by blue-chip venture capitalists and private equity firms, led
by a leader whose natural charisma and cult of personality created a
kind of reality distortion field for investors and employees.

During the company's ascent, WeWork's outlandish pitch was irre-
sistible to its backers. WeWork wasn't just a coworking office rental
startup, according to Adam Neumann, the company's magnetic
founder and chief executive. No, the company was a "worldwide platform that supports growth, shared experiences and true success," he
wrote in the paperwork the company filed for an initial public offering.
WeWork was on a mission to "elevate the world's consciousness" with
"the energy of We." Investors, including many who previously invested
in Uber and Travis Kalanick, were sold on the idea; they wanted to
fund the next Uber.

But while Uber actually provided millions of people with a service they valued, WeWork's value proposition was shakier. Neumann

spent millions acquiring smaller startups that didn't seem to fit with WeWork's core business of shared workspace office rentals. Rebekah Neumann, a founding partner of the company and Adam's life partner, launched an education startup for five-year-olds, linked to WeWork. The endeavor, WeGrow, was, according to Rebekah, meant to help Generation We "understand their superpowers." As WeWork and its side projects swelled in size—and the cash burn rate increased—investors never questioned the dream that WeWork would become a globe-spanning workspace for the next generation of employees. At its height, WeWork was valued at more than $47 billion, bolstered in large part by enormous investments from SoftBank CEO Masayoshi Son.

However, things didn't go as planned. In the wake of Uber's unraveling, reporters began to look more aggressively into Neumann's history of erratic, unorthodox behavior as an executive—his propensity to smoke weed during business travel, or the time he flew across international borders with a chunk of marijuana stuffed into a cereal box for safekeeping. The financial maneuvers Neumann inserted into the IPO paperwork resulted in "egregious self-dealing" to the tune of millions, one former tech CEO said at the time. And beyond Neumann's personal shortcomings, there was the ever-present fact that WeWork was losing mountains of money, fast. With Uber in the back of their minds, and a failing WeWork IPO in the foreground, investors clued in to the fact that Neumann's vision of WeWork was illusory, an idea schemed up by a fast-talking huckster who sold the money men something they wanted to see.

And just like Uber, everything came to a crashing halt. WeWork's board of directors, as if awakening from a coma, revolted. Neumann held the majority of voting shares—sound familiar?—but a series of public leaks and the directors leaning on him ultimately resulted in his stepping down as CEO. He took with him shares initially valued at more than a billion dollars.

Uber and WeWork were infamous, but they represented the ethos of a generation of VC–backed startups. For years, the balance sheets of VC–backed startups were awash in red ink, buoyed by founders' promises that expanding to great size at even greater expense would eventually produce new money, rather than burn it by the truckload. Scale, they argued, was key.

But the consecutive collapses of Uber and WeWork brought a sea change to Silicon Valley, a cultural shift to rival the dot-com bust at the turn of the millennium. Before its IPO, Uber investors believed the company to be worth more than $120 billion. The public markets brought a grimmer reality; investors saw a company losing money, with nebulous notions of how it might turn a profit and a long, difficult road to get there. By the beginning of 2020, Uber's market cap was half that amount. On April Fools' Day of that year, it was down to one-third of its proposed initial valuation. WeWork's financials were even bloodier; Son admitted that he ultimately regretted the multi-billion-dollar stake SoftBank took in WeWork. From a peak of $47 billion, WeWork's valuation was slashed by more than 80 percent, settling below $5 billion. The reality distortion field had malfunctioned.

The malaise spread to other money-losing companies. All of Son's big bets from Vision Fund—his $100 billion slush pile earmarked for technology investment, which used to park billions into companies like Uber, WeWork, DoorDash, and Compass, among dozens of others—were struggling. Many were starting to look like bad investments. Flexport, an automated trucking startup, laid off dozens of employees. Wag.com, a dog-walking startup, shuffled its CEO out the door as the business floundered. Even Zume, the robot pizza startup, couldn't be saved from ousting half of its employees. "My own investment judgment was really bad. I regret it in many ways," Son told investors last year.

As Silicon Valley was going through its great retrenching, something happened that no one could have predicted: the spread of COVID-19, better known as the coronavirus. As the United States government scrambled to contain the spread of the virus in early 2020, the tech industry saw a near immediate impact. The greater Seattle area was hit hard by early infections. The virus quickly spread to northern California. Tech companies shut down their offices, while employees were asked to work remotely. State governors and city mayors began ordering citizens to "shelter in place," barring people from leaving their homes almost entirely. Only "essential businesses" were allowed to operate, which brought upon the near complete collapse of sectors like retail, hospitality, and transportation. Millions filed for unemployment insurance as companies folded overnight. Uber drivers who braved the streets to pick up the few remaining fares wore face masks behind the wheel. At the time of this writing, hospitals were filling with thousands of sick patients, and Wall Street was in utter disarray. The economic gains of the Trump presidency had been erased in the span of a few weeks as a recession even greater than the financial collapse of 2008 loomed; beyond the horror of killing people, COVID-19 threatened to halt the global economy.

In Silicon Valley, one venture capital firm sounded the alarm. On March 3, Sequoia Capital, the most storied and well-regarded firm in the history of technology investment, sent out an open letter to the founders leading the firm's portfolio companies. The firm called coronavirus "the black swan of 2020," a term used for a major event of unforeseen nature, something that is near impossible to predict yet can have an outsized and perhaps fatal impact on an industry. The writing was on the wall for investors, and they wanted to prepare their founders for the reckoning to come. Turning a profit, practicing conservatism in balance sheets, reducing headcount—these were the things companies should focus on.

Sequoia's message to the Valley was a new one: The days of nondiscretionary spending and frivolity were over. Hard choices would have to be made. And not everyone was going to make it.

"In some ways, business mirrors biology," the Sequoia partners told their founders. "As Darwin surmised, those who survive are 'not the strongest . . . nor the most intelligent, [but the] most adaptable to change.' "

And change has indeed come to the Valley. With the fading of characters like Neumann and Kalanick the backslapping bro culture of Silicon Valley's aughts era is now seen as passé. Tech employees are self-aware enough not to be as openly misogynistic at their workplaces as some of Uber's employees were. Even the seven-day bacchanals to Sin City have subsided. (Or, at least they are carried out more covertly—and without company branding.)

But despite moves toward austerity, today's founders' sensibilities remained largely the same. Like the managers of Uber and others before then, entrepreneurs proved willing to sidestep rules and take shortcuts trying to build the next potentially world-changing company.

Take Zoom, for instance: a once-boring teleconferencing enterprise software company, whose usage soared in the wake of the coronavirus outbreak. From stay-at-home office workers to teens locked in their rooms, people congregated online in greater numbers than ever before, and loved the ease and simplicity of doing so with the seamless software. What wasn't as user-friendly were the myriad privacy violations security researchers discovered in the wake of the app's popularity, including lazy loopholes, egregious data-sharing practices, and poorly regarded encryption methods. The company pledged to do better in the wake of the media backlash, of course. But being in the right place at the right time and moving aggressively to seize market share seemed to pay off; by April 2020, Zoom's market cap was north of $35 billion, weathering one of the worst bear markets in American history.

Furthermore, the utopian, world-saving ethos that inspired the previous generation of founders may not be misplaced—after all, companies like Zoom have created tools that let people communicate in ways that would not have been possible when Uber was founded. The next generation of Travis Kalanicks may not be pledging aloud that they'll "always be hustlin',"—Uber cultural value number one. But it doesn't have to. The ethos is implicit in the culture. In the wake of the "techlash" and intensified scrutiny of the industry, the tech world has only grown more insular, more defensive, perhaps even less self-reflective. Critics in the press are seen as uninformed "haters," and the staunchest defenders of startup land have grown more protective of young founders so that they do not fall prey to the cynicism of naysayers.

Tech's hustle culture, then, remains alive and well. Travis Kalanick built his company out of the ruins of the dot-com bubble and the emergence of a new mobile world. The catastrophic upheaval of 2020 has surely already shaped the next generation of tech unicorns—the only question is how.

ACKNOWLEDGMENTS

WHILE ONLY MY NAME IS ON THE COVER, THIS BOOK COULD NOT HAVE been written without the help of dozens of people who supported it over the past two years.

I have to thank Tom Mayer, my editor at W. W. Norton, for his expertise in bringing my ideas and prose to life in such a masterful way. Few editors I've worked with throughout my career are as talented as Tom, and his touch elevated the narrative. I'm a better writer for working with him.

A large team at W. W. Norton, additionally, made this book a reality: Will Scarlett, Dassi Zeidel, Becky Homiski, Beth Steidle, Anna Oler, Nneoma Amadi-obi, Steven Pace, Brendan Curry, Nicola DeRobertis-Theye, Elisabeth Kerr, Meredith McGinnis, and many others. Everyone hustled to make this book a success. Their support and efforts have been invaluable.

I'm grateful to Daniel Greenberg, my agent at Levine Greenberg Rostan, for meeting me in 2014 when I first thought an Uber book might be a good idea. And for his continued persistence despite my decision to ghost him for three years after that initial meeting.

I wouldn't be half the reporter I am today were it not for Pui-Wing Tam, my editor at the *New York Times*, whom I've worked with for years. Pui-Wing is an excellent mentor, and she worked closely with me through every step of Uber's dramatic 2017. I couldn't have penetrated the company without her excellent advice. (Apologies, Pui-Wing, for the many times I've called you after hours with a request to file yet another Uber story.)

And to that end, I owe a great deal of thanks to everyone at the *Times* who supported me in writing this book. Dean Baquet, Joe Kahn, Rebecca Blumenstein, and Ellen Pollock were particularly kind in letting me take leave to write it. My colleagues on the tech desk picked up my slack, and for that I'm super appreciative. And A. G. Sulzberger's kind notes on my coverage were inspiring, especially when I was going insane trying to make sense of all of my reporting.

Sean Lavery was my fact checker and armchair psychologist during more difficult moments. Simone Stolzoff's early research and support was a boon as well; I can't thank either of them enough.

Sam Dolnick and Stephanie Preiss, as always, are my wonderful collaborators at the *Times*.

I would be remiss not to thank all the people who spoke with me, some at great personal risk, for the reporting of this book. Everyone has a motivation to speak with a reporter, but many of my best sources felt they were doing the right thing by coming forward and telling their story in hopes it would help people to better understand the story of Uber. I want to express my thanks to all of you here: I truly could not have done it without you.

Many thanks as well go to the writers and friends who counseled and supported me through the process. Kevin Roose, B. J. Novak, Nick Bilton, and Anna Wiener gave brilliant feedback, while Tristan Lewis, Emily Silverman, and Hana Metzger provided much needed respite from my writing desk.

And finally, to my family—Michael, Lorraine, Joe, and especially Sarah Emerson and Bruna—all of whom have managed to deal with my craziness, long days, and even longer nights of reporting and writing this book. For that I am forever grateful, and this book is dedicated to you.

A NOTE ON SOURCES

THE CONTENTS OF THIS BOOK ARE WRITTEN BASED ON HUNDREDS OF interviews with more than two hundred people over the five years I've been reporting on Uber, and painstaking review of hundreds of never-before-seen documents.

All of the events that occur in the narrative are based on information taken from primary and secondary sources, either from firsthand accounts of the parties involved, or from two or more parties with direct knowledge of the matter. Every scene has been corroborated by multiple people.

All of the dialogue that appears in the book was taken from video recordings, audio files, or transcripts, or recounted by people involved in or with direct knowledge of the situations. Emails or text messages that appear in the book were viewed by or described to the author.

My utmost concern is for the safety and security of my sources. I truly appreciate their help in telling this story.

NOTES

PROLOGUE

xiii **Hales had promised:** Karen Weise, "This Is How Uber Takes Over a City," Bloomberg Businessweek, June 23, 2015, https://www.bloomberg.com/news/features/2015-06-23/this-is-how-uber-takes-over-a-city.

xiv **built into the cement floor:** Max Chafkin, "What Makes Uber Run," Fast Company, September 8, 2015, https://www.fastcompany.com/3050250/what-makes-uber-run.

xv **would later tell reporters:** Weise, "This Is How Uber Takes Over a City."

xiv **the advertisements said:** Alyson Shontell, "10 Ads That Show What A Circus the War Between Uber and Lyft Has Become," Business Insider, August 13, 2014, https://www.businessinsider.com/10-uber-lyft-war-ads-2014-8#heres-a-similar-ad-that-suggests-ubers-are-better-than-taxis-9.

Chapter 1: X TO THE X

3 **according to a letter:** Kara Swisher and Johana Bhuiyan, "Uber CEO Kalanick Advised Employees on Sex Rules for a Company Celebration in 2013 'Miami Letter,'" Recode, June 8, 2017, https://www.recode.net/2017/6/8/15765514/2013-miami-letter-uber-ceo-kalanick-employees-sex-rules-company-celebration.

4 **"fast-growing", "pugnacious", a "juggernaut":** Kara Swisher, "Man and Uber Man," *Vanity Fair,* November 5, 2014, https://www.vanityfair.com/news/2014/12/uber-travis-kalanick-controversy.

6 **a noun coined in 2013:** Aileen Lee, "Welcome to the Unicorn Club: Learning From Billion-Dollar Startups," TechCrunch, October 31, 2013, https://techcrunch.com/2013/11/02/welcome-to-the-unicorn-club/.

6 **under fire for emails:** Sam Biddle, "'Fuck Bitches Get Leid,' the Sleazy Frat Emails of Snapchat's CEO," Valleywag, May 28, 2014, http://valleywag.gawker.com/fuck-bitches-get-leid-the-sleazy-frat-emails-of-snap-1582604137.

6 **Dropbox and Airbnb:** Jack Morse, "Bros Attempt to Kick Kids Off Mission Soccer Field," Uptown Almanac, October 9, 2014, https://uptownalmanac.com/2014/10/bros-try-kick-kids-soccer-field.

11 **"philosophy of work":** Brad Stone, *The Upstarts: How Uber, Airbnb, and the Killer Companies of the New Silicon Valley Are Changing the World* (New York: Little Brown, 2017).

11 **Fourteen core leadership principles:** "Leadership Priciples," Amazon, https://www.amazon.jobs/principles.

13 **employee explained the term:** Alyson Shontell, "A Leaked Internal Uber Presentation Shows What the Company Really Values in Its Employees," Business Insider, November 19, 2014, https://www.businessinsider.com/uber-employee-competencies-fierceness-and-super-pumpedness-2014-11.

Chapter 2: THE MAKING OF A FOUNDER

16 **a former co-worker, said:** Elizabeth Chou, "Bonnie Kalanick, Mother of Uber Founder, Remembered Fondly by Former Daily News Coworkers," *Los Angeles Daily News*, August 28, 2017, https://www.dailynews.com/2017/05/28/bonnie-kalanick-mother-of-uber-founder-remembered-fondly-by-former-daily-news-coworkers/.

17 **an inherent competitive spirit:** Chou, "Bonnie Kalanick."

17 **Travis later said:** Travis Kalanick, "Dad is getting much better in last 48 hours," Facebook, June 1, 2017, https://www.facebook.com/permalink.php?story_fbid=10155147475255944&id=564055943.

17 **in an interview in 2014:** Kara Swisher, "Bonnie Kalanick, the Mother of Uber's CEO, Has Died in a Boating Accident," Recode, May 27, 2017, https://www.recode.net/2017/5/27/15705290/bonnie-kalanick-mother-uber-ceo-dies-boating-accident.

18 **positive relationship with his ex-wife:** Taylor Pittman, "Uber CEO Travis Kalanick and His Dad Open Up on Life, Love and Dropping Out of School," Huffington Post, April 11, 2016, https://www.huffingtonpost.com/entry/uber-travis-kalanick-talk-to-me_us_57040082e4b0daf53af126a9.

18 **built an electrical transformer:** Swisher, "Bonnie Kalanick."

18 **Donald later told a reporter:** Pittman, "Uber CEO Travis Kalanick."

19 **Travis was a top seller:** Adam Lashinsky, *Wild Ride: Inside Uber's Quest for World Domination* (New York: Portfolio/Penguin, 2017), 40.

19 **His prize: An enormous trophy:** Jesse Barkin, "Valley Conference Basketball Honors Top Students," *Los Angeles Daily News*, March 30, 1988, Z10.

20 **$20,000 in knives:** Chris Raymond, "Travis Kalanick: 'You Can Either Do What They Say or You Can Fight for What You Believe,'" Success, February 13, 2017, https://www.success.com/article/travis-kalanick-you-can-either-do-what-they-say-or-you-can-fight-for-what-you-believe.

20 **his commissions growing larger:** Sarah E. Needleman, "A Cutco Sales Rep's Story," *Wall Street Journal*, August 6, 2008, https://www.wsj.com/articles/SB121788532632911239.

20 **recalled Sean Stanton:** Interview with author, 2017.

21 **it occurred to some of them:** TechCo Media, "Travis Kalanick Startup Lessons from the Jam Pad—Tech Cocktail Startup Mixology," YouTube video, 38:34, May 5, 2011, https://www.youtube.com/watch?v=VMvdvP02f-Y.

22 **where he "worked, ate and slept":** Stone, *Upstarts*.

22 **growth was paramount:** John Borland, "Well-Scrubbed Business Plan Not Enough for

Scour," CNET, January 11, 2002, https://www.cnet.com/news/well-scrubbed-business
-plan-not-enough-for-scour/.

22 **"traffic was going through the roof"**: BAMM.TV, "FailCon 2011—Uber Case Study," You-Tube video, 26:18, November 3, 2011, https://www.youtube.com/watch?v=2QrX5jsiico&t=2s.

23 **"we need to go find money"**: BAMM.TV, "FailCon 2011."

23 **"really litigious hardcore dude"**: BAMM.TV, "FailCon 2011."

24 **sued Napster for $20 billion**: Rich Menta, "RIAA Sues Music Startup Napster for $20
Billion," MP3newsire.net, December 9, 1999, http://www.mp3newswire.net/stories/
napster.html.

24 **a lawsuit against Scour**: Matt Richtel, "Movie and Record Companies Sue a Film Trad-ing Site," *New York Times*, July 21, 2000, http://www.nytimes.com/2000/07/21/business/
movie-and-record-companies-sue-a-film-trading-site.html.

24 **Ovitz had sent letters**: Richtel, "Movie and Record Companies Sue."

Chapter 3: POST-POP DEPRESSION

27 **a starting price of $25,000**: "Where Are They Now: 17 Dot-Com Bubble Companies
and Their Founders," CB Insights, September 14, 2016, https://www.cbinsights.com/
research/dot-com-bubble-companies/.

28 **One fifth of all office space**: Matt Richtel, "A City Takes a Breath After the Dot-Com
Crash; San Francisco's Economy Is Slowing," *New York Times*, July 24, 2001.

29 **"Are you frickin' kidding me?"**: BAMM.TV, "FailCon 2011."

30 **"VC's ain't shit but hos and tricks"**: BAMM.TV, "FailCon 2011."

31 **he sold Red Swoosh**: Liz Gannes, "Uber CEO Travis Kalanick on How He Failed
and Lived to Tell the Tale," D: All Things Digital, November 8, 2011, http://allthingsd
.com/20111108/uber-ceo-travis-kalanick-on-how-he-failed-and-lived-to-tell-the-tale/.

32 **"It is in the VC's nature"**: TechCo Media, "Travis Kalanick, Founder & CEO of Uber—Tech Cocktail Startup Mixology," YouTube video, 34:35, June 14, 2012, https://www
.youtube.com/watch?v=Lrp0me9iJ_U.

Chapter 4: A NEW ECONOMY

33 **seized control of Fannie Mae**: Stephen Labaton and Edmund L. Andrews, "In Rescue
to Stabilize Lending, U.S. Takes Over Mortgage Finance Titans," *New York Times*, Sep-tember 7, 2008, https://www.nytimes.com/2008/09/08/business/08fannie.html.

34 **75 percent of American households**: U.S. Bureau of Labor Statistics, "More than 75
Percent of American Households Own Computers," *Beyond the Numbers* 1, no 4 (2010),
https://www.bls.gov/opub/btn/archive/more-than-75-percent-of-american-households
-own-computers.pdf.

34 **more than half of American adults**: John B. Horrigan, "Home Broadband 2008," Pew
Research Center, July 2, 2008, http://www.pewinternet.org/2008/07/02/home-broadband
-2008/.

36 **"nearly killed our company"**: John Doerr, interview with the author, April 3, 2018.

37 **his vision for the iPhone:** Rene Ritchie, "The Secret History of iPhone," iMore, January 22, 2019, https://www.imore.com/history-iphone-original.

39 **coders like Steve Demeter:** Brian X. Chen, "iPhone Developers Go from Rags to Riches," Wired, September 19, 2008, https://www.wired.com/2008/09/indie-developer/.

40 **"the appification of the economy":** Interview with author, April 3, 2018.

Chapter 5: UPWARDLY IMMOBILE

44 ***Casino Royale,* the 2006 reboot:** Brad Stone, "Uber: The App That Changed How the World Hails a Taxi," *Guardian*, January 29, 2017, https://www.theguardian.com/technology/2017/jan/29/uber-app-changed-how-world-hails-a-taxi-brad-stone.

44 **flourish on Bond's cell phone:** Stone, *Upstarts*.

46 **his personal blog:** Travis Kalanick, "Expensify Launching at TC50!!," *Swooshing* (blog), September 17, 2008, https://swooshing.wordpress.com/2008/09/17/expensify-launching-at-tc50/.

46 **a roomful of young engineers:** TechCo Media, "Travis Kalanick, Founder & CEO of Uber."

47 **"VC's are trying to axe the founder":** TechCo Media, "Travis Kalanick Startup Lessons."

48 **"musings and often controversial aphorisms":** https://twitter.com/konatbone.

49 **"respectable clientele":** Garrett Camp, "The Beginning of Uber," Medium, August 22, 2017, https://medium.com/@gc/the-beginning-of-uber-7fb17e544851.

Chapter 6: "LET BUILDERS BUILD"

54 **"Looking 4 entrepreneurial product":** Travis Kalanick (@travisk), "Looking 4 entrepreneurial product mgr/biz-dev killer 4 a location based service.. pre-launch, BIG equity, big peeps involved—ANY TIPS??," Twitter, January 5, 2010, 8:14 p.m., https://twitter.com/travisk/status/7422828552.

54 **"heres a tip. email me":** Ryan Graves (@ryangraves), "@KonaTbone heres a tip. email me :) graves.ryan[at]gmail.com," Twitter, January 5, 2010, 8:17 p.m., https://twitter.com/ryangraves/status/7422940444?lang=en.

54 **found him near Ocean Beach:** Anita Balakrishnan, "How Ryan Graves became Uber's first CEO," CNBC, May 14, 2017, https://www.cnbc.com/2017/05/14/profile-of-ubers-ryan-graves.html.

54 **Graves's Tumblr:** ryangraves, Tumblr, http://ryangraves.tumblr.com/.

54 **One personal favorite:** ryangraves, Tumblr, http://ryangraves.tumblr.com/post/516416119/via-fuckyeahjay-z.

55 **thirty new customers:** Brian Lund, "From Dead-End Job to Uber Billionaire: Meet Ryan Graves," *DailyFinance*, July 3, 2014, https://web.archive.org/web/20140707042902/http://www.dailyfinance.com/on/uber-billionaire-ryan-graves/.

55 **small, metallic statue of an ape-man:** ryangraves, Tumblr, http://ryangraves.tumblr.com/post/336093270/dpstyles-crunchie-closeup-aka-the-heisman-of.

56 **Graves posted to his Facebook:** Ryan Graves, "Into the Infinite Abyss of the Startup

Adventure," Facebook, February 14, 2010, https://www.facebook.com/note.php?note_
id=476991565402.

58 **"Uber CEO 'Super Pumped' ":** Michael Arrington, "Uber CEO 'Super Pumped' about
Being Replaced by Founder," TechCrunch, https://techcrunch.com/2010/12/22/uber-ceo
-super-pumped-about-being-replaced-by-founder/.

59 **posed for an Instagram snapshot:** Uber HQ (@sweenzor), Instagram Photo, September
18, 2013, https://www.instagram.com/p/eatIa-juEa/?taken-by=sweenzor.

59 **hailed UberCab's model:** Leena Rao, "UberCab Takes the Hassle Out of Booking a Car
Service," TechCrunch, https://techcrunch.com/2010/07/05/ubercab-takes-the-hassle-out
-of-booking-a-car-service/.

59 **one *TechCrunch* article by Arrington said:** Michael Arrington, "What If UberCab Pulls
an Airbnb? Taxi Business Could (Finally) Get Some Disruption," TechCrunch, https://
techcrunch.com/2010/08/31/what-if-ubercab-pulls-an-airbnb-taxi-business-could-finally
-get-some-disruption/.

Chapter 7: **THE TALLEST MAN IN VENTURE CAPITAL**

65 **"It's magic":** GigaOm, "Bill Gurley, Benchmark Capital (full version)," YouTube video,
32:48, December 14, 2012, https://www.youtube.com/watch?v=dBaYsK_62EY.

65 **fund returned $250 million:** John Markoff, "Internet Analyst Joins Venture Capital
Firm," *New York Times*, July 14, 1997, https://www.nytimes.com/1997/07/14/business/
internet-analyst-joins-venture-capital-firm.html.

66 **worked at the Johnson Space Center:** Marissa Barnett, "Former Resident Donates $1M
to Dickinson," *Galveston County Daily News*, September 6, 2017, http://www.galvnews
.com/news/article_7c163944-63ee-5499-8964-fec7ef7e0540.html.

66 **Lucia spent her spare time:** Bill Gurley, "Thinking of Home: Dickinson, Texas," *Above
the Crowd* (blog), September 6, 2017, http://abovethecrowd.com/2017/09/06/thinking-of
-home-dickinson-texas/.

66 **one of the first relatively inexpensive:** "Commodore VIC-20," Steve's Old Computer
Museum, http://oldcomputers.net.

66 **he mostly rode the bench:** Eric Johnson, "Full Transcript: Benchmark General Partner Bill Gurley on Recode Decode," Recode, September 28, 2016, https://www.recode
.net/2016/9/28/13095682/bill-gurley-benchmark-bubble-uber-recode-decode-podcast
-transcript.

66 **He played for one minute:** "Bill Gurley," Sports Reference, College Basketball (CBB),
https://www.sports-reference.com/cbb/players/bill-gurley-1.html and "Bill Gurley Season
Game Log," Sports Reference, College Basketball (CBB), https://www.sports-reference
.com/cbb/players/bill-gurley-1/gamelog/1988/.

67 **he was infatuated:** Gabrielle Saveri, "Bill Gurley Venture Capitalist, Hummer Winblad
Venture Partners," Bloomberg, August 25, 1997, https://www.bloomberg.com/news/
articles/1997-08-24/bill-gurley-venture-capitalist-hummer-winblad-venture-partners.

70 **"He's kind of an animal":** Stross, Randall E., *EBoys: The First Inside Account of Venture Capitalists at Work* (Crown Publishers, 2000).

71 **Gurley wrote:** Bill Gurley, "Benchmark Capital: Open for Business," *Above the Crowd*

(blog), December 1, 2008, https://abovethecrowd.com/2008/12/01/benchmark-capital
-open-for-business/.

Chapter 8: PAS DE DEUX

72   **Roughly one third:** Artturi Tarjanne, "Why VC's Seek 10x Returns," *Activist VC Blog*
       (blog), Nexit Adventures, January 12, 2018, http://www.nexitventures.com/blog/vcs-seek
       -10x-returns/.

74   **Kalanick frequently compared:** Amir Efrati, "Uber Group's Visit to Seoul Escort Bar
       Sparked HR Complaint," The Information, March 24, 2017, https://www.theinformation
       .com/articles/uber-groups-visit-to-seoul-escort-bar-sparked-hr-complaint.

75   **"Software is eating the world":** Andreessen Horowitz, Software Is Eating the World,
       https://a16z.com/.

75   **deals increased by 73 percent:** Richard Florida and Ian Hathaway, "How the Geography
       of Startups and Innovation Is Changing," *Harvard Business Review*, November 27, 2018,
       https://hbr.org/2018/11/how-the-geography-of-startups-and-innovation-is-changing.

75   **billions invested post-2010:** Center for American Entrepreneurship, "Rise of the Global
       Startup City," Startup Revolution, http://startupsusa.org/global-startup-cities/.

75   **emerged as the world's epicenter:** Center for American Entrepreneurship, "Rise of the
       Global Startup City."

76   **"to organize the world's information":** "From the Garage to the Googleplex," About,
       Google, https://www.google.com/about/our-story/.

76   **controversial financial instrument:** "The Effects of Dual-Class Ownership on Ordinary
       Shareholders," Knowledge@Wharton, June 30, 2004, http://knowledge.wharton.upenn
       .edu/article/the-effects-of-dual-class-ownership-on-ordinary-shareholders/.

77   **"An Owner's Manual For Google Investors":** Larry Page and Sergey Brin, "2004
       Founders' IPO Letter," Alphabet Investor Relations, https://abc.xyz/investor/founders
       -letters/2004/ipo-letter.html.

77   **$3.5-billion acquisition:** "Snapchat Spurned $3 Billion Acquisition Offer from Face-
       book," *Digits* (blog), *Wall Street Journal*, November 13, 2013, https://blogs.wsj.com/
       digits/2013/11/13/snapchat-spurned-3-billion-acquisition-offer-from-facebook/.

Chapter 9: CHAMPION'S MINDSET

82   **Kalanick once said onstage:** Liz Gannes, "Travis Kalanick: Uber Is Raising More Money
       to Fight Lyft and the 'Asshole' Taxi Industry," Recode, May 28, 2014, https://www.recode
       .net/2014/5/28/11627354/travis-kalanick-uber-is-raising-more-money-to-fight-lyft-and-the.

82   **"There's been so much corruption":** Andy Kessler, "Travis Kalanick: The Transporta-
       tion Trustbuster," *Wall Street Journal*, January 25, 2013, https://www.wsj.com/articles/
       SB10001424127887324235104578244231122376480.

83   **"we're crushing it":** Alexia Tsotsis, "Spotted! Secret Ubers on the Streets of Seattle,"
       TechCrunch, https://techcrunch.com/2011/07/25/uber-seattle/.

84   **"incredibly hot chicks":** Adam Withnall, "Uber France Apologises for Sexist Promotion
       Offering Men Free Rides with 'Incredibly Hot Chicks' as Drivers," Independent, Octo-

ber 23, 2014, https://www.independent.co.uk/life-style/gadgets-and-tech/uber-france
-apologises-for-sexist-promotion-offering-men-free-rides-with-incredibly-hot-chicks-as
-9813087.html.

87 **"Could Uber reach a point"**: Bill Gurley, "How to Miss by a Mile: An Alternative
Look at Uber's Potential Market Size," *Above the Crowd* (blog), July 11, 2014, http://
abovethecrowd.com/2014/07/11/how-to-miss-by-a-mile-an-alternative-look-at-ubers
-potential-market-size/.

87 **In a policy paper published**: Travis Kalanick, "Principled Innovation: Addressing the
Regulatory Ambiguity Ridesharing Apps," April 12, 2013, http://www.benedelman.org/
uber/uber-policy-whitepaper.pdf.

87 **"nervous breakdowns"**: Swisher, "Bonnie Kalanick."

88 **Kalanick would tweet**: Travis Kalanick (@travisk), "@johnzimmer you've got a lot of
catching up to do . . . #clone," Twitter, March 19, 2013, 2.22 p.m., https://twitter.com/
travisk/status/314079323478962176?lang=en.

88 **"It was intense"**: Interview with former Uber executive who worked closely alongside
Kalanick.

88 **Kalanick once admitted**: Swisher, "Man and Uber Man."

89 **"Shave the 'Stache"**: Liz Gannes, "Uber's Travis Kalanick on Numbers, Competition
and Ambition (Everything but Funding)," D: All Things Digital, June 27, 2013, http://
allthingsd.com/20130627/ubers-travis-kalanick-on-numbers-competition-and-ambition
-everything-but-funding/.

90 **told friends of the strategy**: Background interview with early senior employee, San Francisco, 2018.

Chapter 10: THE HOMESHOW

92 **a home-run sale**: Eric Jackson, "Tellme Is One of the Best Silicon Valley Companies
Most People Have Never Heard Of," CNBC, October 23, 2017, https://www.cnbc
.com/2017/10/23/tellme-is-the-best-tech-company-most-have-never-heard-of.html.

95 **"negative churn"**: Tomasz Tunguz, "Why Negative Churn is Such a Powerful Growth
Mechanism," November 18, 2014, https://tomtunguz.com/negative-churn/.

Chapter 11: BIG BROTHER AND LITTLE BROTHER

106 **"like big brother and little brother"**: Jillian D'Onfro, "Google and Uber were like 'Big
Brother and Little Brother'—Until it All Went Wrong," CNBC, February 7, 2018, https://
www.cnbc.com/2018/02/07/travis-kalanick-on-google-uber-relationship.html.

107 **built one of his first robots**: Jack Nicas and Tim Higgins, "Google vs. Uber: How One
Engineer Sparked a War," *Wall Street Journal*, May 23, 2017, https://www.wsj.com/
articles/how-a-star-engineer-sparked-a-war-between-google-and-uber-1495556308.

109 **millions into self-driving research**: Charles Duhigg, "Did Uber Steal Google's Intellectual Property?," *New Yorker*, October 22, 2018, https://www.newyorker.com/
magazine/2018/10/22/did-uber-steal-googles-intellectual-property.

109 **depriving dozens of his colleagues**: Nicas and Higgins, "Google vs. Uber."

110 **head of the self-driving division:** Max Chafkin and Mark Bergen, "Fury Road: Did Uber Steal the Driverless Future from Google?," Bloomberg, March 16, 2017, https://www.bloomberg.com/news/features/2017-03-16/fury-road-did-uber-steal-the-driverless-future-from-google.

Chapter 12: GROWTH

112 **$1 million apiece:** Felix Salmon, "Why Taxi Medallions Cost $1 Million," Reuters, October 21, 2011, http://blogs.reuters.com/felix-salmon/2011/10/21/why-taxi-medallions-cost-1-million/.

112 **one fire-sale auction:** Winnie Hu, "Taxi Medallions, Once a Safe Investment, Now Drag Owners Into Debt," New York Times, September 10, 2017, https://www.nytimes.com/2017/09/10/nyregion/new-york-taxi-medallions-uber.html.

113 **pulled the trigger:** Ginia Bellafante, "A Driver's Suicide Reveals the Dark Side of the Gig Economy," New York Times, February 6, 2018, https://www.nytimes.com/2018/02/06/nyregion/livery-driver-taxi-uber.html.

113 **"I am not a Slave and I refuse to be one.":** Doug Schifter, Facebook, https://www.facebook.com/people/Doug-Schifter/100009072541151.

113 **more than a dozen:** Nikita Stewart and Luis Ferré-Sadurní, "Another Taxi Driver in Debt Takes His Life. That's 5 in 5 Months.," New York Times, May 27, 2018, https://www.nytimes.com/2018/05/27/nyregion/taxi-driver-suicide-nyc.html.

113 **took their own lives:** Emma G. Fitzsimmons, "A Taxi Driver Took His Own Life. His Family Blames Uber's Influence.," New York Times, May 1, 2018, https://www.nytimes.com/2018/05/01/nyregion/a-taxi-driver-took-his-own-life-his-family-blames-ubers-influence.html.

114 **Milan's transportation chief:** Stephanie Kirchgaessner, "Threatening Sign Hung Near Home of Italian Uber Boss," The Guardian, February 12, 2015, https://www.theguardian.com/technology/2015/feb/12/threatening-sign-italian-uber-boss.

114 **120,000 violations:** Andrew Maykuth, "Uber pays $3.5M fine to settle fight with Pa. taxi regulators," Philadelphia Inquirer, April 6, 2017, https://www.philly.com/philly/business/energy/Uber-fine-PA-PUC.html.

114 **"Uber-ON!":** Text message provided to author by an Uber source.

115 **hired lobbyists to get laws rewritten:** Mike Isaac, "Uber's System for Screening Drivers Draws Scrutiny," New York Times, December 9, 2014, https://www.nytimes.com/2014/12/10/technology/ubers-system-for-screening-drivers-comes-under-scrutiny.html.

115 **Amazon, Microsoft, and Walmart combined:** Borkholder, Montgomery, Saika Chen, Smith, "Uber State Interference."

116 **a small, pop-up notification:** Fitz Tepper, "Uber Launches 'De Blasio's Uber' Feature in NYC with 25 Minute Wait Times," TechCrunch, https://techcrunch.com/2015/07/16/uber-launches-de-blasios-uber-feature-in-nyc-with-25-minute-wait-times/.

117 **as many as seven per second:** Rosalind S. Helderman, "Uber Pressures Regulators by Mobilizing Riders and Hiring Vast Lobbying Network," Washington Post, December 13, 2014, https://www.washingtonpost.com/politics/uber-pressures-regulators-by

-mobilizing-riders-and-hiring-vast-lobbying-network/2014/12/13/3f4395c6-7f2a-11e4
-9f38-95a187e4c1f7_story.html?utm_term=.4a82cfdcaccd.

117 **file-folder storage boxes:** Anthony Kiekow, "Uber Makes a Delivery to MTC with
Hopes of Operating in St. Louis," Fox2now: St. Louis, July 7, 2015, https://fox2now
.com/2015/07/07/uber-says-water-bottles-were-symbolic-of-petitions-for-service-in-st
-louis/.

117 **"make your voice heard":** Alison Griswold, "Uber Won New York," Slate, November 18,
2015, http://www.slate.com/articles/business/moneybox/2015/11/uber_won_new_york_
city_it_only_took_five_years.html.

Chapter 13: THE CHARM OFFENSIVE

119 **"asshole culture":** Sarah Lacy, "The Horrific Trickle Down of Asshole Culture: Why
I've Just Deleted Uber from My Phone," Pando, October 22, 2014, https://pando
.com/2014/10/22/the-horrific-trickle-down-of-asshole-culture-at-a-company-like-uber/.

119 **a caricature of a "bro":** Mickey Rapkin, "Uber Cab Confessions," *GQ*, February 27,
2014, https://www.gq.com/story/uber-cab-confessions?currentPage=1.

119 **"face like a fist":** Swisher, "Man and Uber Man."

120 **the impending launch of "Uberpool":** "Announcing Uberpool," Uber (blog), https://
web.archive.org/web/20140816060039/http://blog.uber.com/uberpool.

120 **their corporate blog:** "Introducing Lyft Line, Your Daily Ride," Lyft (blog), August 6,
2014, https://blog.lyft.com/posts/introducing-lyft-line.

121 **Lacy tweeted:** Sarah Lacy (@sarahcuda), "it troubles me that Uber is so OK with lying," Twitter, August 20, 2014, 7:01 p.m., https://twitter.com/sarahcuda/status/502228907068641280.

121 **tone-deaf response:** "Statement On New Year's Eve Accident," Uber (blog), https://web
.archive.org/web/20140103020522/http://blog.uber.com/2014/01/01/statement-on-new
-years-eve-accident/.

122 **read the headline:** Lacy, "The Horrific Trickle Down of Asshole Culture.

122 **Kalanick's obsession with China:** Erik Gordon, "Uber's Didi Deal Dispels Chinese 'El
Dorado' Myth Once and For All," The Conversation, http://theconversation.com/ubers
-didi-deal-dispels-chinese-el-dorado-myth-once-and-for-all-63624.

124 **Most Influential Women in Bay Area Business:** American Bar, "Salle Yoo," https://
www.americanbar.org/content/dam/aba/administrative/science_technology/2016/salle_
yoo.authcheckdam.pdf.

126 **"no fear in Silicon Valley":** Mike Isaac, "Silicon Valley Investor Warns of Bubble at
SXSW," *Bits* (blog), *New York Times*, March 15, 2015, https://bits.blogs.nytimes
.com/2015/03/15/silicon-valley-investor-says-the-end-is-near/.

127 **Manhattan's Flatiron District:** Johana Bhuiyan, "Uber's Travis Kalanick Takes 'Charm
Offensive' To New York City," BuzzFeedNews, November 14, 2014, https://www
.buzzfeednews.com/article/johanabhuiyan/ubers-travis-kalanick-takes-charm-offensive
-to-new-york-city.

128 **hard-hitting news organization:** Mike Isaac, "50 Million New Reasons BuzzFeed Wants to
Take Its Content Far Beyond Lists," *New York Times*, August 10, 2014, https://www.nytimes
.com/2014/08/11/technology/a-move-to-go-beyond-lists-for-content-at-buzzfeed.html.

130 **"Uber's dirt-diggers":** Ben Smith, "Uber Executive Suggests Digging Up Dirt On Journalists," BuzzFeedNews, November 17, 2014, https://www.buzzfeednews.com/article/bensmith/uber-executive-suggests-digging-up-dirt-on-journalists.

Chapter 14: **CULTURE WARS**

132 **"the crazy ones":** http://www.thecrazyones.it/spot-en.html.

132 **went into the technology sector:** Natalie Kitroeff and Patrick Clark, "Silicon Valley May Want MBAs More Than Wall Street Does," Bloomberg Businessweek, March 17, 2016, https://www.bloomberg.com/news/articles/2016-03-17/silicon-valley-mba-destination.

132 **nearly a quarter of them:** Gina Hall, "MBAs are Increasingly Finding a Home in Silicon Valley," Silicon Valley Business Learning, March 18, 2016, https://www.bizjournals.com/sanjose/news/2016/03/18/mbas-are-increasingly-finding-a-home-in-silicon.html.

133 **"In a meritocracy":** Uber's list of 14 values, obtained by author.

134 **Mohrer tweeted:** Winston Mohrer (@WinnTheDog), "#Shittybike #lyft," Twitter, July 11, 2018, 7:21 a.m., https://twitter.com/WinnTheDog/status/1017005971107909633.

135 **designed an algorithm:** Caroline O'Donovan and Priya Anand, "How Uber's Hard-Charging Corporate Culture Left Employees Drained," BuzzFeedNews, July 17, 2017, https://www.buzzfeednews.com/article/carolineodonovan/how-ubers-hard-charging-corporate-culture-left-employees#.wpdMljap9.

135 **"This Safe Rides Fee supports":** "What Is the Safe Rides Fee?," Uber, https://web.archive.org/web/20140420053019/http://support.uber.com/hc/en-us/articles/201950566.

136 **"We live in Uber's world now":** Bradley Voytek, "Rides of Glory," Uber (blog), March 26, 2012, https://web.archive.org/web/20141118192805/http:/blog.uber.com/ridesofglory.

Chapter 15: **EMPIRE BUILDING**

140 **Nearly one-third of that population:** Joshua Lu and Anita Yiu, "The Asian Consumer: Chinese Millenials," Goldman Sachs Global Investment Research, September 8, 2015, http://xqdoc.imedao.com/14fcc41218a6163fed2098e2.pdf.

140 **Nearly 97 percent of Chinese:** Po Hou and Roger Chung, "2014 Deloitte State of the Media Democracy China Survey: New Media Explosion Ignited," Deloitte, November 2014, https://www2.deloitte.com/content/dam/Deloitte/cn/Documents/technology-media-telecommunications/deloitte-cn-tmt-newmediaexplosionignited-en-041114.pdf.

141 **nine of them were Chinese:** Sally French, "China Has 9 of the World's 20 Biggest Tech Companies," Market Watch, May 31, 2018, https://www.marketwatch.com/story/china-has-9-of-the-worlds-20-biggest-tech-companies-2018-05-31.

142 **"I get excited":** Jessica E. Lessin, "Zuckerberg and Kalanick in China: Two Approaches," The Information, March 25, 2016, https://www.theinformation.com/articles/zuckerberg-and-kalanick-in-china-two-approaches.

143 **When press started sniffing:** Amir Efrati, "Inside Uber's Mission Impossible in China," The Information, January 11, 2016, https://www.theinformation.com/articles/inside-ubers-mission-impossible-in-china.

143 **"This kind of growth":** Travis Kalanick, "Uber-successful in China," http://im.ft-static
.com/content/images/b11657c0-1079-11e5-b4dc-00144feabdc0.pdf.

144 **sailing out amid the rough waters:** Octavio Blanco, "How this Vietnamese Refugee
Became Uber's CTO," CNN Money, August 12, 2016, https://money.cnn.com/2016/08/12/
news/economy/thuan-pham-refugee-uber/index.html.

145 **They developed their own codified language:** Leslie Hook, "Uber's Battle for China,"
Financial Times Weekend Magazine, June 2016, https://ig.ft.com/sites/uber-in-china/.

149 **An angry mob of drivers:** Sanjay Rawat, "Hyderabad Uber Driver Suicide Adds Fuel
to Protests for Better Pay," Outlook, February 13, 2017, https://www.outlookindia
.com/website/story/hyderabad-uber-driver-suicide-adds-fuel-to-protests-for-better
-pay/297923.

150 **pending an investigation:** Ellen Barry and Suhasini Raj, "Uber Banned in India's Capital After Rape Accusation," *New York Times*, December 8, 2014, https://www.nytimes
.com/2014/12/09/world/asia/new-delhi-bans-uber-after-driver-is-accused-of-rape.html.

Chapter 16: **THE APPLE PROBLEM**

156 *BuzzFeed* **ran its story:** Ben Smith, "Uber Executive Suggests Digging Up Dirt on Journalists," BuzzFeedNews, November 17, 2014, https://www.buzzfeednews.com/article/
bensmith/uber-executive-suggests-digging-up-dirt-on-journalists.

156 **an enterprising young hacker:** Average Joe, "What the Hell Uber? Uncool Bro.," *Giron-
sec* (blog), November 25, 2014, https://www.gironsec.com/blog/2014/11/what-the-hell
-uber-uncool-bro/.

156 **it landed on** *Hacker News:* "Permissions Asked for by Uber Android App," Y Combi-
nator, November 25, 2014, https://news.ycombinator.com/item?id=8660336.

Chapter 17: **"THE BEST DEFENSE . . ."**

165 **Kalanick also held court over "Hell":** Amir Efrati, "Uber's Top Secret 'Hell' Program Exploited Lyft's Vulnerability," The Information, April 12, 2017, https://www
.theinformation.com/articles/ubers-top-secret-hell-program-exploited-lyfts-vulnerability.

166 **Those programs fell under:** Kate Conger, "Uber's Massive Scraping Program Collected
Data About Competitors Around The World." Gizmodo. December 12, 2017, https://
gizmodo.com/ubers-massive-scraping-program-collected-data-about-com-1820887947/.

167 **more than 50,000 Uber drivers:** Colleen Taylor, "Uber Database Breach Exposed Information of 50,000 Drivers, Company Confirms." TechCrunch. February 27, 2015, https://
techcrunch.com/2015/02/27/uber-database-breach-exposed-information-of-50000
-drivers-company-confirms/

168 **"rebelled" against his hippie parents:** Kashmir Hill, "Facebook's Top Cop: Joe Sullivan," *Forbes*, February 22, 2012, https://www.forbes.com/sites/kashmirhill/2012/02/22/
facebooks-top-cop-joe-sullivan/.

171 **"A lot of companies stop":** Hill, "Facebook's Top Cop: Joe Sullivan."

173 **"We are not going to leave them alone":** Emilio Fernández, "En Edomex *Cazan* al

Servicio Privado," *El Universal*, May 28, 2015, http://archivo.eluniversal.com.mx/ciudad
-metropoli/2015/impreso/en-edomex-cazan-al-servicio-privado-132301.html.

174 **After stabbing Modolo repeatedly:** Stephen Eisenhammer and Brad Haynes, "Mur-
ders, Robberies of Drivers in Brazil Force Uber to Rethink Cash Strategy," Reuters, February 14, 2017, https://www.reuters.com/article/uber-tech-brazil-repeat-insight-pix-tv-g
-idUSL1N1FZ03V.

Chapter 18: CLASH OF THE SELF-DRIVING CARS

176 **"The reason I'm excited":** James Temple, "Brin's Best Bits from the Code Conference
(Video)," Recode, May 28, 2014, https://www.recode.net/2014/5/28/11627304/brins-best
-bits-from-the-code-conference-video.

178 **"We get stuff like this":** Biz Carson, "New Emails Show How Mistrust and Suspicions
Blew Up the Relationship Between Uber's Travis Kalanick and Google's Larry Page,"
Business Insider, July 6, 2017, https://www.businessinsider.com/emails-uber-wanted-to
-partner-with-google-on-self-driving-cars-2017-7.

181 **Trucking was an enormous industry:** American Trucking Associations, "News
and Information Reports, Industry Data," https://www.trucking.org/News_and_
Information_Reports_Industry_Data.aspx.

181 **Trucks drive 5.6 percent:** National Highway Traffic and Safety Administration,
"USDOT Releases 2016 Fatal Traffic Crash Data," https://www.nhtsa.gov/press-releases/
usdot-releases-2016-fatal-traffic-crash-data.

182 **"I want to be in the driver seat":** Duhigg, "Did Uber Steal Google's Intellectual
Property?"

182 **autonomous vehicle engineering specialists:** John Markoff, "Want to Buy a Self-Driving
Car? Big-Rig Trucks May Come First," *New York Times*, May 17, 2016, https://www
.nytimes.com/2016/05/17/technology/want-to-buy-a-self-driving-car-trucks-may-come
-first.html.

182 **Levandowski ignored them:** Mark Harris, "How Otto Defied Nevada and Scored a $60
Million Payout from Uber," Wired, November 28, 2016, https://www.wired.com/2016/11/
how-otto-defied-nevada-and-scored-a-680-million-payout-from-uber/#.67khcq4w5.

183 **"Safety Third":** Chafkin and Bergen, "Fury Road."

183 **"brother from another mother":** Chafkin and Bergen, "Fury Road."

184 **"Super Duper" version of Uber:** From Waymo LLC v. Uber Technologies, 3:17-cv-
00939-WHA, and Paayal Zaveri and Jillian D'Onfro, "Travis Kalanick Takes the Stand
to Explain Why Uber Wanted to Poach Google Self-Driving Engineer," CNBC, February 6, 2018, https://www.cnbc.com/2018/02/06/travis-kalanick-reveals-why-he-wanted
-googles-anthony-levandowski.html.

184 **an entire center in Pittsburgh:** Mike Isaac, "Uber to Open Center for Research on Self-
Driving Cars," *Bits* (blog), *New York Times*, February 2, 2015, https://bits.blogs.nytimes
.com/2015/02/02/uber-to-open-center-for-research-on-self-driving-cars/.

185 **"a pound of flesh":** Zaveri and D'Onfro, "Travis Kalanick Takes the Stand."

185 **"The golden time is over":** Alyssa Newcomb, "Former Uber CEO Steals the Show
with 'Bro-cabulary' In Trade Secrets Trial," NBC News, February 7, 2018, https://www

.nbcnews.com/tech/tech-news/former-uber-ceo-steals-show-court-trade-secrets-bro
-cabulary-n845541.

Chapter 19: SMOOTH SAILING

187 **the public investment arm of Saudi Arabia:** Mike Isaac and Michael J. de la Merced,
     "Uber Turns to Saudi Arabia for $3.5 Billion Cash Infusion," *New York Times*, June 1, 2016,
     https://www.nytimes.com/2016/06/02/technology/uber-investment-saudi-arabia.html.

187 **Kalanick directed his dealmaking stewards:** Eric Newcomer, "The Inside Story of How
     Uber Got into Business with the Saudi Arabian Government," Bloomberg, November
     3, 2018, https://www.bloomberg.com/news/articles/2018-11-03/the-inside-story-of-how
     -uber-got-into-business-with-the-saudi-arabian-government.

189 *The Art of War* **had been his bible:** Sun Tzu's Art of War, "6. Weak Points and Strong,"
     no. 30, https://suntzusaid.com/book/6/30.

191 **Graves was pushed aside:** Greg Bensinger and Khadeeja Safdar, "Uber Hires Target
     Executive as President," *Wall Street Journal*, August 30, 2016, https://www.wsj.com/
     articles/uber-hires-target-executive-as-president-1472578656.

191 **"What are you going to do":** Ryan Felton, "Uber Drivers Ask 'Where Are the Answers?'
     In Shitshow Q&A," Jalopnik, February 16, 2017, https://jalopnik.com/uber-drivers-ask
     -where-are-the-answers-in-shitshow-q-a-1792461050.

191 **"You made it crystal clear":** Felton, "Uber Drivers Ask 'Where Are the Answers?'"

193 **"he was holding the pony's leash":** Emily Chang, "Uber Investor Shervin Pishevar
     Accused of Sexual Misconduct by Multiple Women," Bloomberg, November 30, 2017,
     https://www.bloomberg.com/news/articles/2017-12-01/uber-investor-shervin-pishevar
     -accused-of-sexual-misconduct-by-multiple-women.

196 **"Reflecting On One Very, Very Strange Year At Uber":** Susan Fowler, "Reflecting on
     One Very, Very Strange Year at Uber," *Susan J. Fowler* (blog), February 19, 2017, https://
     www.susanjfowler.com/blog/2017/2/19/reflecting-on-one-very-strange-year-at-uber.

Chapter 20: THREE MONTHS PRIOR

200 **Facebook embedded its own employees:** Sheera Frenkel, "The Biggest Spender of Polit-
     ical Ads on Facebook? President Trump," *New York Times*, July 17, 2018, https://www
     .nytimes.com/2018/07/17/technology/political-ads-facebook-trump.html.

200 **"Facebook enabled a Trump victory":** Max Read, "Donald Trump Won Because of
     Facebook," Intelligencer, November 9, 2016, http://nymag.com/intelligencer/2016/11/
     donald-trump-won-because-of-facebook.html.

200 **Even inside of Facebook:** Mike Isaac, "Facebook, in Cross Hairs After Election, Is Said
     to Question Its Influence," *New York Times*, November 12, 2016, https://www.nytimes
     .com/2016/11/14/technology/facebook-is-said-to-question-its-influence-in-election.html.

200 **Trump had banked more:** Nicholas Confessore and Karen Yourish, "$2 Billion Worth of
     Free Media for Donald Trump," *New York Times*, March 15, 2016, https://www.nytimes
     .com/2016/03/16/upshot/measuring-donald-trumps-mammoth-advantage-in-free-media.html.

201 **"a huge step backward":** Biz Carson, "'I Do Not Accept Him As My Leader'—Uber CTO's Explosive Anti-Trump Email Reveals Growing Internal Tensions," Business Insider, January 24, 2017, https://www.businessinsider.com/uber-cto-internal-email-donald-trump-deplorable-2017-1.

202 **Uber's $3.5-billion funding round:** Isaac and de la Merced, "Uber Turns to Saudi Arabia for $3.5 Billion Cash Infusion."

202 **Some of Uber's institutional shareholders:** Alex Barinka, Eric Newcomer, and Lulu Yilun Chen, "Uber Backers Said to Push for Didi Truce in Costly China War," Bloomberg, July 20, 2016, https://www.bloomberg.com/news/articles/2016-07-20/uber-investors-said-to-push-for-didi-truce-in-costly-china-fight.

202 **Uber conceded the fight:** Paul Mozur and Mike Isaac, "Uber to Sell to Rival Didi Chuxing and Create New Business in China," New York Times, August 1, 2016, https://www.nytimes.com/2016/08/02/business/dealbook/china-uber-didi-chuxing.html.

202 **For investors, it was a win:** https://www.bloomberg.com/news/articles/2016-07-20/uber-investors-said-to-push-for-didi-truce-in-costly-china-fight.

203 **top tech CEOs were called:** David Streitfeld, "'I'm Here to Help,' Trump Tells Tech Executives at Meeting," New York Times, December 14, 2016, https://www.nytimes.com/2016/12/14/technology/trump-tech-summit.html?module=inline.

Chapter 21: #DELETEUBER

205 **"We don't want them here":** Michael D. Shear and Helene Cooper, "Trump Bars Refugees and Citizens of 7 Muslim Countries," New York Times, January 27, 2017, https://www.nytimes.com/2017/01/27/us/politics/trump-syrian-refugees.html.

206 **he called for a complete restriction:** Patrick Healy and Michael Barbaro, "Donald Trump Calls for Barring Muslims From Entering U.S.," New York Times, December 7, 2015, https://www.nytimes.com/politics/first-draft/2015/12/07/donald-trump-calls-for-banning-muslims-from-entering-u-s/.

206 **Thousands of lawyers arrived:** Jonah Engel Bromwich, "Lawyers Mobilize at Nation's Airports After Trump's Order," New York Times, January 29, 2017, https://www.nytimes.com/2017/01/29/us/lawyers-trump-muslim-ban-immigration.html.

206 **"NO PICKUPS @ JFK Airport":** NY Taxi Workers (@NYTWA), "NO PICKUPS @ JFK Airport 6 PM to 7 PM today. Drivers stand in solidarity with thousands protesting inhumane & unconstitutional #MuslimBan.," Twitter, January 28, 2017, 5:01 p.m., https://twitter.com/NYTWA/status/825463758709518337.

207 **a half-delirious meditation:** Dan O'Sullivan, "Vengeance Is Mine," Jacobin, https://www.jacobinmag.com/2016/11/donald-trump-election-hillary-clinton-election-night-inequality-republicans-trumpism/.

208 **"eat shit and die":** Dan O'Sullivan (@Bro_Pair), "congrats to @Uber_NYC on breaking a strike to profit off of refugees being consigned to Hell. eat shit and die," Twitter, January 28, 2017, 8:38 p.m., https://twitter.com/Bro_Pair/status/825518408682860544.

208 **"Don't like @Uber's exploitative anti-labor":** Dan O'Sullivan (@Bro_Pair), "#deleteuber," Twitter, January 28, 2017, 9:25 p.m., https://twitter.com/Bro_Pair/status/825530250952114177.

209 **"You're fascist colluding scabs":** The Goldar Standard @Trev0000r), "done," Twitter, January 28, 2017, 10:50 p.m., https://twitter.com/Trev0000r/status/825551578824396800.

209 **"Taking advantage of the taxi strike":** Simeon Benit (@simeonbenit), Twitter, January 28, 2017 11:04 p.m., https://twitter.com/simeonbenit/status/825555284428988416.

209 **"Catch a rideshare to hell":** _m_(@MM_schwartz), "@uber Hope I'm not too late to the party #deleteUber," Twitter, January 28, 2017, 11:33 p.m., https://twitter.com/MM_schwartz/status/825562459088023552.

210 **attempted a mealy-mouthed apology:** Travis Kalanick, "Standing Up for What's Right," Uber Newsroom, https://www.uber.com/newsroom/standing-up-for-whats-right-3.

210 **"many people internally and externally":** Travis Kalanick, "Standing Up for What's Right."

211 **a well-executed PR stunt:** Rhett Jones, "As #DeleteUber Trends, Lyft Pledges $1 Million to ACLU," Gizmodo, January 29, 2017, https://gizmodo.com/as-deleteuber-trends-lyft-pledges-1-million-to-aclu-1791750060.

211 **Two different engineers asked:** Mike Isaac, "Uber C.E.O. to Leave Trump Advisory Council After Criticism," *New York Times*, February 2, 2017, https://www.nytimes.com/2017/02/02/technology/uber-ceo-travis-kalanick-trump-advisory-council.html?_r=1.

Chapter 22: **"ONE VERY, VERY STRANGE YEAR AT UBER . . ."**

213 **85 percent of Uber engineers were men:** Johana Bhuiyan, "Uber has Published Its Much Sought After Diversity Numbers For the First Time," Recode, March 28, 2017, https://www.recode.net/2017/3/28/15087184/uber-diversity-numbers-first-three-million.

213 **Like sailing "over the moon":** Maureen Dowd, "She's 26, and Brought Down Uber's C.E.O. What's Next?," *New York Times*, October 21, 2017, https://www.nytimes.com/2017/10/21/style/susan-fowler-uber.html.

215 **"I don't use Uber":** Dowd, "She's 26, and Brought Down Uber's C.E.O."

215 **"SREs worked to keep the service online":** Chris Adams, "How Uber Thinks About Site Reliability Engineering," Uber Engineering, March 3, 2016, https://eng.uber.com/sre-talks-feb-2016/.

217 **Performance reviews were little more:** Megan Rose Dickey, "Inside Uber's New Approach to Employee Performance Reviews," TechCrunch, https://techcrunch.com/2017/08/01/inside-ubers-new-approach-to-employee-performance-reviews/.

219 **losing upwards of $9,000 per vehicle:** Greg Bensinger, "Uber Shutting Down U.S. Car-Leasing Business," *Wall Street Journal*, September 27, 2017, https://www.wsj.com/articles/uber-confirms-it-is-shutting-down-u-s-car-leasing-business-1506531990.

219 **"probably just an innocent mistake":** Fowler, "Reflecting On One Very, Very Strange Year at Uber."

220 **recalled a director boasting:** Fowler, "Reflecting On One Very, Very Strange Year at Uber."

220 **"complete, unrelenting chaos":** Fowler, "Reflecting On One Very, Very Strange Year at Uber."

221 **"If we women really wanted equality":** Fowler, "Reflecting On One Very, Very Strange Year at Uber."

Chapter 23: . . . **THE HARDER THEY FALL**

223   **"This is outrageous and awful":** Chris Messina (@chrismessina), "This is outrageous
      and awful. My experience with Uber HR was similarly callous & unsupportive; in Susan's
      case, it was reprehensible. [angry face and thumbs-down emojis]," Twitter, February 19,
      2017, 6:44 p.m., https://twitter.com/chrismessina/status/833462385872498688.

227   **"@travisk showing me his super cool app":** Arianna Huffington (@ariannahuff),
      "@travisk showing me his super cool app, Uber: everyone's private driver uber.com," Twit-
      ter, May 30, 2012, 3:23 p.m., https://twitter.com/ariannahuff/status/207915187846656001.

228   **an unfaithful journalist:** Vanessa Grigoriadis, "Maharishi Arianna," *New York*,
      November 20, 2011, http://nymag.com/news/media/arianna-huffington-2011-11.

228   **a warm, intelligent woman:** Lauren Collins, "The Oracle: The Many Lives of Ari-
      anna Huffington," *New Yorker*, October 13, 2008, https://www.newyorker.com/
      magazine/2008/10/13/the-oracle-lauren-collins.

228   **"Your dowry is your education":** Collins, "The Oracle."

229   **"a kind of liberal foil":** Collins, "The Oracle."

230   **"a unified theory of Arianna":** Meghan O'Rourke, "The Accidental Feminist," Slate,
      September 22, 2006, https://slate.com/news-and-politics/2006/09/arianna-huffington-the
      -accidental-feminist.html.

230   **"There are two schools of thought":** Maureen Orth, "Arianna's Virtual Candidate,"
      Vanity Fair, November 1, 1994, https://www.vanityfair.com/culture/1994/11/huffington
      -199411.

230   **Her political career:** https://www.vanityfair.com/culture/1994/11/huffington-199411.

Chapter 24: **NO ONE STEALS FROM LARRY PAGE**

232   **first consumer-ready version:** John Markoff, "No Longer a Dream: Silicon Val-
      ley Takes on the Flying Car," *New York Times*, April 24, 2017, https://www.nytimes
      .com/2017/04/24/technology/flying-car-technology.html.

233   **"a new way forward in mobility":** Daisuke Wakabayashi, "Google Parent Company
      Spins Off Self-Driving Car Business," *New York Times*, December 13, 2016, https://
      www.nytimes.com/2016/12/13/technology/google-parent-company-spins-off-waymo
      -self-driving-car-business.html.

233   **suing him months ago:** Biz Carson, "Google Secretly Sought Arbitration Against Its
      Former Self-Driving Guru Months Before the Uber Lawsuit," Business Insider, March
      29, 2017, https://www.businessinsider.com/google-filed-against-anthony-levandowski-in
      -arbitration-before-uber-lawsuit-2017-3.

233   **old Google workplace accounts:** Waymo LLC v. Uber Technologies.

234   **downloaded proprietary information:** Waymo LLC v. Uber Technologies.

234   **"Otto and Uber have taken":** Daisuke Wakabayashi and Mike Isaac, "Google Self-
      Driving Car Unit Accuses Uber of Using Stolen Technology," *New York Times*, February
      23, 2017, https://www.nytimes.com/2017/02/23/technology/google-self-driving-waymo
      -uber-otto-lawsuit.html.

236   **Kalanick had announced his hire:** Mike Isaac and Daisuke Wakabayashi, "Uber
      Hires Google's Former Head of Search, Stoking a Rivalry," *New York Times*, January

20, 2017, https://www.nytimes.com/2017/01/20/technology/uber-amit-singhal-google .html?module=inline.

236 **Singhal was pushed out of Google:** Mike Isaac and Daisuke Wakabayashi, "Amit Singhal, Uber Executive Linked to Old Harassment Claim, Resigns," *New York Times*, February 27, 2017, https://www.nytimes.com/2017/02/27/technology/uber-sexual -harassment-amit-singhal-resign.html.

237 **they had just posted a story:** Eric Newcomer, "In Video, Uber CEO Argues with Driver Over Falling Fares," Bloomberg, February 28, 2017, https://www.bloomberg.com/news/ articles/2017-02-28/in-video-uber-ceo-argues-with-driver-over-falling-fares.

238 **"This is bad":** Eric Newcomer and Brad Stone, "The Fall of Travis Kalanick Was a Lot Weirder and Darker Than You Thought," Bloomberg Businessweek, January 18, 2018, https://www.bloomberg.com/news/features/2018-01-18/the-fall-of-travis-kalanick-was-a -lot-weirder-and-darker-than-you-thought.

240 **"I want to profoundly apologize":** Travis Kalanick, "A Profound Apology," Uber Newsroom, March 1, 2017, https://www.uber.com/newsroom/a-profound-apology.

Chapter 25: GREYBALL

241 **"my name is Bob":** I've changed my source's name and any specific details about their identity to protect their anonymity.

241 **A few days prior:** Mike Isaac, "Insider Uber's Aggressive, Unrestrained Workplace Culture," *New York Times*, February 22, 2017, https://www.nytimes.com/2017/02/22/ technology/uber-workplace-culture.html.

243 **"Keep your Uber phone off":** Email redacted for source protection.

243 **"have a wonderful day!":** Documents held by author.

243 **As we watched the video:** The Oregonian, "Portland vs. Uber: City Code Officers Try to Ticket Drivers," YouTube video, December 5, 2014, 1:53, https://www.youtube.com/ watch?v=TS0NuV-zLZE.

244 **"We *will* impound the vehicle":** Victor Fiorillo, "Uber Launches UberX In Philadelphia, but PPA Says 'Not So Fast,'" *Philadelphia*, October 25, 2014, https://www.phillymag .com/news/2014/10/25/uber-launches-uberx-philadelphia/.

244 **"UBERX: REMINDER":** Documents held by author.

245 **a behavior engineers called "eyeballing":** Mike Isaac, "How Uber Deceives the Authorities Worldwide," *New York Times*, March 3, 2017, https://www.nytimes.com/2017/03/03/ technology/uber-greyball-program-evade-authorities.html.

247 **"Uber has for years used":** Isaac, "How Uber Deceives the Authorities Worldwide."

247 **Uber's security chief, prohibited employees:** Daisuke Wakabayashi, "Uber Seeks to Prevent Use of Greyball to Thwart Regulators," *New York Times*, March 8, 2017, https:// www.nytimes.com/2017/03/08/business/uber-regulators-police-greyball.html.

247 **Department of Justice opened a probe:** Mike Isaac, "Uber Faces Federal Inquiry Over Use of Greyball Tool to Evade Authorities," *New York Times*, May 4, 2017, https://www .nytimes.com/2017/05/04/technology/uber-federal-inquiry-software-greyball.html.

247 **the inquiry widened to Philadelphia:** Mike Isaac, "Justice Department Expands Its Inquiry into Uber's Greyball Tool," *New York Times*, May 5, 2017, https://www.nytimes .com/2017/05/05/technology/uber-greyball-investigation-expands.html.

248 **He called it *The Rideshare Guy*:** Harry Campbell, "About the Rideshare Guy: Harry Campbell," *The Rideshare Guy* (blog), https://therideshareguy.com/about-the-rideshare-guy/.

248 **directly due to the string of controversies:** Kara Swisher and Johana Bhuiyan, "Uber President Jeff Jones Is Quitting, Citing Differences Over 'Beliefs and Approach to Leadership,'" Recode, March 19, 2017, https://www.recode.net/2017/3/19/14976110/uber-president-jeff-jones-quits.

250 **"there's just a bunch of models":** Emily Peck, "Travis Kalanick's Ex Reveals New Details About Uber's Sexist Culture," Huffington Post, March 29, 2017, https://www.huffingtonpost.com/entry/travis-kalanick-gabi-holzwarth-uber_us_58da7341e4b018c4606b8ec9.

253 **"I am so sorry for being cold":** Amir Efrati, "Uber Group's Visit to Seoul Escort Bar Sparked HR Complaint," The Information, March 24, 2017, https://www.theinformation.com/articles/uber-groups-visit-to-seoul-escort-bar-sparked-hr-complaint.

253 **reporter's cell phone number:** Efrati, "Uber Group's Visit to Seoul Escort Bar."

Chapter 26: FATAL ERRORS

254 **who called the maneuver illegal:** Mike Isaac, "Uber Expands Self-Driving Car Service to San Francisco. D.M.V. Says It's Illegal.," *New York Times*, December 14, 2016, https://www.nytimes.com/2016/12/14/technology/uber-self-driving-car-san-francisco.html.

254 **Uber issued a statement:** Isaac, "Uber Expands Self-Driving Car Service to San Francisco."

255 **Uber's narrative was false:** Mike Isaac and Daisuke Wakabayashi, "A Lawsuit Against Uber Highlights the Rush to Conquer Driverless Cars," *New York Times*, February 24, 2017, https://www.nytimes.com/2017/02/24/technology/anthony-levandowski-waymo-uber-google-lawsuit.html.

255 **Levandowski was unceremoniously terminated:** Mike Isaac and Daisuke Wakabayashi, "Uber Fires Former Google Engineer at Heart of Self-Driving Dispute," *New York Times*, May 30, 2017, https://www.nytimes.com/2017/05/30/technology/uber-anthony-levandowski.html.

256 **"possible theft of trade secrets":** Aarian Marshall, "Google's Fight Against Uber Takes a Turn for the Criminal," Wired, May 12, 2017, https://www.wired.com/2017/05/googles-fight-uber-takes-turn-criminal/.

256 **expressed contrition in a press interview:** Mike Isaac, "Uber Releases Diversity Report and Repudiates Its 'Hard-Charging Attitude,'" *New York Times*, March 28, 2017, https://www.nytimes.com/2017/03/28/technology/uber-scandal-diversity-report.html.

257 **existence of Uber's program "Hell":** Efrati, "Uber's Top Secret 'Hell' Program."

257 **The team kept tabs:** Kate Conger, "Uber's Massive Scraping Program Collected Data About Competitors Around the World," Gizmodo, December 11, 2017, https://gizmodo.com/ubers-massive-scraping-program-collected-data-about-com-1820887947.

257 **recorded private conversations:** Paayal Zaveri, "Unsealed Letter in Uber-Waymo Case Details How Uber Employees Allegedly Stole Trade Secrets," CNBC, December 15, 2017, https://www.cnbc.com/2017/12/15/jacobs-letter-in-uber-waymo-case-says-uber-staff-stole-trade-secrets.html.

261 **personal, private medical files:** Kara Swisher and Johana Bhuiyan, "A Top Uber Executive, Who Obtained the Medical Records of a Customer Who Was a Rape Victim, Has

Been Fired," Recode, June 7, 2017, https://www.recode.net/2017/6/7/15754316/uber
-executive-india-assault-rape-medical-records.

262 **it was over for Eric Alexander:** Mike Isaac, "Uber Fires Executive Over Handling
of Rape Investigation in India," *New York Times*, June 7, 2017, https://www.nytimes
.com/2017/06/07/technology/uber-fires-executive.html.

262 **Kalanick accepted her resignation:** Mike Isaac, "Executive Who Steered Uber Through
Scandals Joins Exodus," *New York Times*, April 11, 2017, https://www.nytimes
.com/2017/04/11/technology/ubers-head-of-policy-and-communications-joins-executive
-exodus.html.

263 **"The last note I got from her":** Kalanick, "Dad is getting much better in last 48 hours."

265 **"Over the last seven years":** Unpublished letter, obtained by author. The original letter
is over 4,000 words long.

Chapter 27: **THE HOLDER REPORT**

269 **already fired twenty people:** Mike Isaac, "Uber Fires 20 Amid Investigation into Workplace Culture," *New York Times*, June 6, 2017, https://www.nytimes.com/2017/06/06/
technology/uber-fired.html.

270 **"He definitely has my confidence":** Anita Balakrishnan, "Uber Board Member Arianna
Huffington Says She's Been Emailing Ex-Engineer About Harassment Claims," CNBC,
March 3, 2017, https://www.cnbc.com/2017/03/03/arianna-huffington-travis-kalanick
-confidence-emailing-susan-fowler.html.

273 **"Uber has a long way to go":** Emil Michael, "Email from Departing Uber Executive,"
New York Times, June 12, 2017, https://www.nytimes.com/interactive/2017/06/12/
technology/document-Email-From-Departing-Uber-Executive.html.

274 ***"For the last eight years":*** Entrepreneur Staff, "Read Travis Kalanick's Full Letter to Staff: I Need to Work on Travis 2.0," Entrepreneur, June 13, 2017, https://www
.entrepreneur.com/article/295780.

278 **Bonderman was a shuffling giant:** Henny Sender, "Breakfast with the FT: David Bonderman," Financial Times, June 20, 2008, https://www.ft.com/content/569a70ae-3e64
-11dd-b16d-0000779fd2ac.

278 **239th richest man in the world:** "#667 David Bonderman," Forbes, https://www.forbes
.com/profile/david-bonderman/#27d33dd32fce.

280 **the entire contents of the presentation:** JP Mangalindan, "LEAKED AUDIO: Uber's
All-Hands Meeting Had Some Uncomfortable Moments," Yahoo! Finance, June 13, 2017,
https://finance.yahoo.com/news/inside-ubers-hands-meeting-travis-194232221.html.

280 **"Today at Uber's all-hands meeting":** Comment received in email to author, June 13,
2017.

Chapter 28: **THE SYNDICATE**

284 **"we have hit a dead end":** Mitch and Freada Kapor, "An Open Letter to The Uber Board
and Investors," Medium, February 23, 2017, https://medium.com/kapor-the-bridge/an
-open-letter-to-the-uber-board-and-investors-2dc0c48c3a7.

285 **Lake was sexually harassed:** Dan Primack, "How Lightspeed Responded to Cald-beck's Alleged Behavior," *Axios*, June 27, 2017, https://www.axios.com/how-lightspeed -responded-to-caldbecks-alleged-behavior-1513303291-797b3d44-6b7d-4cd1-89ef -7e35782a32e6.html.

287 **through an internal repurchase program:** Katie Benner, "How Uber's Chief Is Gaining Even More Clout in the Company," *New York Times*, June 12, 2017 https://www.nytimes .com/2017/06/12/technology/uber-chief-travis-kalanick-stock-buyback.html.

288 **the two rarely spoke afterwards:** Alex Konrad, "How Super Angel Chris Sacca Made Billions, Burned Bridges and Crafted the Best Seed Portfolio Ever," *Forbes*, April 13, 2015, https://www.forbes.com/sites/alexkonrad/2015/03/25/how-venture-cowboy-chris -sacca-made-billions/#17b4e9866597.

Chapter 29: **REVENGE OF THE VENTURE CAPITALISTS**

297 **eighteen months in the making:** Lori Rackl, "Get A First Look at the 'New' Ritz-Carlton Chicago, $100 Million Later," *Chicago Tribune*, July 19, 2017, https://www.chicagotribune .com/lifestyles/travel/ct-ritz-carlton-chicago-renovation-travel-0730-20170718-story.html.

303 **she claimed on live television:** Sara Ashley O'Brien, "Arianna Huffington: Sexual Harassment Isn't a 'Systemic Problem,' At Uber," CNN Business, March 23, 2017, https:// money.cnn.com/2017/03/20/technology/arianna-huffington-uber-quest-means-business/ index.html.

306 **Seeing the tick-tock of events:** Mike Isaac, "Inside Travis Kalanick's Resignation as Uber's C.E.O.," *New York Times*, June 21, 2017, https://www.nytimes.com/2017/06/21/ technology/uber-travis-kalanick-final-hours.html.

Chapter 30: **DOWN BUT NOT OUT**

309 **"Dear Board of Directors":** Letter obtained by author.

313 **story was about to be published:** Eric Newcomer, "Uber's New CEO Short List Is Said to Include HPE's Meg Whitman," Bloomberg, July 25, 2017, https://www.bloomberg.com/ news/articles/2017-07-25/uber-s-new-ceo-short-list-is-said-to-include-hpe-s-meg-whitman.

314 **a top candidate for Uber CEO:** Eric Newcomer, "GE's Jeffrey Immelt Is on Uber's CEO Shortlist," Bloomberg, July 27, 2017, https://www.bloomberg.com/news/articles/2017-07 -27/ge-s-jeffrey-immelt-is-said-to-be-on-uber-ceo-shortlist.

314 **Kalanick . . . pushing Immelt:** Mike Isaac, "Uber's Search for New C.E.O. Ham-pered by Deep Split on Board," *New York Times*, July 30, 2017, https://www.nytimes .com/2017/07/30/technology/uber-search-for-new-ceo-kalanick-huffington-whitman .html.

314 **three brief tweets to the world:** Mike Isaac, "Uber's Next C.E.O.? Meg Whitman Says It Won't Be Her," *New York Times*, July 27, 2017, https://www.nytimes.com/2017/07/27/ technology/ubers-next-ceo-meg-whitman-says-it-wont-be-her.html.

314 **"Normally I do not comment":** Meg Whitman (@MegWhitman), "(1/3) Normally I do not comment on rumors, but the speculation about my future and Uber has become

a distraction.," Twitter, July 27, 2017, 10:04 p.m., https://twitter.com/megwhitman/status/890754773456220161.

314 **"So let me make this as clear":** Meg Whitman (@MegWhitman), "(2/3) So let me make this as clear as I can. I am fully committed to HPE and plan to remain the company's CEO.," Twitter, July 27, 2017, 10:04 p.m., https://twitter.com/MegWhitman/status/890754854632787969.

314 **"Uber's CEO will not be Meg Whitman":** Meg Whitman (@MegWhitman), "(3/3) We have a lot of work still to do at HPE and I am not going anywhere. Uber's CEO will not be Meg Whitman." Twitter, July 27, 2017, 10:05 p.m., https://twitter.com/megwhitman/status/890754932990763008.

315 **the firm filed a lawsuit:** Mike Isaac, "Uber Investor Sues Travis Kalanick for Fraud," *New York Times*, August 10, 2017, https://www.nytimes.com/2017/08/10/technology/travis-kalanick-uber-lawsuit-benchmark-capital.html.

315 **"We do not feel it was either prudent":** Mike Isaac, "Kalanick Loyalists Move to Force Benchmark Off Uber's Board," *New York Times*, August 11, 2017, https://www.nytimes.com/2017/08/11/technology/uber-benchmark-pishevar.html.

317 **He bankrolled his college years:** Cyrus Farivar, "How Sprint's New Boss Lost $70 Billion of His Own Cash (and Still Stayed Rich)," Ars Technica, October 16, 2012, https://arstechnica.com/information-technology/2012/10/how-sprints-new-boss-lost-70-billion-of-his-own-cash-and-still-stayed-rich/.

317 **He returned to Japan:** Andrew Ross Sorkin, "A Key Figure in the Future of Yahoo," Dealbook, *New York Times*, December 13, 2010, https://dealbook.nytimes.com/2010/12/13/a-key-figure-in-the-future-of-yahoo/.

317 **"crazy guy who bet on the future":** Walter Sim, "SoftBank's Masayoshi Son, the 'Crazy Guy Who Bet on the Future,'" *Straits Times*, December 12, 2016, https://www.straitstimes.com/asia/east-asia/softbanks-masayoshi-son-the-crazy-guy-who-bet-on-the-future.

317 **He designed the investment vehicle for speed:** Dana Olsen, "Vision Fund 101: Inside SoftBank's $98B Vehicle," PitchBook, August 2, 2017, https://pitchbook.com/news/articles/vision-fund-101-inside-softbanks-93b-vehicle.

Chapter 31: **THE GRAND BARGAIN**

319 **The storied corporation had lost:** Steve Blank, "Why GE's Jeff Immelt Lost His Job: Disruption and Activist Investors," *Harvard Business Review*, October 30, 2017, https://hbr.org/2017/10/why-ges-jeff-immelt-lost-his-job-disruption-and-activist-investors.

320 **"There's this chip you have":** Sheelah Kolhatkar, "At Uber, A New C.E.O. Shifts Gears," *The New Yorker*, April 9, 2018, https://www.newyorker.com/magazine/2018/04/09/at-uber-a-new-ceo-shifts-gears.

320 **After Allen & Company:** https://www.newyorker.com/magazine/2018/04/09/at-uber-a-new-ceo-shifts-gears.

324 **"I have decided not to pursue":** Jeff Immelt (@JeffImmelt), "I have decided not to pursue a leadership position at Uber. I have immense respect for the company & founders – Travis, Garrett and Ryan." Twitter, August 27, 2017, 11:43 a.m., https://twitter.com/JeffImmelt/status/901832519913537540.

324 **People close to Immelt:** Kara Swisher, "Former GE CEO Jeff Immelt Says He Is No Longer Vying to Be Uber CEO," Recode, August 27, 2017, https://www.recode .net/2017/8/27/16211254/former-ge-ceo-jeff-immelt-out-uber-ceo.

325 **Shortly after five o'clock:** Mike Isaac, "Uber Chooses Expedia's Chief as C.E.O., Ending Contentious Search," *New York Times*, August 27, 2017, https://www.nytimes .com/2017/08/27/technology/uber-ceo-search.html.

326 **Son reached a deal:** Mike Isaac, "Uber Sells Stake to SoftBank, Valuing Ride-Hailing Giant at $48 Billion," *New York Times*, December 28, 2017, https://www.nytimes .com/2017/12/28/technology/uber-softbank-stake.html.

327 **It was a pure preemptive strike:** Katie Benner and Mike Isaac, "In Power Move at Uber, Travis Kalanick Appoints 2 to Board," *New York Times*, September 29, 2017, https:// www.nytimes.com/2017/09/29/technology/uber-travis-kalanick-board.html.

327 **"one share, one vote":** Katie Benner and Mike Isaac, "Uber's Board Approves Changes to Reshape Company's Power Balance," *New York Times*, October 3, 2017, https://www .nytimes.com/2017/10/03/technology/ubers-board-approves-changes-to-reshape-power -balance.html.

EPILOGUE

331 **"180 days of change":** Rachel Holt and Aaron Schildkrout, "180 Days: You Asked, and We're Answering," Uber, https://pages.et.uber.com/180-days/.

332 **"We do the right thing. Period.":** Dara Khosrowshahi, "Uber's New Cultural Norms," LinkedIn, November 7, 2017, https://www.linkedin.com/pulse/ubers-new-cultural-norms -dara-khosrowshahi/.

332 **a full-throated repudiation:** Mike Isaac, "Uber's New Mantra: 'We Do the Right Thing. Period.,'" *New York Times*, November 7, 2017, https://www.nytimes.com/2017/11/07/ technology/uber-dara-khosrowshahi.html.

332 **dubbed the new "Dad":** Geoffrey A. Fowler, "I Was Team #DeleteUber. Can Uber's New Boss Change My Mind?," *Washington Post*, May 11, 2018, https://www.washingtonpost .com/news/the-switch/wp/2018/05/11/i-was-team-deleteuber-can-ubers-new-boss -change-my-mind/?utm_term=.affb048f5b91.

332 **Uber blanketed media with ads:** Priya Anand, "Uber to Spend Up to $500 Million on Ad Campaign," The Information, June 5, 2018, https://www.theinformation.com/articles/ uber-to-spend-up-to-500-million-on-ad-campaign.

333 **"Khosrowshahi has been perfecting his brand":** Jessi Hempel, "One Year In, The Real Work Begins For Uber's CEO," Wired, September 6, 2018, https://www.wired.com/story/ dara-khosrowshahi-uber-ceo-problems-lyft/.

333 **"He's back?":** Anthony Levandowski, "Pronto Means Ready," Medium, December 18, 2018, https://medium.com/pronto-ai/pronto-means-ready-e885bc8ec9e9.

333 **"Way of the Future":** Mark Harris, "Inside the First Church of Artificial Intelligence," Wired, November 15, 2017, https://www.wired.com/story/anthony-levandowski-artificial -intelligence-religion/.

338 **Larry Page grew increasingly "unpumped":** Daisuke Wakabayashi, "Why Google's Bosses Became 'Unpumped' About Uber," *New York Times*, February 7, 2018, https:// www.nytimes.com/2018/02/07/technology/uber-waymo-lawsuit.html.

338 **"He answered every question":** Eric Newcomer, "Inside the Abrupt End of Silicon Valley's Biggest Trial," Bloomberg, February 9, 2018, https://www.bloomberg.com/news/articles/2018-02-09/inside-the-abrupt-end-of-silicon-valley-s-biggest-trial.

338 **As part of the terms of the settlement:** Daisuke Wakabayashi, "Uber and Waymo Settle Trade Secrets Suit Over Driverless Cars," *New York Times*, February 9, 2018, https://www.nytimes.com/2018/02/09/technology/uber-waymo-lawsuit-driverless.html.

INDEX

Page numbers followed by *n* refer to footnotes.